Black Atlantic Religion

Black Atlantic Religion

TRADITION, TRANSNATIONALISM, AND MATRIARCHY IN THE AFRO-BRAZILIAN CANDOMBLÉ

J. Lorand Matory

PRINCETON UNIVERSITY PRESS

PRINCETON AND OXFORD

Copyright © 2005 by Princeton Unviersity Press
Published by Princeton Unviersity Press,
41 William Street, Princeton, New Jersey 08540
In the United Kingdom: Princeton University Press,
3 Market Place, Woodstock, Oxfordshire OX20 1SY

ISBN: 0-691-05943-8
ISBN (pbk.): 0-691-05944-6

LIBRARY OF CONGRESS CATALOGING-IN-PUBLICATION DATA
Matory, James Lorand.
   Black Atlantic religion : tradition, transnationalism, and matriarchy in the
   Afro-Brazilian Candomblé / J. Lorand Matory.
      p. cm.
   Includes bibliographical references and index.
   ISBN 0-691-05943-8 (cl : alk. paper)—ISBN 0-691-05944-6 (pbk. : alk. paper)
      1. Candomblé (Religion) I. Title.
   BL2592.C35M38 2005
   299.6'73—dc22   2004057273

British Library Cataloging-in-Publication Data is available

This book has been composed in Sabon

Printed on acid-free paper. ∞

pup.princeton.edu

Printed in the United States of America

10 9 8 7 6 5 4 3 2 1

Contents

List of Illustrations

Introduction

THIS IS A STORY OF AFRICA in the Americas. But it is just as much a story about the Americas in Africa, in defiance of the outmoded supposition that internal integration and the isomorphism of cultures with local populations are the normal conditions of social life. This story suggests that lifeways, traditions, and the social boundaries they substantiate endure not *despite* their involvement in translocal dialogues but *because* of it. Candomblé (pronounced cahn-dome-BLEH, with a final vowel sound resembling the *e* in "pet") is one such lifeway and tradition, which is both the product and one of the greatest producers of a transoceanic culture and political economy known as the "black Atlantic."

Candomblé is an Afro-Brazilian religion of divination, sacrifice, healing, music, dance, and spirit possession. The only rival to its beauty is its complexity. Though this religion is headquartered in the coastal Brazilian state of Bahia, it has counterparts and offshoots all over urban Brazil. Believers attribute miraculous powers and exemplary flaws to gods known variously as *orixás, voduns, inquices,* and *caboclos,* depending on the Candomblé denomination. The adventures, personalities, and kinship relations of these superhuman beings are described in an extensive mythology and body of oracular wisdom, which also serve to explain the personalities and fates of their human worshipers, as well as the worldly relations among those worshipers. Through blood sacrifice and lavish ceremonies of spirit possession, the gods are persuaded to intervene beneficently in the lives of their worshipers and to keep the foes of those worshipers at bay.

The Candomblé temple, or "house," also serves the social and economic needs of its class-diverse and largely urban membership. It is usually the primary residence of the chief priest, some of his or her lieutenants, and their wards, as well as a temporary shelter for fugitives from police persecution, domestic crises, and poverty. The temple is also often a conduit of bourgeois largesse, a source of job contacts, an employer in its own right, and a major port of call for politicians. Yet priests and practitioners, no less than the social scientists and politicians who seek to speak for them, tend to emphasize the ancientness and fixity of Candomblé and of the ritual "tradition" that both constitutes its ultimate purpose and shapes the deepest part of community life.

This book is a sequel to my *Sex and the Empire That Is No More: Gender and the Politics of Metaphor in Ọ̀yọ́-Yorùbá Religion* (1994; also forthcoming b), and shares in the intent of that book to understand supposedly local and primordial "folk" cultures and "primitive" religions not primarily in terms of their roots in a pristine past but in the context of the dynamic politics, economics, and long-distance communication that are the lived realities of the "folk." The five-hundred-year-old black Atlantic is five hundred years older in its translocalism and more than two hundred years older in its disruption of nation-state boundaries than the epoch now fashionably described as "transnationalism," giving rise to the suspicion that the chief exponents of this term have chosen to disregard Africa or to disregard the past generally. Yet my aim is not to nominate a new date for the beginnings of transnationalism so that it encompasses African history but, instead, to embrace the truths that this term actually highlights about both the African diaspora and the entire course of human cultural history: the isolation of local cultural units has long been the exception rather than the rule, and territorially bounded social groups have never monopolized the loyalty of their members.

The 15th- to 19th-century Atlantic slave trade and European colonialism were the founding conditions of the black Atlantic but were hardly the first moments of dynamism or translocalism in African cultural history. Commerce and migration across the Sahara, across the Indian Ocean, and across the diverse regions of sub-Saharan Africa are among the forces that have long made Africa a cosmopolitan and ever-changing place. Hence, the focus of this project on the past two hundred years is intended to demonstrate not that we can no longer treat geographical isolation as a condition of cultural reproduction in Africa but that we never could, in Africa or anyplace else.

Moreover, far from emerging from the death throes of the nation-state, translocalism long predated nationalism everywhere in the world. Indeed, translocalism was a founding condition of nationalism. I regard transnationalism less as a sudden thirty-, forty-, or fifty-year-old challenge to the nation-state than an analytic focus on how important translocalism *remains*, even at moments when the rulers of a few exceptionally powerful nation-states claim to believe that their territorial jurisdictions are culturally, economically, and politically autonomous. What is most worthy about the recent literature on transnationalism is what it ultimately implies about a very old phenomenon: territorial jurisdictions might command great loyalty from their citizens and subjects, and they might impose significant constraints on their conduct; however, territorial jurisdictions have never monopolized the loyalty of the citizens and subjects that they claim, and they are never the sole founts of authority or agents of constraint in such people's lives. Such is true of the nation-state now, as it was and is of kingdoms, empires, religions, acephalous republics, and fiefdoms.

By way of illustration, this book documents a series of transnational dialogues—involving West African, Afro-Brazilian, and Afro-Cuban priests alongside European and American slave traders, European imperialists, postcolonial Latin American and African nationalists, black trans-Atlantic merchants, and an international community of ethnographers—in the absence of which the massive changes in the ethnic identities, sacred values, and gendered leadership associated with Candomblé over the past century and a half would have been difficult to explain. Hence, this story defies not only the current fashion to describe transnationalism as a recent phenomenon but also the old chronotrope (Fabian 1983) that homelands are to their diasporas as the past is to the present and future. The irony at the core of this story is that diasporas create their homelands. The circum-Atlantic forces that have produced the Afro-Brazilian Candomblé, which range from the slave trade to the return migration of Afro-Latin Americans to Africa, as well as Boasian anthropology itself, have also produced a range of novel West African ethnic identities.

I have chosen to tell a story about the Afro-Brazilian religion of Candomblé not simply because it presents some of the most beautiful spectacles of black divinity in the world but also because it illustrates black ingenuity under duress, an ingenuity that created its transnational, transimperial, and transoceanic networks before the word "transnationalism" was ever known. It is an ingenuity that refuses to be written out of history. Here I will tell a new tale of Candomblé's old past and its dynamic present, a present of which I myself am a part.

A CHALLENGE TO ETHNOGRAPHY

What would a culture look like, and how would an ethnography look, if we attended consciously to the transnational processes that have constantly informed the meanings and motives of its participants?

The genre of anthropological ethnography began as an effort to understand the deeply local character of meaning and the local institutional context of collective human endeavors. The conventions of the genre consciously demoted history because the only forms of history available for the Pacific island, African, and Native American societies in which anthropologists specialized seemed to be diffusionist speculations about where a local "trait" had come from or evolutionist constructions of how "high" a society had risen in some putatively universal trajectory toward the lifeways best exemplified by northwest European white men. The lack of long-term documentation of these societies' histories is more easily resolved now than it once was—social into the late 19th century, such black Atlantic nations brought their citizens together in work crews, manumission societies, Catholic lay brotherhoods, and rebel armies. Today they are held together—often with tremendous success—by obedience to shared gods, shared ritual standards, shared language, and, in some sense, a shared leadership.

With counterparts in Nigeria, the People's Republic of Bénin, Trinidad, Cuba, and everyplace where Cuban exiles have lived since 1959, Candomblé attracts many nonbelievers with its festivals, where the beautifully clothed gods and goddesses dance before drum orchestras and bring blessings to their earthly devotees. Its African–inspired gods and Native American–inspired *caboclo* spirits are associated with historical characters (kings, queens, knights, and Indian chiefs), with "forces of nature" (oceans, rivers, trees, hunting, rainbows, snakes, and storms), and with particular human personality and body types. The god a person worships personifies elements of his or her character and destiny and prescribes the worshiper's potential roles in the ritual life of the temple. A worshiper also obeys his or her god's food and behavioral taboos. Knowledge of other peoples' gods is taken to illuminate virtually any social situation.

When people are afflicted or in trouble, divination sometimes identifies the source of the problem as a neglected tutelary god, a malevolent spirit sent by an enemy, other people's jealousy, or a harmful "influence" picked up from the street. Candomblé priests command the technology to purify bodies, houses, and other vessels of unwanted influences and to insert, or secure the presence of, the divinities in the bodies and altars of their devotees. The ritual control of what enters and what stays out of people's bodies dramatizes people's situational and lifelong belonging in trans-Atlantic communities that crosscut multiple territorial nations.

Among the instruments of this technology are pots and soup tureens, which, at the time of initiation, are invested with shells, stones, and metal objects representing the divinities (Matory 1986). These and other objects representing a divinity and embodying its protective force must be in the color, material, shape, and number emblematic of the divinity they represent. Moreover, they have been bathed in the herbs and in the blood of animals consecrated to that divinity. Each divinity also demands specific species and a particular sex of animal, as well as cooked vegetable foods specific to that divinity. Well cared for, a divinity guarantees the maximum possible protection, prosperity, mental health, and physical well-being for its sons and daughters. The protocols of Candomblé also create "houses" or "homes" (*casas* or *ilês*) where their members and wards find an alternative to, and sometimes refuge from, the nation-state.

I am not the first non-Brazilian to be swept up in Candomblé's vivid imagery of identity and utility (see, e.g., Harding 2000; Carelli 1993; Wafer 1991; Omari 1984; Bastide 1978[1960]; Cossard-Binon 1970). Yet, even with its paradigmatic status in the study of "African survivals" and of Afro-Atlantic religions, the Candomblé boasts no surfeit of book-length histories or ethnographies in English. What follows is a historical ethnography of the Brazilian Candomblé and of the multifarious "transnational" phenomena that have shaped and been shaped by it.

Geographically, Brazil is now the fifth-largest country in the world, and it is by far the largest country in "Latin America." In population, it is the sixth largest in the world, with nearly 176 million people.[4] Judging by North American racial definitions, only Nigeria's population of more than 110 million outnumbers Brazil's black population. Because of Brazil's famously ambiguous definitions of race, estimates of its population of African descent range from 58 million to 132 million and from 33 to 75 percent of the national population (Minority Rights Group 1995). By contrast, only about 34.7 million, or 12.3 percent, of the United States' population declares itself "Black," thus constituting only the *second*-largest population of sub-Saharan African descent outside of Africa (U.S. Census 2000). (In the United States, the capitalization of the term is meant to convey more than a shared, externally perceived phenotype; it denotes an internal sense of nationality shared by dark- and light-skinned people of African descent.) However, this triumvirate of black Atlantic powers—Nigeria, Brazil, and the United States—is linked by more than just ancestry. They have long been bound together in an intensely transnational dialogue over race, gender, the environment, and the fate of nations.

The volume to which this is a sequel, *Sex and the Empire That Is No More* (1994; forthcoming a), documents two centuries of changes in the gender conceptions and political uses of òrìṣà worship among the Ọ̀yọ́-Yorùbá, who live in what is now Nigeria. Over a similar period, the present volume traces the activities of an itinerant, trans-Atlantic religious and commercial elite whose lifeways are no more reducible to the conditions of Brazilian enslavement and racial oppression than to some primordial African "origins." Chiefly since the mid–19th century, this elite has been active in renaming, reshaping, and redefining political identities and formations on both sides of the Atlantic. These itinerant Afro-Brazilians have been pivotal actors in the simultaneous formation of the Brazilian, Nigerian, Beninese, Nagô, and Jeje national identities. By their example, I am concerned to restore the Afro-Brazilian subject, or agent, to the narrative of Candomblé's remarkable elaboration and spread over the past two centuries. Although I am interested in the *fact* of Candomblé's impressive continuities with African linguistic and ritual forms, I am more interested here in the *historical processes and activities* that have created such continuities, privileged some continuities over others, and officially masked numerous discontinuities.

I wish to show that, at the hands of Afro-Brazilian merchants and priests, the remaking of Candomblé's trans-Atlantic nations has paralleled and interacted with the similarly trans-Atlantic processes that made the Brazilian territorial nation.

Transnationalism and Its Neglected Forebears

The global spread of the nation-state idea reached its apogee in the mid–20th century, when scores of formerly colonized countries in Asia, Africa, and the Caribbean followed the model of late 18th-century and 19th-century Europe and the continental Americas in declaring independence from sprawling European-centered empires. In theory, the outward movement of European people and ideas fixed an inexorable, if approximate, trajectory for global social change (Anderson 1991[1983]).

After the 1960s, however, the sizable reverse flow of formerly colonized, or "third world," populations into the "first world" even took the academy by surprise. The facility with which migrants and multinational corporations participated simultaneously in the politics, economics, and cultural formations of multiple nation-states sent scholars rushing to identify new analytic units and to name the seemingly novel processes afoot. Since the 1980s, various disciplines in the social sciences and the humanities have produced a sizable bibliography under the rubric of "transnationalism," as well as a series of new journals—*Diaspora, Public Culture, Positions,* and so forth—that establish the integrity of studying dispersed and supraterritorial sociopolitical units (see Lie 1995; Kearney 1995). This literature teaches a great many lessons about the global interconnectedness of late 20th-century and early 21st-century life, but some of its expositors rely on arbitrary exclusions to emphasize their point. First, translocalisms that preceded the nation-state are usually ignored or diminished rather than analyzed comparatively. Second, the undeniable growth in transportation and communication technology is treated as a recent sea change rather than a millennial trend; before one or another recent historical moment, all changes in the technology or organization of translocal movement are deemed unworthy of notice. Third, momentous transborder communities structured by forces other than "capitalism"—such as Islam and international socialism—are cast outside of this new monocausal grand narrative. Finally, for many theorists, it has become a truism that transnationalism spells the end of the nation-state. They thus define as irrelevant those forms of translocalism that have either produced the nation-state or subsidized it from the very beginning.

The premise of this book is that such exclusions are more definitional than empirical or instructive. The "transnationalism" rubric invites a useful but neglected range of comparison cases—the Islamic world; the Jewish diaspora and Zionism; the commercial networks of the Indian Ocean; the dispersion of Chinese speakers across Southeast Asia; of South Asians to the South Pacific, East and southern Africa, the Caribbean, and South America; of Irish to the Caribbean and the United States; of Syro-Lebanese to West Africa and South America; and, of course, of Europeans and Africans to the Americas.

Apparently, this last case has so far been too obvious or too overwhelming to be fruitfully compared to what is now conventionally called "transnationalism." Or, perhaps, it bears characteristics too disruptive of the current theoretical fashion to be convenient. At the time of the Euro-African colonization of the Americas, the nation-state was not a precondition to be disrupted by transnationalism, as the usual story of transnationalism goes; rather, the nation-state was an epiphenomenon of that early colonial dispersion of Europeans and Africans. The late 18th and the 19th centuries were in fact a time when both American territorial units (such as Brazil, Cuba, and British colonial North America) and transoceanically dispersed black ethnic groups (such as the Nagô, Jeje, Angola, and Congo) were becoming "nations" for themselves. Overlapping in the same geographical space and engaged in mutually implicating dialogues, these white Creole territorial nations and these dispersed black Creole nations were simultaneously in gestation.

It is true that a now-outmoded type of ethnography once imagined local communities as though they were bounded. And, in today's world, any such ethnography silences major dimensions of how the community under study works. However, inattention by much of the older ethnography to translocal forces should not be mistaken, as it is in much current theorizing, for proof that before the late 20th century, most societies functioned in an entirely local field, that every present-day society is equally enmeshed in transnational forces, or that the entire world is becoming a boundary-free whole. While some theorists regard amorphousness, limitless extension, instantaneous long-distance communication, and boundarylessness as the distinguishing features of today's transnationalism, I submit that earlier flows were no more or less socially and geographically patterned than today's and that, even today, talk of simultaneity in long-distance communication is much exaggerated. We must be careful not to allow our enthusiastic description of the present to reduce the past to a one-dimensional foil.

Nor must we ignore the ways in which old patterns continue to co-exist with new ones. Indeed, they sometimes complement each other. If we included this range of old translocalisms and transnationalisms in our analyses, what would we learn about the genesis of nation-states and "transnations" (Appadurai 1996), about the long-term relationships between them, about the motives behind these contrary "imaginations" of community, and about the broader sociogeographical contexts in which they operate?

Braudel (1992[1949]) offers a foundational model for such geographical contexts in his treatment of the "Mediterranean and the Mediterranean World," one such regional system of movement, exchange, and culture formation. A similar regional system has coincided with, subsidized, and overlapped with the spread of the nation-state itself. Philip Curtin and coauthors' discussion of the "South Atlantic System" (1978), Robert Farris Thompson's "black Atlantic" world (1983), Paul Gilroy's "black Atlantic" (1993), John Thornton's "Atlantic World" (1992), and Joseph Roach's "circum-Atlantic world" (1996) name the ways in which cultural and political developments at any given locale on the Atlantic perimeter flow from what I call an ongoing "dialogue" with multiple other locales—and nation-states—on that same perimeter.

These processes deserve an ethnographic and historical description specific to the Atlantic perimeter region, which is the main aim of *Black Atlantic Religion*, but the diasporic dialogue around the Atlantic perimeter is arguably the foundation of world capitalism, of the industrial revolution, and of the emergence of the European, American, African, and Asian nation-state. As scholars debate whether "transnationalism" spells the imminent death of territorial nationhood (see, e.g., Glick Schiller, Basch, Szanton-Blanc 1992; Appadurai 1996), we would do well to consider the long record of how diasporas created nation-states and how the two have continuously reshaped each other.

African or Not?

This book also proposes a new kind of solution to a quandary that has been central to the study of New World black populations throughout the time that their worthiness of citizenship in the nation-state has been debated. The question of whether New World blacks are culturally African and the worthiness of their cultures' integration into the nation-state were debated cross-racially and transnationally not only in movements ranging from abolitionism to pseudoscientific racism and eugenics but also at the transnational intersection among the United States–based Harlem Renaissance, Havana-based *Afro-Cubanismo*, the cultural nationalism of the Haitian Bureau d'Éthnologie, the pan-Francophone

Négritude movement, São Paulo–based Modernismo, and Northeastern Brazilian Regionalismo. These debates co-occurred and overlapped with various bourgeois *indigenismos*, which rethought the Native American presence in the nation-state and flourished particularly in Mexico from the 1930s onward.

Whatever the peculiarities of the Brazilian case, it quickly assumed the status of a locus classicus in the debate over the nature of African-diaspora, or African-American, cultures generally. Its exemplary form, the Herskovits-Frazier debate, concerned whether African-American lifeways were African-inspired or not. Melville J. Herskovits (e.g., 1958[1941]; 1966[1937], 1966[1930], 1966[1945]) argued that African Americans in virtually every American nation-state "retained" some greater or lesser legacy of the African cultural past. Assuming that the cultures of the West African Fòn, Yorùbá, and Ashanti (as reconstructed in the "ethnographic present") represented the extant "base line," or starting point, of African-American cultural history, Herskovits's "ethnohistorical" method, or "social laboratory," posited that less acculturated African-American groups revealed the stages and intermediate forms through which African cultural traditions had been "transmuted" into their counterparts among more highly acculturated African-American groups, such as the blacks of the United States. Thus, as one such intermediate form, spirit possession in the Afro-Brazilian Candomblé religion could be taken to demonstrate the African derivation of "shouting," or the behavior of those "filled with the Holy Spirit," in Black North American churches (Herskovits 1958[1941]:220–21). Dozens of scholars have usefully employed similar reasoning in the study of African-American cultural reproduction.[5] The problem, however, is that Herskovits's descriptions tend to represent the distinctive qualities of African-American cultures—with their enduring "deep-seated drives," "bents," "underlying patterns," and tenacious "personality characteristics of the African"—as instinctual, giving little attention to their meaning in the eyes of the African-American actors or to the proximate mechanisms of their transmission from one generation to the next.

The eminent sociologist E. Franklin Frazier, who was African-American, sharply condemned Herskovits's "Africanisms" thesis, generating one of the most central debates in African-American studies. While recognizing the complexity of West African religion and the "fusion of Christian beliefs and practices with African religious ideas . . . in the Candomblé in Brazil," Frazier emphasizes the processes of capture, abuse, and surveillance whereby "the Negro was stripped of his social heritage," *foremost* in the United States. In his view, Blacks in the United States retained nothing of African language, social organization, or religion. Frazier summarizes these developments as "The Break with the African

Background" (Frazier 1974[1964]: esp. 10, 15). In important ways, Frazier recapitulated the optimistic narrative of the North American "melting pot," but he did not always restrict its implications to the North American case. He and Herskovits also clashed over the relevance of African polygyny in explaining the proclivity of Afro-Brazilian men to father children by multiple women (Frazier 1942, 1943; Herskovits 1943).

Twentieth-century scholarship on Afro-Brazilian religions is structured by a similar debate. Many Brazilian and non-Brazilian scholars have found Afro-Brazilian religions interesting insofar as they demonstrated continuity with their African past or proved the capacity of Africans and their descendants in the New World to "preserve" or "remember" African culture (e.g., Herskovits, 1958[1941]; Verger 1970[1957], 1981; Bastide 1978[1960]; Elbein dos Santos 1976; R. N. Rodrigues 1935[1900/1896], 1945[1905]; Ramos 1940[1934]; Pierson 1942; Landes 1947; Carneiro 1986[1948]).

For Bastide, the Candomblé and Xango religions are transplants from Africa whose Africanness owes something to the ongoing traffic between Africa and Brazil, but he adds that their survival in Brazil depended on the social role they have played in redressing urban anomie, joblessness, police persecution, and the absence of other forms of social leadership (Bastide 1978[1960]:171). His conclusion that African religion endures only insofar as it remains "communalistic" and escapes the effects of class and ethnic competition is quite the opposite of the conclusion that I reach here. However, his insightful argument that African culture endures in the Americas only insofar as it finds or creates a social "niche," in which it can regularly be practiced by organized groups of people, foreshadows the widely cited work of Mintz and Price (1992[1976]). No longer could African culture in the Americas be conceived of merely as a set of "bents" that survive, as if atavistically, without the aid of enactment by living and embattled communities.

Mintz and Price (1992[1976]) have added to Herskovits's legacy by grasping the importance of the social conditions and processes shaping the selective reproduction and transformation of African cultural dispositions in the Americas. For them, the enslavement of Africans and their arrival in any given American country with ethnically heterogeneous "crowds" of other Africans meant that the African slaves could collectively reinstitute not whole African cultures or concrete African practices but, at most, African "cognitive orientations." Against Herskovits's diffusionist hypothesis, Mintz and Price argued that the African origins of any given practice are less important than its role in an African-American community's process of institution-building and its creation of social solidarity, autonomy from the white world, and meaning. Mintz and Price argue that the processes by which African-Americans cultures took form and endured or changed over time must be studied within the specificity of particular American locales, rather than being inferred through speculative devices, such as Herskovits's "social laboratory."

However, Mintz and Price report several recurrent patterns in their comparative study of African-American cultures. First, both African culture and the cultural heterogeneity of American slave populations encouraged "additivity," or the tendency to borrow creatively from numerous cultural traditions rather than to preserve any particular cultural tradition exclusively or purely. Second, the stripping of status that resulted from enslavement encouraged individual expressiveness and the ever-changing creativity of African-American cultures. Third, local African-American cultural systems tended to form soon after the first arrivals from Africa. Thereafter, they remained highly resistant to change under the influence of arrivals from new African locales and of other exogenous events.

Mintz and Price posit that African-American cultures originated specifically in the "slave sector," that social space in which the white masters, who otherwise monopolized power, allowed the black slaves to act autonomously. By giving priority to the "slave sector" as the baseline of African-American culture and the source of enduring "collective memory" and by focusing on African-American creativity, Mintz and Price have inspired or foreshadowed a great deal of exciting scholarly work (e.g., S. Price and Price 1999; R. Price 1983, 1990; Dayan 1995; Gilroy 1993; Palmié 2002).

Though much inspired by Mintz and Price's socially contextualizing approach, Stephan Palmié (2002) shows that even in ethnically heterogeneous black Cuba, the names of Afro-Cuban religious denominations reflect enormous cultural debts to specific parts of Africa. However, Palmié supports Mintz and Price's aim to explain enduringly African names and practices within the more important context of their use within American sociocultural "systems." In Cuba, the religious practices identified with diverse African origins have taken on new and contrasting moral valences specific to their New World context. Yorùbá-identified Lucumí practices and West-Central African–identified Palo Mayombe draw their primary meaning not only from their respective African cultural precedents but from the moral contrasts between them as they are perceived in Cuba— that is, the *contrast* between, on the one hand, a morally laudable Lucumí ethos of "reciprocal interchange and divine initiative" and, on the other hand, an emically amoral or immoral logic of "wage labor and payment, dominance, and potential revolt" characterizing Cuban Congo practices (Palmié 2002:25). This defining contrast is American, not African.

A related Brazilian scholarly trend so emphasizes the constraints imposed by a white-dominated system that it credits contemporary local

Euro-American elites with the power to "invent tradition" in black Brazil. In this model, African-American cultural debts to Africa and agency in the construction of their own traditions are either marginalized or denied, and the power of Euro-Brazilian sponsors to reshape Afro-Brazilian religious practice in their own oppressive interest is emphasized (Motta 1994, 1992; Dantas 1982, 1988; D. D. Brown and Bick 1987; Wafer and Santana 1990; Frigerio 1983; Eco 1986:103–12).

Scholars who study African culture and seek to understand what it has to do with the cultural creativity of oppressed African Americans ignore these Brazilian lessons at their peril. However, these models of white elite domination provoke us to ask when and how people torn from their natal societies, stripped of their accustomed social statuses, cast in with heterogeneous crowds of other Africans, and culturally committed to individual expressiveness also *construct self-conscious traditions and canons*. And what has been the breadth of their freedom to do so in their own interests? In numerous cases, African-American groups have chosen to anchor themselves in nostalgic conceptions of and reconnaissance about Africa that are purist and, in effect, anti-"additive" in their logics. In particular, Mintz and Price's choice to focus on the culture of rural enslaved people provokes us to ask how urban people, free people, and people with more cosmopolitan social connections are constrained and empowered by their social circumstances. African Americans have long been diverse in their statuses and socioeconomic circumstances. In sum, what social contexts and nonprimordially African "cognitive orientations" have motivated the purist and Africanizing trends that also appear in African-American cultures? Analytic models that represent African-American cultures as encompassed in and constrained by local Euro-American power structures (Mintz and Price 1992[1976]), as reactions against the failed promises of Western modernity (Gilroy 1993), or as locally specific antitheses of Western modernity (Palmié 2002) lay the groundwork for an answer to this question but leave open an important space for Africanists to enter the dialogue.

John Thornton (1992), while implicitly embracing Mintz and Price's demand for the nonspeculative study of specific social contexts, reaches a different conclusion about the proper scope of that context. For Thornton, the Atlantic world and the commercial geographies that united specific supplying and receiving regions in the slave trade become the relevant sociopolitical context. With the benefit of greater historical documentation of this context, Thornton points out that the captives arriving in any given American locale were seldom as heterogeneous as Mintz and Price had imagined. Slaveholders in any given region and sphere of the economy often preferred one African "nation" over others, and, even where local slave populations were ethnically heterogeneous, people of the same "nation" could often visit each other or join the same social organizations. Hence, Thornton argues that the endurance of African cultures in the Americas was much more widespread and powerful than Mintz and Price suggest.

However, there are points of agreement among these scholars as well. Along with Herskovits and Mintz and Price, Thornton embraces the view that sizable regions of Africa share sufficiently similar cultural principles or cognitive orientations that such principles have endured as a major element of African-American cultures. Like Mintz and Price, Thornton emphasizes the internal dynamism of African-American cultures, relating it to African logics of communion with the divine—that is, the ability of spirit mediums to convey unprecedented messages from the divine. This tradition of discontinuous "revelation" has shaped African-American religions and promoted their tendency to change continually.

However, I am less committed than Herskovits, Mintz and Price, or Thornton to the view that bents, cognitive orientations, and underlying logics are what objectively constitutes the Africanness of African-American cultures. Such Africanness is also constituted by a genealogy of interested claims and practices, available for selective invocation as precedents. The naturalization of these claims and practices is situational and impermanent. In other words, both the explicit, formal practices and the underlying logics of African and African-American cultures can, in my view, change vastly without making those cultures objectively un-African. Rather, African and African-American actors—along with the white scholars who study them, the white nationalists who appeal to their loyalty, and the white slaveholders and officials who oppress them—have regularly debated the meaning of Africanness, claimed and/or renounced it in ways that no African or African-American people has been able to escape. This debate has powerfully shaped the history of African culture around the Atlantic perimeter.

The slave plantation, identified by Mintz and Price as the foundation of African-American culture, is probably the most useful context in which to regard African-American cultures as locally bounded and internally integrated. This context is, no doubt, important, even for scholars who believe that Africa before the slave trade was an equally important foundation. The present study supplements these foundational logics with a third sort of contextualization. The ongoing 19th- to 21st-century dialogue among the massive urban black populations of the Atlantic perimeter has, to my mind, done as much to constitute the Africanity and the creativity of these populations as has any ancestral African or plantation culture. The social contexts of not only Candomblé but also Dahomean/Beninese Vodun, Cuban Ocha, West African and Cuban Ifá divination, Rastafarianism, North American jazz, and black

Protestantisms all over the Anglophone Americas (to name just a few famous instances of Afro-Atlantic "folk" culture) have always had important supralocal, interethnic and cross-class dimensions. In all these traditions, African-American practitioners borrowed from, studied, and communicated with Africa (and strategically manipulated Africa's image) as they institutionalized their own African-American forms of solidarity and social hierarchy. An African-Americanist cultural history need not assume, even in the context of plantation slavery, that African Americans lacked a means of access to Africa. And they never lacked their own strategic priorities.

At stake in the Frazier-Herskovits debate is just such a strategic question. Were African Americans *dignified* by proof that their cultural legacy links them to a distant place with legitimate cultures of its own, or was evidence of African culture in the Americas also proof of African-Americans' *inability* to learn the dominant culture? Is proof of the absence of African culture in the Americas proof that Africans and, therefore, African-Americans have no legitimate cultures of their own, or proof that African-Americans are just as American and, therefore, just as entitled to the rights of citizenship as are the descendants of European immigrants (see Jackson 1986)?

At stake in the Brazilian debate is a slightly different political issue. Who is empowered to define what is "African" in these religions, and who stands to benefit from any given definition thereof? A point typically missed in these debates is the role of African-American (including African-Brazilian) agency in creating these forms of cultural representation and self-representation. Something greater than "collective memory," the endurance of African "logical principles," white-dominated local contexts, or the passive reception of ideas from the "dominant" race or class has shaped black Atlantic religion—that is, a transnational, Afro-Atlantic dialogue. In the chapters to come, I detail the lives of the African-American travelers, scholars, writers, pilgrims, merchants, and priests who have, since the mid–19th century, profoundly shaped these "folk" religions, who have done so in defense of their own interests, and who, in doing so, have sagaciously interpreted forces far beyond any single country of origin or country of destination.

Visual Portraits of Candomblé and Its Ọ̀yọ́ Counterpart

Contemporary Ọ̀yọ́ religion and Bahian Candomblé clearly share many of the same West African precedents. Yet these contemporary forms differ in fascinating ways. *Sex and the Empire That Is No More* (1994; forthcoming a) details how a changing translocal political economy and the strategic responses of West Africans have reshaped the forms and meanings of òrìṣà worship in the Ọ̀yọ́-ethnic town of Ìgbòho. The present volume charts the historical transformation of Bahian Candomblé. Candomblé combines Ọ̀yọ́ precedents with European, Native American, and African precedents from places other than Ọ̀yọ́. No less important, Afro-Brazilians have invented forms and logics of worship all their own. And all these dimensions of cultural reproduction have been constrained and motivated by the political economy of Brazil and of the broader black Atlantic.

The following visual portraits lay the groundwork for recognizing what is useful about these already well-trodden theoretical paths and extends the path toward the more controversial argument of this book: the political economy, iconographic vocabulary, and interpretive discourses that Candomblé has produced and been produced by have never been Brazilian alone, African alone, or Brazilian and African alone. They have always been radically transnational and, particularly, circum-Atlantic, even in the middle of the 19th century.

This case calls for us not to overlook what once seemed clear enough to explain in terms of African "survivals," "retentions," "deep-seated drives," "cognitive orientations," and "logical principles," or in terms of African-American "reinterpretations," "syncretism," "collective memories," and "creolization." Rather, mine is a call to recognize the more encompassing geographical frames, political hierarchies, and networks of long-distance communication that have long made it impossible for cultures to reproduce themselves within the closure of a bounded, self-defining set of social relationships and meanings. Within such a frame, the involuntarism of cultural "survivals" and "cognitive orientations" often gives way to the *strategic* assertion of Africanness in a trans-Atlantic commercial, cultural, and political arena.

Let me offer a few vivid illustrations of the similarities and differences between Brazilian Candomblé and èsìn ìbílẹ̀, or "traditional religion," in a rural Ọ̀yọ́-ethnic town in West Africa. Figures 1 through 4 represent a present-day religion of spirit possession, blood sacrifice, divination, and healing in the Nigerian town of Ìgbòho, which lies in the historic heartland of the Ọ̀yọ́ kingdom and ethnic group. At the end of the 16th century, on the eve of Ọ̀yọ́'s growth into the largest royal empire in the precolonial past of the Yorùbá, Ìgbòho had been the capital of the Ọ̀yọ́ kingdom. Elevated atop earthenware pots, the two calabashes on the altar in figure 1 contain the river stones and shells that represent the river goddess Yemoja. Next to them sits an earthenware pot containing the thunderstones of her son Ṣàngó, god of thunder and lightning. Priestesses present food offerings and the blood of sacrificial chickens to these vessels to strengthen the gods and guarantee protection for the devotees.

Figure 1. The shrine of the goddess Yemọja in an Ọ̀yọ́-Yorùbá palace. The two calabashes contain the sacred emblems of the goddess, while the stones in the neighboring pot (right foreground) are consecrated to Ṣàngó. Photograph by the author (1988).

Priestesses use kola nuts and, during the New Yam Festival, quartered yam tips to divine whether the goddess is satisfied with the offerings and with the conduct of her devotees. Similar calabashes, earthenware pots, and wooden vessels containing stones and sacred emblems represent other gods—like the river goddess Ọ̀ṣun, the lord of gestation Ọbàtálá, and the farming god Òrìṣà Oko—on Ọ̀yọ́ altars.

Most possession priests in Ìgbòho and elsewhere in Yorùbáland are women; male possession priests wear hairstyles, jewelry, cosmetics, and clothing that suggest they are like women, in that they are "wives of the god" (Matory 1994b). Whereas newly initiated possession priests of Ṣàngó, like the one in figure 2, wear women's blouses, wrap skirts, baby slings, jewelry, and cosmetics, senior Ṣàngó possession priests like the one posing in figure 3, my friend Adeniran, braid their hair like women and, during festivals, wear skirts of embroidered or appliquéd panels called *wàbì*. Figure 4 shows the Ṣàngó of a priest named Ogundiran parading in classic *wàbì* during the festival. In figure 3, by contrast, Adeniran models *wàbì* of modern, machine-embroidered velvet.

Counterparts of Ṣàngó and Yemọja are the most widely and publicly worshiped "African" gods in Brazil. Yet certain features of the West African mythology of these West African gods would surprise Brazilians,

Figure 2. A recent initiate, or "bride" (*ìyàwó*), of the Ṣàngó. He wears the blouse (*bùbá*), wrap skirt (*ìró*), and baby sling (*òjá*) of a Yorùbá wife. Photograph by the author (1988).

who tend to see "purity" as a primordial feature of Yorùbá religion. For example, West African priests describe both of these gods as Nupe (a non-Yorùbá people) or as children of local non-Yorùbá Bariba dynasties' intermarriage with Nupe women. Indeed, the mythology and panegyrics of the *òrìṣà*, or gods (pronounced OH-REE-SHAH), report

Figure 3. The rear view of Ṣàngó possession priest Adeniran wearing the *wàbì* skirt. Note also his braided hair. Photograph by the author (1995).

the gods' northern or Near Eastern origins. Ṣàngó himself is said to be a light-skinned Muslim, albeit an iconoclastic one. The savanna region in which Ìgbòho is located has had a significant Muslim and/or Arab population since the 16th century, to which the iconography of Ṣàngó

Figure 4. Computer-digitized video image of Ṣàngó during a festival in Ìgbòho.
Video by the author (1988).

worship and even of the famous Yorùbá Ifá divination system appear to
owe a great deal. Hence, much that is classified as "traditional" (*ìbílè*)
in Yorùbá life is also highly cosmopolitan in its intentional referents and
its historically discoverable sources (Matory 1994a).

The oldest, largest, and wealthiest Bahian Candomblé houses tend to
identify themselves as members of the Quêto, or Nagô, nation. Nowadays, the West African cognates of "Quêto"—"Kétou" or "Kétu"—refer
to a specific Yorùbá-speaking town in the People's Republic of Bénin and
to a kingdom that cuts across the border of Nigeria and the People's
Republic of Bénin. The West African cognates of "Nagô"—"Nagôt,"
"Nagô," or "Ànàgó"—refer either to a specific Beninese Yorùbá group
or, in Beninese parlance, to the Yorùbá speakers as a whole. In Brazil,
too, both "Quêto" and "Nagô" have come to be equated consciously
with the inclusive Yorùbá ethnic group. Cubans identify the same inclusive Yorùbá group, including its Latin American diaspora, as "Lucumí."

Figures 5 and 6 depict important buildings on the grounds, or *roça*, of a
Brazilian Candomblé house called Ilê Axé Opô Afonjá. The building in
figure 5 is dedicated to the Brazilian counterpart of Ṣàngó, called "Xangô,"
and is named in honor of Àfònjá, who, having been a historically important Ọ̀yọ́ royal, is in Brazil regarded as an avatar (*marca*) of Xangô.

Figure 5. Side view of the Xangô house at the Bahian Candomblé temple Ilê Axé Opô Afonjá. Photograph by the author (1987).

Indeed, Johnson identifies Àfọ̀njá as the Ọ̀yọ́ royal military chief whose greed to wear the crown resulted in the collapse of the Ọ̀yọ́ Empire and the departure of hundreds of thousands of Ọ̀yọ́ people for the Americas (Johnson 1921:188–90).

Among Brazilian temples, Ilê Axé Opô Afonjá is said to be the most African or most "traditional"; for the same reason, it is the most respected and envied. Like other Candomblé houses of the Quêto/Nagô nation, Opô Afonjá worships gods generically called *orixás* (oh-ree-SHY-eesh) after the Yorùbá term *òrìṣà*, which are also identified individually by names thoroughly recognizable to any contemporary West African Yorùbá person. Among them are not only the thunder god Xangô (cf. "Ṣàngó") and the sea goddess Iemanjá (cf. "Yemọja," a river goddess in Yorùbá-land) but also the lord of pestilence Omolú (cf. "Ọmọlúayé," a West African praise name for Sọnpọnnọn), the hunter god Oxossi (cf. "Ọṣọ́ọ̀ṣì"), and the god of iron Ogum (cf. "Ògún"). Brazilian priests "seat" (*assentam*) these gods in ceramic, wooden, or calabash vessels containing herbally treated and blood-fed stones and shells, much as West African priests do in Ìgbòho and other Yorùbá towns.

Though most Brazilian houses identify the Catholic saints to one degree or another with the African gods, the head priestess of Opô Afonjá, Mãe

Figure 6. The *barracão*, or festival hall, of Ilê Axé Opô Afonjá.

Stella, is famous for her public denial that the Catholic saints have any legitimate role in Candomblé. She also denies that the Brazilian Indian spirits called *caboclos* (kah-BOH-kloosh) have any place in her house. In figure 7, she is seen on the occasion of her most widely cited pronouncements against "syncretism," at the Third International Congress of Orisha Tradition and Culture in New York City (Azevedo 1986). At the far left of the photo stands my dear friend Gracinha of Oxum. A further index of Opô Afonjá's "African purity," by Bahian standards, is that the head priestess forbids the filming of sacred festivals (*festas*) in the house. Thus I had to find other houses that would allow me to film a festival. More anon.

In Bahia, the dying Jeje nation (whose worship of the *vodum* (voh-DOO[n]) gods identifies it with the Ewè, Gèn, Ajá, and Fòn speakers neighboring Yorùbáland) is closely associated with the Quêto/Nagô nation. The ritual protocols, songs, and ritual language of the Jeje and Quêto/Nagô nations are so profoundly indebted to each other that local ethnographers describe them as one "Jeje-Nagô" ritual complex (e.g., Costa Lima 1977). There are only a few Jeje houses in Salvador, and even the most respected among them lack the size, name recognition, and national influence of the

Figure 7. Mãe Stella of Ilê Axé Opô Afonjá at a conference of *òrìṣà* priests in New York City. Photography by the author (1986).

so-called great Quêto/Nagô houses. Ironically, insofar as the Jeje temples remain "purely African" (which also entails the idea of their resisting assimilation to more populous nations), they also guarantee the eventual death of their own nation. Distinctively Jeje songs are said to be so obscure and complex that normal audiences cannot answer them in the way that proper worship for the gods and a lively festival atmosphere require. Thus, Jeje houses tend to perform large numbers of the highly recognizable songs of the popular Quêto/Nagô nation, thus progressively renouncing their distinctness from the larger Quêto/Nagô nation.

The third major nation in contemporary Bahia is Angola, which has largely absorbed the Congo nation. It claims origins among West-Central African Bantu speakers, who outnumbered West Africans among the captives taken to Brazil in general but were far outnumbered in the state of Bahia by captives from the West African Bight of Benin (Eltis and Richardson 1995:31; see also Curtin 1969; Verger 1976[1968]). Houses of the Quêto/Nagô nation are widely recognized as "more African" and more worthy of respect than are the Congo-Angola houses.

Therefore, even soi-disant Angola houses regularly imitate Nagô ritual practices and must, in order to make themselves understood by the general public, define the *inquice* (cf. the KiKongo *nkisi*, meaning "god" or repository of power) as counterparts of one or another Nagô *orixá* (see, e.g., Paixão 1940). Insofar as it performs its songs and key rituals according to standards regarded as faithful to "Angolan," or supposedly West-Central African, prototypes, even an Angola house might be recognized as "purely African." But virtually all the sacred songs of the Angola repertoire are full of Portuguese phrases. Hence, it is precisely the houses that are least preoccupied with "African purity" that are most likely to call themselves "Angola" and to allow filming. Plenty of smaller and geographically marginal but, by local standards, well-administered and ritually orderly houses allow filming, but upward mobility in the world of the "great houses" (*grandes casas*) with name recognition and high-bourgeois clienteles tends to involve prohibitions against filming and culminate in decampment from the Angola to the Quêto/Nagô nation.

Figure 8 is a computer-digitized image of a festival for Xangô's wife— the goddess of wind and storm, Iansã—in the small but elegant and

Figure 8. The goddesses Oxum (left) and Iansã (right) dancing at a festival in Mãe Bebé's house. Note their billowy skirts and the hybrid style of their crowns. Video by the author (1987).

category of spirits so named in Candomblé includes an ethnic array ranging from Tupi Indians to mixed-race cowboys, Turks, and Gypsies. Though defiant in her public celebration of her *caboclo*, Dona Maura makes her house no exception to the pattern of *anagonização* (anagonization)—or the juggernaut expansion of Nagô influence—that we have already seen in the prevalence of Nagô ritual forms and mythical references in the Angola and Jeje nations. Dona Maura classifies her house as Quêto, and even her *caboclo* asked me to bring the priestess an entire outfit of the costly Yorùbá woven cloth—"cloth from the Coast"—when I returned from my next trip to Africa. In Brazil, "cloth from the Coast" is conventionally the preserve of senior priestesses (*ebomins*) of the Quêto nation, and such cloth is worn only around the waist. I gladly complied with Pena Branca's request.

As they grow in size and/or seek recognition in the world of the "great" Quêto houses, upwardly mobile Candomblé houses tend to decrease sharply the public attention they devote to their *caboclos*. For one, *caboclos* lack the authority of being "African" or "traditional" and therefore respectable according to the definitions of the ranking Quêto houses, the newspapers, the tourist agencies, and the public authorities that in recent years have periodically funded temple renovations. But, more important in the life of the average Candomblé house, the *caboclos* represent the kind of charismatic leadership that is useful in the hiving off, founding, and early expansion of a house. On the other hand, such independent leadership is threatening to the routinized and institutionalized authority of the senior priests in a well-established house or centralized family of houses.

In Bahia, the difference between *caboclo* Indian spirits and the African *orixás* is summed up in the dictum that "*caboclos* have no mother or father," which quality is nearly sacrilegious in those houses where the ranking authorities are themselves addressed as "Mother" (Mãe) and "Father" (Pai). The ranking authorities are "mothers" and "fathers" in the sense that they have given birth ritually to their followers through the act of initiation, which then confers upon them absolute authority over their "children" (*filhos*). Hence, the initiates of a priest are called his or her "children-in-saint" (*filhos-de-santo*), and the initiating priest is called their "mother-in-saint" (*mãe-de-santo*) or "father-in-saint" (*pai-de-santo*). Like only some *orixás*, *caboclos* start possessing their mediums spontaneously. Yet *orixás* expect their mediums eventually to be initiated at the hands of a "mother-" or "father-in-saint." *Orixás* will severely punish a medium or a called person who delays initiation for too long. On the other hand, the mediums of *caboclos* require no initiations at all. They dare to open up a house and practice their profitable healing craft without authorization by anyone. Unlike *orixás, caboclos* require no training to dance properly or to speak the sacred words of their craft. The language in their land of origin is but a modified "backwoods Portuguese" (Wafer 1991:65–68).

Orixá festivals are full of gestures of obeisance and gestural indications of rank. Whereas the *orixás* typically *follow* the directions of the drums and of the senior priests, looking lost and on the verge of keeling over when the drums stop, the *caboclos* obey no one. They run the show. Thus, without the elaborate paraphernalia and costly drum ensembles characteristic of *orixá* festivals, *caboclos* can host sessions every week and draw the same loyal and ever-expanding crowds. *Orixás* are infinitely more costly to feed, dress, and celebrate. Their worship relies not only on interclass patronage but also on an elaborate commerce that is coextensive with the typical professional engagements of the "Africans" in the Bahia of the 1930s—butchery, the vending of "African" food and herbal medicine, dressmaking, laundering, and the importation of commodities (such as palm oil and kola nuts) from Africa. Candomblé relied on and subsidized the personal fortunes of many an "African" merchant and craftsperson during the century culminating in the 1930s.

In a move recalling the apologetics of the Lagosian tract writers of the 1890s, Mãe Aninha told a visiting anthropologist in the late 1930s that her religion was much like Old Testament religion, that it was a more authentic form of Christianity on the grounds that Candomblé-members "worship nature" rather than the human-made images used in Roman Catholicism (Pierson 1942:294). Decades later, sympathetic outsiders and insider spokespeople have expanded and centralized this message, summarizing Candomblé and its African antecedents as "the worship of nature," which, to my mind, is more or less like describing Christianity as "the worship of the sky," since, after all, the Christian high god and his son are said to reside there.[6] Such summations delete or contradict much of the content of these religions, despite their selective truthfulness. Though many of the chief gods of the Candomblé are associated with rivers, the ocean, lightning, the rainbow, or the wind, for example, some equally important gods are associated chiefly with very human, social, or technological phenomena, such as war, revolution, and iron, the hunt, or medicine. And *all* the gods are simultaneously represented as royals and nobles of the most civilized kind. They are arrayed in the crowns, swords, jewels, money, and sumptuous clothing associated with their aristocratic and cultured social class. These gods are also all associated with human "nature" (also *natureza* in Portuguese)—that is, the diverse personality and somatic types with which each person has been born.

The summation that Candomblé is the "worship of nature," however, most effectively evokes environmentalist sympathy and the Western

Enlightenment logic that "savages," "primitives," and lower-income Westerners are close to "nature" in their often admirable and sometimes abhorrent ways. This is not to deny that Candomblé is the "worship of nature"; the phrase has been repeated so often in and around the Candomblé that it is part of the emic reality. I am saying, however, that this gloss is too specifically ideological a claim to suffice as an analytic or ethnographic description.

A more comprehensive summation of Candomblé's rituals and motifs cannot fail to mention its themes of spirit possession, animal sacrifice and food offerings, cowry-shell divination, herbal medicine, and metaphysical healing. Underlying these operations is a repeated logic of purification—of removing impurities and heat from the body (and simultaneously from the ritual vessels that metaphorically represent the body), and of inserting or developing contents that associate the person more purely with the gods, with the temple, with the family of temples that ensures its legitimacy, and, ultimately, with its diasporic African "nation." Food and sexuality are the major idioms in which the purity of the person and of the sacred lineage, or *axé*, is dramatized.

My point is not to disprove that many ritual forms and underlying ritual logics have "survived" the trans-Atlantic slave trade and slavery itself. Rather, it is to prove that central features of local linguistic and ritual practice, as well as the meanings and motives that believers invest in them, resulted from a long-distance dialogue with colonial Africa and with other American locales, much of which took place after both the slave trade and slavery had ended.

This book is dedicated to Pai Francisco Agnaldo Santos of Bahia, Brazil—high priest, or *doté*, of the Jeje nation. He has been my friend and mentor since we met in August 1987 and is likely to be blamed for any possible error or indiscretion appearing in the following pages. It is therefore important to point out that my sources of information have been numerous. My interpretation of Candomblé is not Pai Francisco's. I have done my best to follow my sincerest understanding of what my friends and teachers thought should be kept secret. However, it has been my scholarly obligation to comment on some such matters that have already been published by others. I am moved by matters far more important than formality to assure the reader that any errors and indiscretions here are my own, and that Pai Francisco is chiefly responsible for whatever virtue lies between these covers.

I owe an additional and profound debt of gratitude to the vast networks of people, funding agencies, government bureaucracies, travel agencies, airlines, postal services, telephone companies, international committees, religious alliances, and universities that have made possible our complex and enduring transnational relationship. For their generous hospitality and willingness to instruct me in their rich legacy, I thank Ẹlẹ́gùn Àgbà Adeniran, Ẹlẹ́gùn Ṣangodara, Ẹlẹ́gùn Ṣangobunmi, and Mọ́gbà Ìyá Elégbo, all of Ìgbòho, Nigeria, and Ẹlẹ́gùn Àgbà Ogunnikẹ Mosadogun of Ìdí Arẹ̀rẹ, Ìbàdàn. Equally important are my Brazilian friends and teachers Dona Maura, Pai Xico de Oxum, Pai Amilton, Mãe Sylvia, Pai Cate, Pai Mauro, Dona Elsa, Pedro and Kylla, Rita de Oxum, Claudecir, Marinalva, Lucas, Cida, Gracinha, Dona Detinha, and Nidinha.

I joyfully thank a number of institutions as well: the National Science Foundation, the Committee on Institutional Cooperation, the Social Science Research Council, the W. E. B. Du Bois Institute and the Faculty of Arts and Sciences at Harvard University, and the National Endowment for the Humanities for their generous support of my fourteen months of field research in Brazil—in 1987, 1992 and 1995–96—the best part of which has been my ever-deepening friendship with Pai Francisco and the other devotees of the Candomblé. I am also grateful to Williams College, Brown University, and Harvard University for supporting Pai Francisco's travels to the United States, where he was able to teach my students far more about black Atlantic religion and cultural diversity than I could have, and to those students of mine who have subsequently visited him in Bahia, adding not only continuity but also new dimensions to our shared transoceanic life.

Finally, I must thank a number of people and organizations that facilitated my historical research. Foremost among these is the National Endowment for the Humanities, which funded archival research in Europe, West Africa, and Brazil in 1995 and 1996. I received extensive guidance from J.D.Y. Peel and Elysée Soumonni, as well as apt advice from Sandra Barnes, Kristin Mann, Ọlabiyi Yai, João José Reis, David Eltis, Robin Law, Antônio Risério, Rebecca Scott, Paul Lovejoy, Roland Abiọdun, John Thornton, Toyin Falọla, Phil Zachernuk, Jeferson Bacelar, Manuela Carneiro da Cunha, Roberto Motta, Karin Barber, J. Michael Turner, Michael Herzfeld, Chris Dunn, Matthew Gutman, Ian Baucom, Charles Piot, Engseng Ho, Doris Sommer, Richard Price, Patrick Manning, and Sherry Ortner, as well as the students of my "Afro-Atlantic Religions" course in 2003. I am most grateful for their assistance. I am additionally indebted to Chris Dunn, not only for his invaluable friendship but also for the use of his beautiful photography.

I could have accomplished little without the help of archivists and librarians at the Public Record Office (Kew, U.K.); the British Library (London); the Church Missionary Society Archives at the University of Birmingham (U.K.); the School of Oriental and African Studies at the University of London; the National Library of France (Paris and Versailles); the National Archives of France (CARAN, Paris); the Society of

African Missions (Rome); the National Archives of Nigeria (Ìbàdàn); collections, including the Herbert Macaulay Papers at the University of Ìbàdàn Library, the National Archives and National Library of Brazil (Rio), the General Archives of the City of Rio de Janeiro, the Public Archives of the State of Bahia (Salvador); the archives of the *Tribuna da Bahia*, the *Tarde da Bahia*, the *Estado de São Paulo*, the *Folha de São Paulo*; the archives and library of Bahiatursa (the Bahian state office of tourism) in Salvador; the Secretariat of the Environment (Prefeitura de Salvador); and the Lajuomi Memorial, housing the books, letters, and mementos of the late Iyalaxé Caetana Sowzer Bangboxê. Among the archivists, I am particularly indebted to Bernard Favier of the Society of African Missions, whose vast knowledge is matched only by his kindness. I also wish to thank Alphonse Labitan of the National Archives of the People's Republic of Bénin (Porto-Novo), as well as Anna Amélia Vieira Nascimento and Mercedes Dantas Guerra of the Public Archives of the State of Bahia.

Equally indispensable were the guidance and historical information I received from the living descendants of some great Afro-Atlantic travelers and early priestesses of the Candomblé. My most special thanks go to Dr. George Alakija, Aïr José Sowzer de Jesús (Aizinho de Oxaguiã), Albérico Paiva Ferreira, and Beatriz da Rocha of Salvador, Bahia, as well as Paul Lọla Bamgbọṣe-Martins and Yinka Alli Balogun of Lagos. To preserve the confidentiality of my friends and teachers, I have changed some names here and elsewhere in the book. I hope, however, that all of them will recognize themselves in these few words, which could not possibly capture the depth of my gratitude.

REVISING THE HISTORY OF BLACK ATLANTIC RELIGION AND CULTURE: ON THE CHAPTERS THAT FOLLOW

So how would a culture and an ethnography of it look if we recognized the long history of transnationalism that has shaped them? This book is offered as both a theoretical proposal and an empirical illustration. It is an effort to understand one Afro-Brazilian religion—Candomblé—both ethnographically and historically amid the transnational flows of goods, people, and ideas that it has shaped and by which it has been shaped. Archaeologists have long studied the role of river basins in the genesis of civilizations, such as Egypt, Mesopotamia, and China. Other sorts of navigable open space serve better to explain the cultural and political might of Greek, Roman, Sahelian, Anglo-Saxon, Swahili, and Yorùbá civilizations. Thus, I situate Candomblé in a supralocal geographical context that is more familiar to historians than to anthropologists, a sort of context that I call "circum-oceanic" or "circum-desert" fields. Like the Indian Ocean and the great Afro-Asiatic desert at the center of the Islamic world, the Atlantic has for centuries been the focus of enormous flows that cross it and of a veritably global array of tributary flows beyond it.

Though these flows have hardly been an exchange among equals, I am anxious to refute what I take to be two reasons for which enslaved Africans and their descendants are so often overlooked in current theorizing about transnationalism. First, it has been easy for theorists impressed by the post–World War II flood of Africans, Asians, Caribbeans, and Latin Americans into Europe and the United States to forget that flows among Africa, Asia, and Latin America are centuries old and, in many cases, momentous. We must not ignore South-South transnationalism. Second, Africans and their descendants were neither passive nor marginal to the processes by which the territorial nations were "imagined" into being (Anderson 1991[1983]). Nor, as ambivalent participants in these normally racist natural processes, did Africans submit their imaginations entirely to the images of community propagated by nation-states.

The historical ethnography that follows describes the multiple "imagined communities" that overlap among the devotees of the Afro-Brazilian Candomblé religion, the textual forms in which they are articulated, the commercial forms that make different imaginations of community profitable to different degrees, and the ritual forms through which they are embodied. Like many of my informants and their friends, I have, for decades now, experienced this circum-Atlantic field through written and oral history, newspapers, ritual, and travel. These have been important media of collective life in Candomblé in ways that are here documented from the first half of the 19th century. And they continue to provide the idioms in which life is locally negotiated in that religion. At this locus classicus of African diaspora scholarship, I hope to demonstrate, then, that what has long been seen as the most conspicuous passive "survivals" of African culture in the Americas or as instinctual African "bents" are in fact products of ongoing human agency in a circum-Atlantic field, the overall form of which I call the "Afro-Atlantic dialogue."

The evidence of this dialogue lies not only in the passport records of multiple nations and empires but also in the preeminent print media of national consciousness—the newspapers. Alongside these sources, multiple generations of ethnography and nationalist folklore studies of the Nigeria, Brazil, the Bénin Republic, Cuba, and the United States are used here to reveal that the peoples of the black Atlantic never simply embraced nation-states as sufficient indices of their collective identities. Nor did they suddenly find liberation from them amid a thirty-year-old transnationalism. Rather, they have always made strategic and situational choices about the long-distance and territorial communities in terms of which they imagined themselves.

Overall, the chapters that follow are organized to proceed from the mid–19th century to the early 21st, from the historical to the ethnographic to the biographical, and from macroscopic patterns to the personal experiences of the interlocutors I met in this translocal dialogue. The discursive styles of these chapters will necessarily change apace.

Chapter 1 explains the trans-Atlantic genesis of the allied peoples known as Yorùbá in West Africa, Nagô in Brazil, and Lucumí in Cuba, as well as the role of literacy—and of the African cultural nationalism known as the "Lagosian Cultural Renaissance"—in making them the most prestigious of Afro-Atlantic nations. Diasporas, I argue, are regularly responsible for creating the cultures and communities that they call their homelands. This chapter explores the ongoing function of self-conscious "purity" in a world that students of the postcolonial have seen as dominated by the "hybrid" and the "creole." I will argue that the tension between these idioms of collective identity is structural to social life in a transnational context.

Chapter 2 tacks between two genres: straight historical investigation and theoretical debate. Hence, on the one hand, this chapter reconstructs the historical origins of the mysterious Jeje nation in Bahia, which had not been known in West Africa until after the end of the slave trade. I examine the role of the rivalry between French and British imperialists in the brief success of this identity on the West African "Coast," as well as the apparent role of trans-Atlantic black merchants in resurrecting and reforming a dying nation in the Bahia of the 1920s and 1930s. On the other hand, this chapter confronts the juggernaut conviction in today's academy that transnationalism is qualitatively and entirely new. This chapter both critiques that claim and, by actually setting it alongside a detailed historical analysis of the past, dramatizes what is systematically missing from the conventional wisdom.

Chapter 3 explains how the themes of African racial and cultural "purity" propagated by the literati of the 19th-century Lagosian Cultural Renaissance and by the trans-Atlantic merchants of the early 20th century are now embodied in Bahian lives through Candomblé ritual practice.

Chapter 4 explores the subnational politics of Regionalism that have motivated the Brazilian nation-state's embrace of Candomblé, and particularly of its most transnational forms, as symbols of the Brazilian nation in the sphere of international public opinion.

Chapter 5 argues that "matriarchy" in Candomblé arose not from "African tradition," as is usually supposed, but from the Margaret Mead–like rhetorical strategies of transnational feminist Ruth Landes and from the international embarrassment of Brazilian nationalist culture brokers over Landes's revelation of the local view that most of the male priests were "passive homosexuals." This chapter illustrates how the propagation and protection of such open secrets define the communal boundaries and hierarchies of overlapping imagined communities—including those of the territorial nation, transnational feminism, and the new Nigerian Yorùbá diaspora in the United States.

Rather than relegating the autobiographical sources of my perspective to the introduction, I have chosen to place them where they belong in the chronological order of this cultural history—at the end. However, my birth, my institutional involvements, my profession, and my translocal friendships are the backdrop, the main proof, and the conclusion of the story I tell about a centuries-old black transnationalism. Chapter 6 could therefore be read first or last among the six main chapters. It might be read as a deeply personal narrative of events that led me to reanalyze translocal history as I do, or it might be read as a dense and ultimate illustration of the thematic points made in each of the previous chapters. Chapter 6 reveals, then, how Candomblé's ritual idioms, the towering prestige of the highly literate Nagô/Yorùbá nation, various old and new transnational flows, Brazilian Regionalism, and the progressive feminization of Candomblé's leadership since the time of Landes's intervention have defined the imaginary and material possibilities in two lives—my own and Pai Francisco's.

I conclude the book with chapter 7, which argues for the utility of circum-oceanic or circum-desert "dialogue" as an analytic metaphor, illuminating a whole range of historically important transnationalisms and translocalisms that have been overlooked in the cultural history of the African diaspora, in the ethnography of every world region, and in the theory of globalization generally.

CHAPTER ONE

The English Professors of Brazil

ON THE DIASPORIC ROOTS OF THE YORÙBÁ NATION

"My religion is pure Nagô."
—MÃE ANINHA, founder of the Ilê Opô Afonjá Temple[1]

ARE HOMELANDS TO DIASPORAS as the past is to the present? Is Africa to the black Americas as the past is to the present? Much in the conventional writing of cultural history would suggest that it is. This chapter both revises existing narrations of African-diaspora cultural history and proposes, at the Afro–Latin American locus classicus of Herskovitsian studies, some nonlinear alternatives to Herskovits's and others' visions of diasporas generally. I argue here that slavery and African-American isolation from the major trends of Euro-American culture were probably not the typical conditions under which African-inspired culture flourished in the Americas (see also Matory 2000).

The empirical terrain of this revision is the Jeje-Nagô, or Fòn- and Yorùbá-affiliated, temples of the Brazilian Candomblé religion, and their supposedly West African origins, which have become a locus classicus in the study of African "collective memory," "retention," and "continuity" as the mechanisms of community formation and cultural transmission in the African diaspora. These temples are cited more often and with greater certainty than any other African-American institution as proof that African culture has "survived" in the Americas (e.g., Raboteau 1980[1978]; Bastide 1967; Herskovits 1966[1955]:227). The formal and lexical parallels between the Jeje-Nagô Candomblé and the contemporary religions of the West African Fòn and Yorùbá are indeed impressive, as trans-Atlantic researchers M. J. Herskovits, Roger Bastide, William Bascom, Pierre Verger, Mikelle Smith Omari, Robert Farris Thompson, Margaret Drewal, Deoscóredes and Juana Elbein dos Santos, and I have agreed.

One of the sharpest implicit challenges to the Herskovitsian project has come from a leading figure in cultural studies. Although Paul Gilroy's book *The Black Atlantic* (1993) borrows extensively from the anthropological and Herskovitsian lexicon (employing such terms as "syncretism," "creolisation," "ethnohistory," and the term "black Atlantic" itself), Gilroy overlooks these debts and appears to dismiss the question of the diaspora's cultural and historical connection to Africa as "essentialist." Instead, Gilroy describes the African diaspora primarily in terms of a *discontinuous cultural exchange* among diverse African-*diaspora* populations. Drawing examples primarily from the English-speaking black populations of England, the United States, and the Caribbean, Gilroy's *Black Atlantic* argues that the shared cultural features of African-diaspora groups generally result far less from shared cultural memories of Africa than from these groups' mutually influential responses to their exclusion from the benefits of the Enlightenment legacy of national citizenship and political equality in the West.

Gilroy usefully gives new salience to the role of free black Atlantic travelers and of cultural exchanges among freed or free black populations in creating a shared black Atlantic culture and shared black identities that transcend territorial boundaries. This approach is anticipated in Brazilianists' study of the ongoing, two-way travel and commerce between Brazil and Africa.[2] Yet the Brazilianists who briefly attend to the cultural consequences of that travel and commerce for Brazil have tended to assume their *preservative* or *restorative* effects on the "memory" of an unchanging African past, rather than their *transformative* effects on both Brazil and Africa (see R. N. Rodrigues 1935[1900/1896]:169; Bastide 1971[1967]:130; 1978[1960]:165, Elbein dos Santos and Santos 1981[1967]).[3]

The case to be discussed in this chapter—that is, the historical connections between Africa and what is often described as the most "purely African" religion in the Americas—will demand a reunion of these models of the African diaspora. It suggests that greater space be given to African agency, or deliberate action in making the world (which is neglected in Herskovits's and Gilroy's models), and to African cultural history (which is neglected in Gilroy's alone). Both African agency and African culture have been important in the making of African diaspora culture, but more surprisingly, the African diaspora has at times played a critical role in the making of its own alleged African "base line" as well.[4]

The following revision of diasporic cultural history is based on the premise that Africa is historically "coeval" (Fabian 1983)—or contemporaneous—with its American diaspora. Herskovitsians and their allies distort important realities when they treat African cultures and practices as the "past," the "base line," the "provenience," the "origin," and the prototype of cognate American realities. The further premise of this revision is closer to the original sense in which Fabian employed the term "coeval." The West Africanist ethnographers and folklorists whom African-Americanists have tended to cite for information on the

"African origins" of New World practices are no more external to the politics of statecraft and knowledge in colonial Africa than are African-Americanists from the racial politics of the postslavery Americas. Not only traveling black pilgrims, businesspeople, and writers but traveling white writers and photographers, as well as their publications, have long been vehicles of transformative knowledge in the production of what Thompson (1983) and Gilroy (1993) call the "black Atlantic" culture.

I will argue that, aside from the introduction of the "culture" concept shared among E. B. Tylor, Franz Boas, and Boas's student Melville J. Herskovits, the greatest impulse behind the respectful study of African culture in the Americas occurred at the hands of Africans or, properly speaking, through a dialogue among West Africans and African-American returnees to colonial Lagos, now in Nigeria. I will argue that, to this day, neither African-American lifeways nor the scholarly discussion of them escapes the influence of the Lagosian Cultural Renaissance of the 1890s. Yet few scholars are aware of this movement's impact. Various literatures have masked its principles as characteristics of a primordial African culture, taken them for granted as natural dimensions of cultural memory, or mistaken them for the arbitrary preferences of Euro-American scholars.

THE BLACK ATLANTIC NATIONS: EXPLAINING
THE SUCCESS OF THE YORÙBÁ

Since the 19th century, one Afro-Latin nation has risen above all the rest—preeminent in size, wealth, grandeur, and international prestige. It is studied, written about, and imitated far more than any other, not only by believers but also by anthropologists, art historians, novelists, and literary critics. The origin and homeland of this trans-Atlantic nation is usually identified as "Yorùbáland," which is now divided between southwestern Nigeria and the People's Republic of Bénin on the Gulf of Benin (see maps A and B).[5]

Yet among its other major metropolises are Havana, Miami, Oyotunji (South Carolina), New York City, Chicago, Los Angeles, and Washington, D.C. These are the lands of the Lucumí. However, this trans-Atlantic nation possesses no greater American metropolis than Salvador da Bahia, which one priestess famously called the "Black Rome" (see map A).[6] Located in the Brazilian state of Bahia, this city is usually called "Salvador" or "Bahia" for short and is now the capital of African-inspired religion in Brazil and a pilgrimage site for practitioners and scholars from the United States, the Caribbean, Argentina, France, and

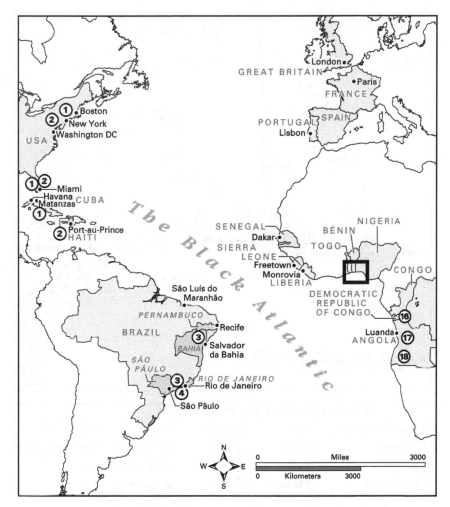

Map A. Cultures and capitals of the Black Atlantic world. Map by Adam Weiskind.

Nigeria itself. In Bahia, numerous Candomblé temples identify themselves as members of the nation they call "Quêto," or "Nagô," but they have integrated enough ritual practice and vocabulary from Brazil's second most prestigious nation—the Jeje—for scholars since Raimundo Nina Rodrigues to describe their *practice* as "Jeje-Nagô" (1988[1905]:215).

Though the equivalence among, for example, the Cuban Lucumís, the Brazilian Nagôs, the Haitian Nagôs, the French West African Nagots, and the British West African Yorùbá was not fully evident in their names, late 19th-century and early 20th-century ethnographers clearly

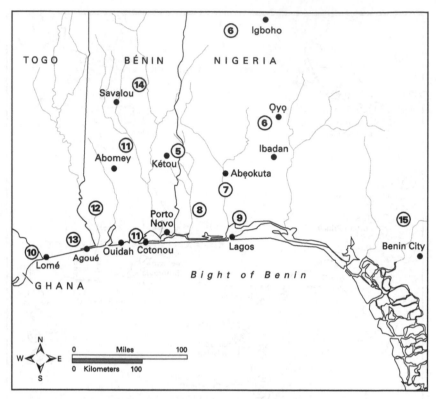

Map B. Inset—The Bight of Benin. Map by Adam Weiskind.

assumed that unity in a way that I will argue was consistent with the interests of a powerful class of black "ethnicity entrepreneurs" (see, e.g., F. Ortiz 1973[1906]:28, 32–33; R. N. Rodrigues 1935[1900/1896]:24; 1945[1905]:175; Ellis 1970[1890]:29–30).[7] The trans-Atlantic vicissitudes of the Jeje and Nagô nations, as well as their citizens' role in reshaping a range of the territorial nations that host them, open this new narration of an old transnationalism.

The *dominant* narration of African religious history in the Americas comes down to us through Melville J. Herskovits and his legion of followers. It is the story of slaves and of their descendants—separated by time, distance, and cruel fate from Africa—heroically remembering and preserving their ancient ancestral culture. Among the most successful were slaves in Bahia, Brazil (see map A), and their descendants, who, in the late 20th century, practiced what partisan observers call "purely African" rituals and sang in what such observers call "perfect Yorùbá." According to the received narrative, the descendants of the Yorùbá—members of the Nagô, or Quêto, nation—were so successful at preserving their primordial heritage that certain "houses" call themselves "purely African" or "purely Nagô."

The apparently extraordinary success of the Brazilian Nagô nation at preserving its African religion is paralleled by the success of the cognate Lucumí nation in Cuba (e.g., Ortiz 1973[1906]:28, 51, 52, 59, 60) and by the success of the Fòn-connected Rada and Mina-Jeje nations in Trinidad (Carr 1989[1955]), Haiti (e.g., Price-Mars 1990[1928]), and São Luís, Brazil (Ferretti 1985). Scholars have conventionally explained the success of the Brazilian Nagô nation in terms of an interaction among multiple factors. However, many of the factors cited rely for their explanatory value on the imputation of general causal mechanisms, nomothetic principles, or inductive patterns that, I will argue, are not borne out by the comparative literature and are not subject to any consensus among the various authors.

First, the Brazilian scholar Raimundo Nina Rodrigues and his followers offered the principal explanation that, at the time of the slave trade, the West African "Nagôs" had possessed a more organized priesthood and a more highly evolved and therefore more complex mythology than had the other equally numerous African peoples taken to Brazil. In Rodrigues's opinion, the Jejes, or Fòn/Ewè , ran a close second in evolutionary complexity (e.g., R. N. Rodrigues 1945[1905]:342; 1988[1905]:230–31).[8]

On the contrary, Robert Farris Thompson and Bunseki Fu-Kiau's discussions of Kongo cosmology suggest that what has come to be called "Yorùbá" religion possessed no monopoly on complexity (e.g., Thompson 1983:101–59). In any case, it is not obvious that *more* complex religions are more attractive to converts than are *less* complex ones. Christianity did not replace Greco-Roman religion because of Christianity's relative mythic complexity or the superior organization of its priesthood. Nor would such an explanation account for the spread of Buddhism and Islam. Whether simpler or more complex than Yorùbá culture, the Central, East, and southern African cultures of the Bantu speakers, including the BaKongo, are the products of a demographic and cultural expansion within Africa that dwarfs the transoceanic influence of the Yorùbá's ancestors. By the eighth century A.D., the Bantu languages had spread from a small nucleus in what is now Nigeria to Zanzibar, off the coast of East Africa (Curtin et al. 1978:25–30). Bantu speakers now dominate virtually the entire southern half of the African continent and have significantly influenced the music, religion, and language of the Americas as well (e.g., Kubik 1979; Vass 1979; Thompson 1983). Bastide 1978[1960]:60) attributes the decline of Bantu religion in Brazil to what he regards as its nearly exclusive focus on the patriarch's veneration of dead ancestors and the irrelevance of such a pursuit under American slavery, where the African family and its internal leadership have been torn asunder. Yet ritual negotiation with the spirits of the dead remains a prominent feature of *caboclo-* and Yorùbá-inspired Egungun worship in Brazil, as well as the Petwo nation of Haiti and the explicitly Bantu-inspired Palo Mayombe religion of Cuba, where the West-Central African–inspired *inkices* have entered into a powerful moral symbiosis with the West African–inspired *orichas* (Palmié 2002). Dealings with the dead, including Bantu-inspired ones, were adaptable to a wide range of African-American sacred projects. There is no systematic contradiction, across American locales, between cults of the dead and the regime of slavery.

Verger spearheaded a less popular, nonevolutionary explanation for Nagô success, attributing the strength of Yorùbá influence to "the recent and massive arrival of this people" in Bahia and to the presence of numerous Yorùbá captives "originating from a high social class, as well as priests conscious of the value of their institutions and firmly attached to the precepts of their religion" (Verger 1976[1968]:1). A related demographic explanation is that the arrival of the Brazilian Nagôs' ancestors in Brazil was indeed concentrated in the 19th century, at the last stage of the slave trade, making them the most recent and therefore least acculturated of major African ethnic groups in Brazil. They also immigrated in huge numbers. However, studies of other African-American cultures have indicated the disproportionate influence not of the last-arriving immigrant groups but of the *earliest*-arriving ones (e.g., Mintz and Price 1992[1976]:48–50; Creel 1987; G. Hall 1995[1992]; Aguirre Beltrán 1989a, 1989b). Moreover, the long-developing preeminence of Yorùbá/Nagô divinities in the Center-South (Rio and São Paulo [see map A]) did not depend on Nagô numerical dominance there.[9] Though the Nagô nation is the most prestigious and imitated in the Center-South, its African ancestors had never predominated among the Africans enslaved there. Likewise, Dahomean divinities and terms prevail in Haitian religion despite the fact that Dahomeans were always a minority among Haitian slaves. Thus, students of the diaspora disagree sharply over the causal relationship between numerical population and cultural predominance among African-American groups.[10]

Even where the Nagôs did predominate numerically, being the most *common* has never guaranteed that a particular subculture would become as prestigious as has Nagô religion. In fact, being "common" is precisely what *excludes* many practices from the canon of elite culture in societies all around the globe.

Other demographic factors might have contributed to the success of Nagô religion in Bahia. In 19th-century Bahia the Nagôs were disproportionately represented among *urban* slaves and among *negros de ganho*—that is, the slaves who freely moved about contracting work for themselves in order, then, to supply their masters with some agreed-upon portion of their income. Their freedom of movement also allowed a certain freedom to organize themselves and to commemorate their ancestral practices beyond the supervision of their masters.[11] It should be noted, however, that these explanations appear to contradict the overall pattern of explanation dominant in the pan-American literature. The Brazilian Nagô case defies the usual Afro-Americanist view that rural isolation and poverty are the normal conditions for the "retention" of African culture.[12]

Such contradictions may have contributed to the endurance of Rodrigues's evolutionary explanation. Rodrigues's sense that social and biological "evolution" had made the pre–slave trade Nagôs superior and thereby allowed their Brazilian descendants to preserve and diffuse their religion and identity was credible to generations of ethnographers in Rodrigues's train, including Arthur Ramos, Édison Carneiro, Ruth Landes, and Roger Bastide, as well as the laypeople who still unwittingly quote them.[13]

Advocates of a third and more recent model expressly deny the explanatory relevance of African cultural history and of the ethnic demographics of Brazilian slavery. Instead, they say, the supremacy of the Nagô nation is the product of arbitrary, local invention by whites since the 1930s. To the once-conventional claims of Nagô purity and superiority, these antiessentialists (Motta 1992, 1994; Henfrey 1981; Dantas 1982, 1988; Fry 1982; Frigerio 1983; D. D. Brown and Bick 1987; Wafer and Santana 1990) replied that these religions are "purely Brazilian," if they are anything pure at all. Brazilian Nagô religion has been reinvented in conformity to the arbitrary preferences of its white or light-skinned Brazilian patrons. Indeed, the prestige attached to the claim of "Nagô purity" is a mere artifact of Euro-Brazilian scholars' consent to protect from the police only those houses that embraced the *scholars'* definition of Africanness, which included a disavowal of "black magic." The added consequences of this "invention of Africa" and factitious "Africanization" of the Candomblé, the argument goes, were the devaluation of allegedly less "pure" Afro-Brazilian religious practices and the contentment of too many Afro-Brazilians with the full rights of citizenship only in some imaginary, otherworldly "Africa."

This new antiessentialist ethnography is persuasive to many right-minded critics of Brazil, where the state now publicly endorses certain Afro-Brazilian religions and performing arts, while colluding in the racist exclusion of blacks from political and economic power. These social constructionists have wrongly posited, however, that this scenario depends on the complete denial of black agency in this process of cultural change. A more carefully drawn history will, in my opinion, reveal the role of Afro-Brazilians themselves in creating and, indeed, inventing the standards of a trans-Atlantic culture, with consequences no less revolutionary in Africa than in Brazil.

VECTORS OF "AFRICAN PURITY": THE ENGLISH PROFESSORS OF BRAZIL

Let us consider, first, one of the more surprising distinctions of the people who were calling themselves "Yorùbá" in late 19th-century and early 20th-century Bahia. Indeed, some of these were also the informants who convinced the early ethnographers of Yorùbá/Nagô superiority and created the image of that nation's extraordinary survival and traditionalism. Many of the towering leaders and perennial informants quoted in the literature on Afro-Bahians from the 1890s to the 1940s were famous for their Anglicized names, English-language competency, or Anglo-Saxon social connections. Their contemporaries identified them, with no intentional irony, both as outstanding priests or exemplars of a "purely Yorùbá" tradition *and* as "professors of English" (e.g., Carneiro 1986[1943]:120; Andrade Lima 1984:7).

The first was Martiniano Eliseu do Bonfim, born on 16 October 1859 to freed African parents in Bahia. In 1875, his father took him to be educated in Lagos, West Africa, where he remained for eleven years. During that time, he once visited his parents in Bahia, and his father once visited him. In Lagos, he attended the Presbyterian Faji School, where all his teachers were Anglophone Africans. Though he never traveled inland, he drank deeply from the emergent Lagosian literature on "Yorùbá traditional religion" and underwent initiation as a *babaláwo*, or Ifá diviner, in Lagos between 1875 and 1886. His contemporary in the 1930s and 1940s, the Bahian journalist Édison Carneiro, reported that Martiniano not only was fluent in Yorùbá but also had visited England and taught English to financially comfortable blacks in Bahia (Carneiro 1986[1948]: 120; also L. Turner 1942; Frazier 1942; Braga 1995:37–55).

The last of the Afro-Bahian *babaláwos* was Felisberto Sowzer, who, like some other Brazilian returnees to Lagos, had Anglicized his originally Portuguese surname, Souza. His descendants tell me that he took pride in "*being* English," since Lagos was a British colony at the time and, in Brazil, he valued the distinction of being able to speak English. Carneiro reports that Felisberto, too, spoke fluent English and Yorùbá (Carneiro 1986[1948]:120).[14]

Like many Lagosian and Bahian *male* elites of his day, Felisberto was a Freemason. Indeed, the first Masonic lodge in Lagos was established in 1868 in the city's Brazilian Quarter. According to Ayandele, that lodge comprised "nearly all the African leaders" of Lagos (Ayandele 1966:268). Today, conspicuous borrowings from the iconography of Freemasonry— such as the unblinking eye, the inverted V, and the compass—unite Haitian Vodou, Cuban Palo Mayombe, the Trinidad Shouters, Nigerian òrìṣà worship, and the Nigerian Reformed Ogbóni Fraternity, which emerged in 1914 from the cultural nationalism of bourgeois Christians in Lagos (see figures 12 through 17).[15]

Felisberto proudly displayed the Masonic sign of his elite, trans-Atlantic affiliations right alongside his "Englishness" and his skills as an Ifá diviner. The side of his house in Bahia bore a seal featuring an Ifá divining board, the Masonic compass, the Yorùbá proverb *Suru ni oogun aiye* (meaning "Patience is the medicine of life"), and two biblical inscriptions in English—"The Lord is my Helper" and "Wait on the Lord and keep his way" (Pierson 1942:259; see my figure 12). Indeed, Sowzer is part of an impressive dynasty of Brazilian-Lagosian travelers and priests, beginning with his diviner grandfather from Ọ̀yọ́, Manoel Rodolfo Bamgboṣe, and ending with numerous priestly grandchildren in Lagos, Bahia, and Rio.[16]

Perhaps the only trans-Atlantic dynasty more famous than the line of Bamgboṣe is the Alakija family. Though they are not noted for any particular connection to the Candomblé religion, they are a central feature of latter-day recollections of the relationship between Bahia and the Africa from which Candomblé is said to have originated. Appearing in figure 13, Brazilian returnee Plasido Alakija (aka Sir Adeyẹmọ Alakija) was a District Grand Master of the Freemasons in Nigeria. His nephew and my main informant in the family, psychiatrist Dr. George Alakija, is also a Freemason. Sir Adeyẹmọ was also the first Lord of Lords, or Olórí Olúwo, of the Reformed Ogbóni Fraternity, an office he occupied until his death in 1953.[17] His nephew, Nigerian Chief Justice Sir Adetokunbọ Adegboyega Ademọla, succeeded him in that office (Anyebe 1989:108–13).

During his research in Bahia in the mid-1930s, Donald Pierson was shown a copy of the *Nigerian Daily Times* of December 1932. One of the articles therein detailed the professional accomplishments of the Alakija family, which included a lawyer, an otolaryngologist, and a civil engineer. Throughout the generations of their 20th- and 21th-century residence in Bahia, the Brazilian branch of the family has borne English

SYMBOLS OF THE SOOTHSAYER'S ART PAINTED ON THE
WALL OF A NEGRO DWELLING IN MATATU

Figure 12. Reproduction of the seal on the side of the home of Bahian diviner
Felisberto Sowzer, Salvador da Bahia, Brazil. From Pierson 1942[1939].

names like George and Maxwell (Pierson 1942:243). Indeed, even the
surname "Alakija" retains the *k* of Anglo-Yorùbá orthography rather
than the *qu* that would normally represent the same sound in Portuguese.

The defensive pride the African-Bahian travelers and their children took in
the Africa they knew is summed up in one remark recorded by Pierson:
"These people here in Bahia think Africans are all barbarous and uncivilized.

Figure 13. Plasido Alakija (aka Sir Adeyẹmọ Alakija) of Nigeria, wearing his Masonic attire (probably 1930s or 1940s). Photograph reproduced by the permission of Dr. George Alakija of Salvador, Bahia.

They won't believe we write our language and that books are printed in it. . . . They don't know that in Lagos there are good schools, better than they've got in Bahia. Look at this [showing a photograph of a school in Lagos]! Is there anything in Bahia as fine as that?" (Pierson 1942:272).

This genre of recollection of Africa was prominent in Bahia during the first half of this century. These are not memories of the pristine African or Yorùbá culture that the victims of American slavery *remembered* well enough to *retain* and *preserve*. Instead, these accounts are based on the experiences, souvenirs, and photographs of a class of literate and well-traveled Africans who, I will argue, helped to bring Yorùbá culture and Yorùbá traditional religion as such into existence, established their prestige around the Atlantic basin, and canonized them as the preeminent classical standard of African culture in the New World.

From the mid–19th century onward, what many Bahians remembered firsthand (or heard about from those who did) was a 19th- and 20th-century West African coast in which English language, Roman script, Masonic temples, and a lively press were the stuff of daily life. Their dialogue with early Brazilian ethnographer Raimundo Nina Rodrigues produced images of a cosmopolitan but, to their minds, still "purely African" culture. It was these images that prompted my own archival and oral historical search in 1995–96. Whatever "Africanness" and "purity" might have meant to these African-Brazilian travelers and their descendants, it was surely not *invented* by the Euro-Brazilian researchers who interviewed them and later advocated their cause.

In truth, we are now closer to understanding the cultural hybridity of the African-Brazilians who propagated Yorùbá culture in Bahia during the late 19th century and the early 20th. But we have yet to see how Afro-Brazilian talk and action helped to generate a so-called Yorùbá culture in West Africa that is in fact *younger than its Brazilian diaspora*. Whence derives the presumed unity of this trans-Atlantic culture? Whence derives the ideology of Yorùbá *superiority* to other African cultures and its persuasiveness to so many New World blacks, browns, and whites?

Thus the final question. If Euro-Brazilian scholars are *not* the source of the idea, whence comes the extreme value placed upon cultural and, at times, racial "purity" in the Brazilian Nagô nation? The Afro-Brazilian nation identified with the Yorùbá has flourished in a 20th-century Brazil dominated, since the 1930s, by the public valorization of cultural and racial *hybridity (mestiçagem)*, and the Brazilian Nagô nation claims descent from a West African religion that, by all evidence, has also long valued cultural borrowing and hybridity (Matory 1994a). Neither of the scholarly histories conventionally given for the emergence of the Candomblé—either as the survival of an ancient African religion or as the invention of the Euro-Brazilian bourgeoisie—would seem to predict the value accorded to racial and cultural purity from the mid–20th century onward. Where *did* that value come from?

"THE COAST": ON THE TRANSNATIONAL GENESIS OF THE YORÙBÁ

Brazilians have long called western Africa "the Coast" *(a Costa), pars pro toto*, and identified it as the classical origin of the finest in Afro-Brazilian culture. Indeed, the region of western Africa most frequented and occupied by Afro-Brazilian returnees was and is the West African seaboard between Freetown and Lagos. Elements of the regional society that emerged there in the 19th century had been precedented among inland political formations, cultural centers, and trade routes. For example, the kingdom of Ifẹ̀ was much admired as a spiritual and cultural capital by various peoples beyond its small domain, including the people of the Benin kingdom to the east and Ewè-Gèn-Ajá-Fòn speakers to the west. The kingdom of Ọ̀yọ́ spread its influence through conquest, government, and trade. However, as a political and cultural identity uniting Ifẹ̀, Ọ̀yọ́, and other groups, "Yorùbá"-ness was *created* in the creole society of the Coast, in a place and in a time that put it in constant dialogue with the nations of the Afro-Latin diaspora.

By the late 18th century, the Ọ̀yọ́ kingdom, whose population the Arabs and the Hausa called "Yarriba" (Clapperton 1829:4; also Law 1996:67–68; 1977:5; Awde 1996:14), had by fits and starts secured control over between a third and a half of what is now called Yorùbáland. The Ọ̀yọ́ Empire hosted a large and prosperous savanna population, which mediated between the trans-Saharan trade to the north and the trans-Atlantic trade to the south. The trade in Ọ̀yọ́ handwoven strip cloth across a wide belt of the West African interior is evidence of the extensive lateral projection of Ọ̀yọ́ cultural values as well (Aderibigbe 1975). Those kingdoms that Ọ̀yọ́ came to dominate militarily also came, in various degrees, to speak dialects highly similar to Ọ̀yọ́'s and to practice forms of worship deeply indebted to Ọ̀yọ́ ritual and mythology. With the fall of the Ọ̀yọ́ Empire around 1830, Ọ̀yọ́-ethnic military leaders in Ìbàdàn came to rule an even larger territory than the antecedent royal empire, and, throughout Ìbàdàn's sphere of influence, Ọ̀yọ́ gods became potent sources of symbolic power (see, e.g., Apter 1992:36, 51).

With its 18th-century capital and heartland more than three hundred miles from the Atlantic coast, Ọ̀yọ́ was not naturally predestined to supply the defining cultural emblems of the Coastal society. In fact, the kingdom of Benin (an ancient monarchy whose capital and current territory lie entirely in Nigeria, and which is largely unconnected to the present-day Republic of Bénin) was once and might have remained the culturally and politically sovereign power in the seaside city of Lagos. Until the 1790s, Lagos had been but a minor, outlying town in the sprawling Benin Empire.[18] Well into the 19th century, Lagos was so firmly under Benin influence and so peripheral to Ọ̀yọ́ imperialism that it had not been "Yorùbá" in the original, Hausa sense of that term or, as we shall see, in the modern sense that would later be introduced by Sàró and Afro-Latin returnees.

From the last quarter of the 18th century until the middle of the 19th, the decline of the Ọ̀yọ́ Empire and the rise of the Dahomean kingdom coincided to produce a tide of captives from Ọ̀yọ́, Ẹ̀gbá, Ẹ̀gbádò, Ilé ṣa, and nearby regions of what would become Yorùbáland, including a small western group called the "Nàgó," or "Ànàgó," which, being especially vulnerable to Dahomean predations, gave its name to the emergent Afro-Brazilian nation. They were captured and forced aboard ships at the Coastal ports of Lagos, Badagry, Porto-Novo, and Ouidah. Particularly large numbers of them reached Cuba and Brazil. As the slave trade grew in the late 18th and early 19th centuries, Lagosian merchants also grew wealthy, powerful, and increasingly independent of the Benin Empire (Aderibigbe 1975). The British colonization of Lagos in 1861 finalized the eclipse of Benin influence, as the Pax Britannica made Lagos the commercial and cultural Mecca of multiple and culturally diverse groups. Ìjèṣà, Ọ̀yọ́, Ẹ̀gbá, and especially Ìjẹ̀bú people flooded into Lagos as its economy grew. Thus, a combination of the slave trade and British colonization turned Lagos from a sleepy, seaside town into a metropolis. Moreover, the character of British colonialism in Lagos made that city into fertile ground for diasporic returnees' articulation of an Ọ̀yọ́-centered, pan-Yorùbá identity.

The role of the British in this cultural convergence was no more predestined than that of Ọ̀yọ́. By the mid–19th century, when, in the interest of suppressing the slave trade, the British first established political dominion over Lagos, several earlier centuries of *Portuguese* commercial activity on the Coast had already established a Portuguese-speaking belt along the West African littoral. That both Lagos and Porto-Novo still bear Portuguese names is symptomatic of the depth of the Portuguese presence. The rush of the British and the French to establish actual dominion over expanding areas of the region in the late 19th century created new forms of unity, as well as divisions, among towns and peoples. But there was no unifying force as great as the "return" of Westernized ex-slaves from Brazil, Cuba, and Sierra Leone to the Coastal towns near their ancestral homes and the *cross-regional* communication that those returnees set up among themselves.

The Ọ̀yọ́, Ẹ̀gbá, Ẹ̀gbádò, Ìjèṣà, Ìjẹ̀bú, and Nàgó captives who reached Brazil in the 18th and 19th centuries came collectively to be called "Nagô." They were initially grouped together not because they had identified with each other but because slave traders regarded them as similar to each other and collected them from the same ports of embarkation.[19] These captives, and especially the Muslims among them, were prominently involved in a series of insurrections and conspiracies in Bahia between 1807 and 1835. In small numbers, some freed people had returned to West Africa from the late 18th century onward, but the number of returnees peaked between 1835 and 1842, following the largest of the insurrections, in 1835 (Verger 1976[1968]; J. M. Turner 1975[1974]; Manuela C. da Cunha 1985; Reis 1987). Bahian officials ordered the expulsion of many suspected participants in the rebellion. Furthermore, an atmosphere of increased repression, head taxes, harsh punishment, curfews, restriction of movement, and so forth inspired many of the other manumitted Africans in Bahia suddenly to act upon their earlier dreams of going home. Some manumitted Africans chartered ships to carry groups of friends and family; others bought passage on commercial passenger or cargo ships (esp. Verger 1976[1968]:317; J. M. Turner 1975[1974]:51; Lindsay 1994:24–25).

For the most part, they returned to Lagos, Porto Novo, and other such ports in what are now western Nigeria, the Republic of Bénin, Togo, and Ghana. From 1820 to 1899, about 8,000 Afro-Brazilians returned to West Africa. Others returned from Cuba (Sarracino 1988), some passing through Brazil en route to Africa.[20] Many of them suffered unspeakable physical abuse and extortionate taxation at the hands of Dahomean and other Coastal African kings. Their lives were everywhere insecure, but far less so, they discovered, under the abolitionist British protectorate in Lagos. In recompense, the Brazilian returnees provided many of the architectural and other technical skills with which colonial Lagos was built. By 1889, even as the general African population of Lagos continued to swell, one in seven Lagosians had lived in Cuba or Brazil—that is, about 5,000 out of 37,458 (Moloney 1889:269; Lindsay 1994:27, 47n31).

Equally crucial to the emergence of Coastal society and its Latin American diaspora was the later return of the "Sàró" to Lagos and its hinterland (see Kopytoff 1965). From the first decade of the 19th century, the British had made efforts to enforce legislation outlawing the slave trade. A pillar of these efforts was the British Royal Navy's capture of slave ships departing from the West African coast. The African captives thus rescued were generally resettled in Freetown, Sierra Leone, where they and their descendants came to be called "Creoles," or "Krios." British missionaries quickly established programs to convert and school them, thus preparing a highly influential class of Western-educated Africans, whose families had originated disproportionately in what *the recaptives* would later help to denominate as "Yorùbáland." Upon their return to Lagos, they themselves would be called "Sàró"— from the local pronunciation of "Sierra Leone."

The sudden proximity of Ọ̀yọ́, Ẹgbá, Ẹgbádò, Ìjèṣà, and so forth in their Sierra Leonean exile had produced a number of effects relevant to our discussion. First, it made them cognizant of their similarities, particularly in the context of their *shared difference* from the populations of the local Sierra Leonean interior. Only then did a united Yorùbá—or, in the Sierra Leonean context, "Aku"—ethnic identity become conceivable. Second, the Krios, and especially the Akus of that time, were disproportionately Western educated, literate, Christian, commercially active, and prosperous. They had access to employment in British government, schools, and missions far out of proportion to their numbers. Krio society produced numerous administrators, teachers, and ministers, whom the British colonial and missionary projects dispersed to various parts of West Africa.

By 1880, the number of Krios, or Sàró, in Lagos roughly equaled that of the Afro-Latin returnees, but the Sàrós' education in English and in British social conventions enabled them to achieve greater commercial success and a greater number of administrative positions in the emergent British colonial government of Lagos. More important, their literacy allowed Sàró Christian ministers to produce the orthographic and lexical standards by which the emergent lingua franca of the Akus would be reduced to writing. Indeed, it was in the process of preaching and translating the Bible into a language that Òyó, Ègbá, Ìjèṣà, Èkìtì, Òndó, Ìjèbú, Ègbádò, and "Nàgó" could all understand that Sàró ministers *produced* for the first time a "standard" language, one that thereby reified the ethnic unity of these peoples and named that unity by a term previously used only by outsiders and reserved for the Òyó—that is, Yorùbá.

Born near Ìṣéyìn, in the Òyó-speaking region, the African missionary Samuel Ajayi Crowther wrote *A Vocabulary of the Yoruba Language* in 1843. In 1844, he preached to the ex-captives from Òyó, Ègbá, Iléṣa, and so forth gathered in Freetown in a hybrid language predominantly Òyó in its morphology and syntax, and predominantly Ègbá in its phonemes, with its lexicon enriched by coinages and the speech of Lagos and the Sàró, Nagô, and Lucumí diasporas (Ajayi 1960; Adetugbo 1967, cited in Peel 1993[1989]:67). Numerous others, such as the Òyó-born and British-educated Church Missionary Society (CMS) missionary Samuel Johnson, helped to establish a Yorùbá literary canon, which included not only a Yorùbá translation of the Bible but grand narrations of the Yorùbás' supposedly shared Òyó legacy. In sum, though the original Yorùbá were Òyó, Lagos in the late 19th century became the capital of a much more broadly conceived Yorùbá people—one that embraced multiple ethnic groups and indeed crosscut continents. Moreover, the privileged position of the Sierra Leonean recaptives and, more generally, of diasporic returnees to Lagos in the British colonial project made Lagos the capital of the emergent Yorùbá nation and furnished the new nation with a reputation for superiority to other Africans.

By the mid–19th century, Sàró returnees of the proto–Yorùbá Ègbá ethnic group had set the stage for white Baptist and CMS missionaries' special admiration of the Yorùbá. After the mid–19th century, CMS and Southern Baptist missionaries greatly amplified claims of Yorùbá superiority to other Africans. Much of their admiration arose from the well-organized character of local agriculture, the grandeur of the local monarchies, and the beauty of local religion and art, which all belied the racist apologetics of American slavery. However, Africa has never been short on magnificent and well-organized civilizations. Two facts may have inspired the *selectiveness* of missionaries' praise. First, whereas white Christian missionaries had repeatedly been turned away and mistreated by the materially and administratively impressive Ashanti and Dahomean kingdoms, the proto-Yorùbá Ẹ̀gbá city of Abẹ̀òkúta welcomed them with open arms (Ayandele 1966). Second, the Ọ̀yọ́-centric reflections of white missionaries like T. J. Bowen suggest that they were influenced specifically by the Ọ̀yọ́-centric and pro-Yorùbá writings of Crowther and other Ọ̀yọ́-born writer-missionaries (see Bowen 1968[1857]). And, third, praising the potential of a particular part of Africa where the missionaries had established missions was one strategy to persuade congregations in the West to continue subsidizing these missions. It was the strategy adopted by Southern Baptist missionaries T. J. Bowen and William H. Clarke, for example, as they declaimed the educational, religious, agricultural, linguistic, and commercial accomplishments and potentials of the local people, as well as the specific but exemplary educational and literary accomplishments of the black Yorùbá missionary Samuel Ajayi Crowther (Bowen 1968[1857]: esp. 288–93; Clarke 1972: esp. 271, 291).

By the 1880s, the laudatory reasoning of African CMS missionaries and their Euro-American contemporaries had profoundly affected British colonial discourses as well. The proto-Yorùbá Ọ̀yọ́ and many neighboring African groups amply shared the British predilection for royalism and tradition. However, British colonialists had further reasons to praise the new Yorùbá collectivity. For example, the governor of Lagos, Alfred Moloney, endeavored to encourage the economic development of the Lagos Colony and its hinterland by inviting the return of more African-Brazilians. So he addressed one gathering of returnees thus: "You are the representative embodiment of the steady annual flow of Africans from Brazil to your own land in some instance[s,] in others to the land of your parents and ancestors, (the grand, rich, and intellectual *Yoruba land*)" (quoted in Verger 1976[1968]:551–54). So committed was Moloney to the reality and dignity of "Yoruba"-ness that he declared it erroneous to call the returnees "Brazilians" at all; they were, properly speaking, "repatriated Yorubas." Years later, he regaled them, anachronistically, for having kept their dignified and ancient "Yorùbá" language alive in Brazil and Cuba (Verger 1976[1968]:551–54). In fact, no language by that name had existed at the time that most of them or their ancestors had arrived as slaves in Brazil. Captain W. A. Ross, the chief colonial officer in the Ọ̀yọ́ kingdom from 1906 to 1931 and a personal friend of the ascending king, became another, even more forceful spokesman of Yorùbá dignity (Atanda 1970; Adas 1989:155–56, 158).

Thus, the reality of "Yorùbá"-ness and the *discourse* of its ancientness and dignity among nations had become an official truth. Whatever the economic advantages British colonial officials wished to use these official words to secure, these officials had now come to reinforce, authorize, validate, and valorize a novel identity borne of African and African-diasporic experience. Moloney and others added official weight to an identity that *crosscut* linguistic, state, and even *maritime* boundaries. Indeed, in 1890 and 1891, the colonial government subsidized the trial voyages of a regular steamship line between Lagos and the cities of the Brazilian coast. Though it ultimately proved too costly to operate, the project demonstrates the willingness of British colonial officials to encourage this trans-Atlantic identity, which repeatedly received similar support from the late 20th-century Brazilian and Nigerian nation-states as well (see chapter 4). Support from missionaries and colonial authorities, far from being circumstantial or epiphenomenal to Yorùbá identity, is a major part of what brought that identity into being from the mid–19th century onward, established its equivalence to a range of New World black identities, and fashioned them into a trans-Atlantic nation of unparalleled prestige.

By the late 19th century, some variant of the Yorùbá language had become a lingua franca in Bahia. Whereas some attribute its status to the sheer number of proto-Yorùbá people in the population of the state (e.g., R. N. Rodrigues 1988[1905]:130), Ramos attributes the dominance of that language to the literacy and the literary accomplishments of its speakers (Ramos 1942:20).

In sum, Ọ̀yọ́ and various existing sociopolitical units around the Gulf of Benin supplied much of the cultural raw material of the Lucumí identity formulated in Cuba, the Nagô identity formulated in Brazil, and the Yorùbá identity that was first articulated in Sierra Leone and then amplified in Lagos. But the Yorùbá identity that embraces the Ọ̀yọ́, the Ẹ̀gbá, the Ẹ̀gbádò, the Ìjẹ̀bú, the Ijèsà, and the Nàgó did not exist before the dispersion of these peoples to Cuba, Brazil, and Sierra Leone. Their novel collective identity was given substance by the new linguistic and literary forms that its advocates created in the 19th century, by its colonial and missionary uses, and, as we shall see, by the religious motifs that set this nation apart from others in the diaspora.

The cognates of terms like "Yorùbá," "Nagô/Nàgó," "Lucumí," and "Kétu/Quêto" had probably predated the slave trade, but their application to all those peoples now called "Yorùbá" is of recent vintage.[21] Equally recent is the idea that they share a common primordial culture. Thus, to call the self-identified Ọ̀yọ́, Ẹ̀gbá, Ẹ̀gbádò, Ìjẹ̀bú, and Èkìtì captives of even the late 19th century "Yorùbá" is, in most cases, as Robin Law (1977; 1993[1984]; 1996) and J.D.Y. Peel (1993[1989]) have demonstrated, an anachronism. Nor was there anything inevitable about these groups' coming together as a nation on the basis of their linguistic and cultural similarities, as the case of the neighboring Gbè speakers and that of the Twi speakers clearly demonstrate. Calling the the Ọ̀yọ́, Ẹ̀gbá, Ẹ̀gbádò, Ìjẹ̀bú, and Èkìtì peoples of the 19th century and their pre-19th century ancestors "Yorùbá" reads a commonsense reality of the late 20th century back onto a period in which that reality was only beginning to be produced by formerly captive populations of Nagôs, Lucumís, and Krios, who "returned" to colonial Lagos and its hinterland.

From Cooperation to Cultural Nationalism

Despite British infrastructural and moral support for the Yorùbá identity, the 1880s and 1890s were hard times for the returnee bourgeoisie in Lagos. First, there was an economic downturn in the 1880s, which large and well-funded European commercial houses survived far more easily than did the smaller, African-owned enterprises. Second, the progress of British colonialism into the Lagos hinterland in the 1890s undermined preexisting forms of cooperation between Europeans and Westernized Africans by authorizing in their stead a policy of white racial privilege.

As European tropical medicine improved, increasing numbers of white British administrators, doctors, and missionaries arrived to take up jobs previously held by Westernized Africans, and the new arrivals were prepared to use racialist discourses to assert their privileged access to the best jobs and the highest pay. Even when race was not offered as the official explanation, it became evident to many Westernized Africans that *racism* was in fact the grounds for many of the exclusions, demotions, and firings that they had begun to suffer in the colonial service and the churches. The young white missionaries' rejection of Bishop Samuel Ajayi Crowther's leadership and his consequent resignation from the Niger Mission in 1890 were but the most famous case so interpreted by the Sàró and Afro-Latin returnees (see, e.g., Omu 1978:109), and it precipitated a major movement toward the formation of independent African churches.

That resignation also contributed to an ethos that the historian J.F.A. Ajayi (1961) and later scholars have described as a "renaissance" and as "cultural nationalism." This turn-of-the-century Lagosian cultural nationalism was a "crusade" or "campaign" carried out mainly by journalists, tract writers, and composers. It sought to stimulate interest in "African history, language, dress, names, family life, religion, dance, drama and art forms" (Omu 1978:107–14). Though it occurred primarily in the 1890s, it had precedents in the 1880s, in the form of articles in the *Lagos Times*. A prominent theme there had been the "preservation" of the "Yorùbá language," since—and I quote one such article—it was considered "one of those important distinguishing national and racial marks that God has given to us" (Omu 1978:107–8).[22]

Thus, in a generation, the Sàró appear to have changed from avid Anglophiles, grateful for their liberation from slavers at British hands, to advocates of various forms of nativist African nationalism—or what was appropriately called "patriotism"—that opposed British cultural hegemony. Though Sàró literacy has made it easiest to document this process among these returnees from Sierra Leone, it is also clear that, as a consciously political gesture, numerous Afro-Latin returnees also publicly adopted what they construed as "African" names, dress, and marital standards during this period (Ajayi 1961; Cole 1975; J. M. Turner 1975[1974]; Lindsay 1994; Mann 1985; Omu 1978).[23] It has now become commonplace for New World òrìṣà devotees in Brazil, Cuba, and the United States to do the same as a mark of their own adopted sacred nationality.

These nationalist-inspired changes in the self-conception of Westernized Africans on the 19th-century Coast are generally called "cultural nationalism" to distinguish them from direct challenges to British political sovereignty and to mark them as mere precursors to the forms of activism that would later lead to the independence of the Nigerian national state. But two further points are curiously absent from most existing discussions of this period, and certainly from the implications of the label "cultural nationalism." First, the people whose oppression concerned these nationalists and whose unity they assumed were characterized as much by their "race" as by their "culture" in the ethnographic sense. Nationalist writers of this period variously assumed that the "Black race" shared certain cultural characteristics separating them from whites or took all the various African cultures that preceded the Westernized Creole culture of the Coast to be more "authentic" and appropriate to members of the "Black (or Negro) race."

Analysts who note the racial essentialism of these discourses tend to attribute them to bourgeois Africans' rather passive adoption of the vocabulary generated by European writers, such as Comte Joseph Arthur de Gobineau and James Hunt, which were then propagated by European colonialists (see, e.g., Okonkwo 1985:2–3; Law 1990:81–99). There are clearly parallels between some West African writers' and many European thinkers' conflation of "race," culture, language, religion, and nation.

However, there is no particular reason to believe that African nationalists' peculiar shuffling of these ideas was determined by any particular European intellectual's theory. Indeed, there is far more in Lagosian

social history than in European intellectual history to inspire local na-
tivists to conflate Yorùbá culture with Black racial unity. To begin with,
racial binarism was, for Lagosians, no mere philosophical postulate.
Lagos from the 1880s to 1915 was rife with racial discrimination. A
Lagos that had in the mid–19th century been divided among European,
Brazilian, Sàró, and indigene was, by 1910, divided into black and
white. Visiting Lagos in 1911, the art historian Leo Frobenius reported
that even black policemen and postal clerks were hostile toward whites
(Frobenius 1968[1913]:38–40; Cole 1975:75, 89). The theories of Arthur
de Gobineau and James Hunt were themselves products of a changing
colonial dialogue rather than blueprints for the Coastal African response.

Indeed, if any particular literary or scholarly articulations of the
"race" idea strongly shaped the African Creole response, they were
those articulations that the African Creoles imported from the African
diaspora. Far more than Lagosians read their experience in terms of
the natural history of Arthur de Gobineau, they read it in terms of the
highly publicized experience of Black North Americans, particularly as
revealed in the news items and editorials reprinted from Black American
newspapers and in the serialization of writings by W.E.B. Du Bois and
Booker T. Washington, not to mention those of the well-traveled Danish
Virgin Islander Edward Wilmot Blyden, who had himself studied and
experienced racial discrimination in the United States. Blyden's flight
from the United States took him in turn to the Black North American returnee colony of Liberia and then to Sierra Leone. This returnee and advocate of racial uplift and liberation even supported British colonialism
as a means to pursue his goal of uniting large parts of Africa into a single state. His thinking enormously influenced generations of Black political thinkers on both sides of the Atlantic.[24]

From the turn of the century well into the second decade of the 20th
century, Lagosian newspapers regularly documented the lynching of
Black North Americans and the racial forms of cooperation mobilized
by them to escape murder and oppression, including "repatriation" to
Liberia. Lagosian newspapers exchanged much correspondence with
Black North American newspapers, not to mention the fact that some of
the foremost Lagosian publishers and political leaders of the day themselves had West Indian, U.S., or Liberian social origins, as did the Jamaican mulatto Robert Campbell, who edited the Anglo-African. The
editor of the Lagos Weekly Record, John P. Jackson, was the son of
Black North American settlers in Liberia. Many Lagosians, such as Orishatukeh Fadumah, a returnee from Sierra Leone, had visited the
United States. Thus, Lagosian "cultural nationalism" was also a racial
nationalism inspired by the social experiences, political activism, and literary self-expression of the African diaspora.

Hence, the further oversight implicit in the term "cultural national-ism" is that any nationalism self-consciously referring to Black North American politics, to Black repatriation to Africa, and to racial unity at the turn of the century can hardly be represented as *pre*-political.

It should be noted, though, that Lagosian nationalism was no more *determined* by Black North American ideology and social conditions than by European ideology. For example, turn-of-the century Lagosian newspapers participated in rather *non*–North American modes of doctoring the bloodlines of Lagosians' heroes. Everyone from Africanus Horton and Alexander Crummell to W.E.B. Du Bois and Booker T. Washington was lauded in the Lagosian press as a "pure-blooded Negro,"[25] paralleling testimony recorded in Bahia between the 1890s and the 1940s by Raimundo Nina Rodrigues and in the late 1930s by Ruth Landes and Donald Pierson. For example, trans-Atlantic traveler Martiniano Eliseu do Bonfim and many of his Afro-Bahian contemporaries were quoted proclaiming the superiority of black or African racial purity, which they appear to have associated with and held on par with African *cultural* purity (Landes 1947: e.g., 23, 28; Pierson 1942:273 292; see also R. N. Rodrigues 1935[1900/1896]:170–71); 1988[1905]:216).

Moreover, the talk of racial purity on the West African Coast surely also has precedents in colonialists' and slave owners' contempt for their culturally and racially mixed subjects (see, e.g., Law 1990:88). Traveler Mary Kingsley's ideas on the dignity of African racial and cultural purity were also widely quoted in the Lagosian press. Even more widely quoted were the similar ideas of E. W. Blyden, who associated the dilution of both African culture and African blood with weakness, disease, and social decay (see Law 1990:88). Such black racial and cultural purism is documented in Bahia precisely from the period of the Lagosian Cultural Renaissance onward. The simultaneous presence of this novel and distinctive ideology in both colonial Lagos and the Bahian Candomblé underlines the strength and the transformative power of the link between the two places.

As we have seen, numerous studies by Brazilianists have credited the Euro-Brazilian intellectuals and social elites themselves with arbitrarily inventing notions of Yorùbá superiority and Nagô purity. However, despite the tandem claim that earlier Brazilianist ethnographers saw and endorsed "Nagô purism" in the Candomblé from the 1890s to the 1970s, it is difficult to find references to any such observations or endorsements in the early ethnography of Bahia. While Raimundo Nina Rodrigues, Arthur Ramos, and Èdison Carneiro clearly approve of the Nagôs' preeminent dignity and relative "preservation" of their traditions, these early ethnographers also describe the complex interethnic and Afro-Catholic "syncretism" found in all the Brazilian Candomblé temples they observed (e.g., R. N. Rodrigues 1935[1900/1896], 1945[1905]; Ramos 1940[1937]: 35, 137–38; Carneiro 1967:263–64). The earliest and most influential of these ethnographers, Raimundo Nina Rodrigues, focuses his arguments on the evolutionary inablity of blacks to embrace pure monotheistic Christianity, not on the purity of any given African practice. Strikingly, most of the explicit claims and endorsements of racial and cultural purity in these early ethnographies appear in *direct quotes* from the African-Brazilian priests and travelers themselves (e.g., Landes 1947:23, 28, 80, 155, 169, 196–97, 200; Pierson 1942:273, 292). In sum, there is no reason to think that the early ethnographers—or any members of the Euro-Brazilian bourgeoisie—taught Afro-Brazilians this value. The value attached to black racial and cultural purity, and the belief in Yorùbá superiority have a complex contextual genesis and a cosmopolitan set of precedents—in the Lagosian Renaissance and its dialogue with the Anglophone black Americas in an age of Jim Crow. The chief advocates of Nagô-centrism and African purism in Brazil were African-Brazilian priests and travelers themselves.

"Traditional Religion" and the Invention of the Yorùbá

Though literate Christians and missionaries did much to create the Yorùbá identity, "traditional religion" (*èsìn ibílè*) remains one of its chief emblems. For example, the kingships of Yorùbá towns remain attached to rites and mythologies of "traditional" gods, or *òrìsà*. For this and other reasons, the gods Ifá, Odùduà, and Sàngó remain central in any 20th-century discussion of collectively Yorùbá political and cultural tradition. Primers like Daramola and Jeje's *Àwon Àsà àti Òrìsà Ilè Yorùbá (Traditions and Òrìsà of Yorùbáland,* 1967) continue to wed Yorùbá identity and literacy to dignified images of non-Christian religious traditions.

But what accounts for the image in the diaspora that Yorùbá "traditional religion" is superior to that of other African peoples? Anglo-Yorùbá literacy and the returnees' success as colonial intermediaries were certainly persuasive media of Yorùbá prestige. However, the image of Yorùbá superiority was not articulated in Western terms alone. For example, in Freetown, the Aku masking societies modeled on the Egúngún ancestral masquerades of the Òyó and Ègbá became a "cultural diacritic," as Barth (1969) might call it, marking and reinforcing Krio distinctness from other local ethnic groups. Yet these masking societies were so closely associated with Krio power and prestige that they became an object of imitation for the non-Krio migrants from the Sierra Leonean interior.

Hence, the premise of Yorùbá religious superiority to other African ethnic groups appeared at the very 19th-century origins of the pan-Yorùbá ethnic identity (see, e.g., Desribes 1877:318–19, 322, 348) and would remain closely associated with Yorùbá/Nagô/Lucumí religion in the Brazilian and Cuban diasporas, as noted in the testimonies recorded by Raimundo Nina Rodrigues, Donald Pierson, and Ruth Landes in Bahia.

In the late 19th century, African-Brazilian travelers also participated not only in validating the image of Yorùbá superiority but in establishing its foundational role in the emergent scholarly literature on African-American religions. First, African-Brazilian returnees were the chief informants, translators, and hosts of the French Catholic Society of African Missions, affiliates of which penned a range of documents on òrìṣà religion that are cited extensively in the African-Americanist literature, including Fernando Ortiz's seminal Yorùbá-centric account of Afro-Cuban religion (1973[1906]: e.g., 28, 51, 52, 59, 60).[26]

The most influential English-language and Yorùbá-language literature on West Africa entered Afro–Latin American scholarship through the mediation of the Afro-Brazilian travelers. On their frequent return trips to Brazil, these travelers brought knowledge of the Lagosian Renaissance literature. In an intellectual biography of Raimundo Nina Rodrigues (founder of Afro-Brazilian studies and founding scholarly advocate of Yorùbá superiority), Andrade Lima cites the pivotal contribution of an African-Brazilian traveler and friend of Rodrigues: "Lourenço Cardoso, of Lagos, Nigeria, merchant, [Rodrigues's] English professor and Nagô [Yorùbá] translator, who was [in Bahia] for a time and referred Rodrigues to the work of Col. A. B. Ellis" (Andrade Lima 1984:7). Colonel Ellis of the British West India Company is the author of The Yoruba-Speaking Peoples of the Slave Coast of West Africa (1894). For his part, Raimundo Nina Rodrigues's African-Brazilian informant Martiniano do Bonfim provided Rodrigues with a translation and critique of the CMS text Iwe Kika Ẹkẹrin Li Ede Yoruba (The Fourth Primer in Yorùbá Language—see, e.g., R. N. Rodrigues 1935[1900/1896]:42–43; 1945[1905]:356).[27] Both of these texts and their African-Brazilian interpreters are cited prominently in Raimundo Nina Rodrigues's classic works on Afro-Brazilian religions, O Animismo Fetichista dos Negros Bahianos (The Fetishistic Animism of the Bahian Blacks, 1935[1896]) and Os Africanos no Brasil (The Africans in Brazil, 1945[1905], 1988[1905]). Note also that not only Ellis's but also Nina Rodrigues's books are, in turn, prominently cited in the work of early Cuban folklorist Fernando Ortiz (see, e.g., Ortiz 1973[1906]:51) and that the image of Yorùbá preeminence surfaces again among Afro-Cubans and in the Afro-Cubanist textual legacy (e.g., Cabrera 1983[1954]:27; Barnet 1994[1966]:34–35).

Moreover, highly literate informants like Martiniano do Bonfim were actively involved in the establishment of the Bahian Candomblé houses now regarded as the most "purely African" and "purely Nagô" in Brazil. In fact, based on his interpretation of his academic studies in Lagos, Martiniano directly designed a set of liturgical innovations that survive up to this day in the Bahian Ilê Axé Opô Afonjá temple, which is now regarded as *the* single most "purely Nagô" temple in all of Brazil. Yet these were not the only Coastally inspired ritual innovations that made Opô Afonjá "pure." Part of what impressed sociologist Roger Bastide with the "purity" of the Opô Afonjá temple in Bahia was that it "conserves the old lunar calendar, found in a Bible translated into Nagô" (Bastide 1983:263).[28] I am unaware of any precolonial West African precedent for the use of the lunar calendar (from the Bible or any other source) in the conduct of òrìṣà worship. Thus, the African sources of the Bahian Candomblé are complex and often closely related to major cultural developments in Africa that postdated the end of the slave trade.

The most influential work of the Lagosian Renaissance era was *The Yoruba-Speaking Peoples of the Slave Coast of West Africa*, published in 1894 by the commander of the British West India Regiment in Lagos, Colonel A. B. Ellis. It clearly plagiarized Father Baudin's earlier, Brazilian returnee-influenced account of "Yorouba" religion in Porto-Novo. Among the features that Ellis added, however, was a laudatory tone typical of his Lagosian surroundings during the nationalist Renaissance. Citations of Ellis are omnipresent in the early 20th-century literature on Afro-Brazilians (e.g., R. N. Rodrigues 1988[1905]; Ramos 1940[1934]; 1946[1937]; on Afro-Cubans (e.g., Ortiz 1973[1906]); and on Black North Americans (e.g., Puckett 1969[1926]).

It should not be forgotten that Ellis's book continued the most extensive project of apologetics and documentation ever undertaken, up to that date, for any African religion other than ancient Egypt's. *The Yoruba-Speaking Peoples* appeared amid an explosion of lectures, pamphlets, books, and newspaper articles by bourgeois Yorùbá intellectuals, and it effectively conveyed to numerous African-Americanist authors the sort of mythological complexity and organization that they then assumed was absent from other, more sparsely documented African religions. In English and in Yorùbá, turn-of-the century Yorùbá intellectuals constructed, codified, and valorized an ancestral religious legacy. Ellis's publication also followed decades of Edward Wilmot Blyden's and Mary Kingsley's ambiguous campaign of doubts about the suitability of Christianity for "the Negro," as well as the publication of J. Abayọmi Cole's *Revelations of the Secret Orders of Western Africa* (1887) and "Astrological Geomancy in Africa" (1888). Ellis's publication briefly preceded J. O. George's "Historical Notes on the Yoruba Country" (1895); Orishatukeh Fadumah's article "Religious Beliefs of the Yoruba People" (1896); Rev. E. M. Lijadu's book *Ifa* (1897); Rev. James Johnson's *Yoruba Heathenism* (1899); Dr. Abayọmi Cole's lectures on "Yoruba" religion; Rev. Mojọla Agbebi's studies of *òrìṣà* religion in Ọyọ́, Òwu and Ọwọ̀;[29] Oyeṣile Keribo's pamphlet "History of the Gods" (1906); and various series of newspaper articles by Adesọla (aka Rev. E. T. Johnson), by someone writing under the initials F. S., and by editor Chris Johnson in the *Nigerian Chronicle* (1908–12).

Note that these writings by Christians do not necessarily recommend the practice of *òrìṣà* religion. Rather, they advertise its complexity, wisdom, dignity, and ancient pedigree. They claim Yorùbá religion as an ancestral or national heritage and as a natural precursor to the Yorùbá people's conversion to an African form of Christianity (see Barber 1990). Yet, like Baudin and Ellis, these Yorùbá Christian writers advance a project of documentation and codification that has been easily appropriated by literate nationalists on both sides of the Atlantic, in projects validating the Yorùbá national legacy and propagating *òrìṣà* worship itself.

For example, a direct heir to the Lagosian cultural renaissance, the Christian Reverend David Ọnadele Epega, not only wrote several tracts on Ifá early in the 20th century but founded a late 20th-century dynasty of priests of the Yorùbá "traditional" religion. The elder Epega had read the Black nationalist writings Edward Wilmot Blyden and quoted the latter in Epega's own treatises on Yorùbá religion (Epega 1931:19). His grandsons are *òrìṣà* priests and have initiated other priests in Chicago, Florida, São Paulo, and probably elsewhere. They continue their forefathers' publishing legacy as well.[30]

Tradition and Trade in the Trans-Atlantic Yorùbá Nation

Thus, during the African cultural renaissance in turn-of-the-century Lagos, Africans and at least one European wrote numerous books, pamphlets, and newspaper articles describing and dignifying so-called Yorùbá traditional religion. However, the effectiveness of these texts as vehicles of trans-Atlantic Yorùbá identity depended on their mobility and benefited from their service to the commercial interests of the Afro-Brazilian travelers. The publication and availability of these writings coincided with the period when sailors, black businesspeople and English professors, and mail ships frequently traversed the Atlantic. Recall that the government of colonial Lagos itself once subsidized direct steamship service between Lagos and the cities of coastal Brazil.

In the lamentably incomplete Bahian archives of return voyages from Lagos, I have counted dozens of ships and hundreds of free Africans traveling from Lagos *to* Bahia or *through* Bahia to Rio or Pernambuco (see map A) between 1855 and 1898. Journalistic, epistolary, and oral historical evidence reveal the repeated journeys of another score of African-Brazilian travelers up to the 1930s.[31] Many of them carried British passports, and most appear to have been engaged in commerce.

Theirs was a truly transnational identity configuration and commercial endeavor, long before "transnationalism" and "globalization" became the catchwords of an era. These traders appear to have capitalized on their intercontinental nationality—providing West Africans with òrìṣà-related goods and services identified with "the white man's country" and providing Brazilians with òrìṣà-related goods and services uniquely authenticated by their African origin.[32] The African-Brazilians' religious expertise and the relatively low profit margins made religious supplies one sphere of trade exempt from major European competition (see Manuela Carneiro da Cunha 1985). Thus, not only their national identities but their livelihood clearly depended on the construction and defense of Africa's sacredness, of Yorùbá—and, as we shall see, Jeje—superiority, and of the "African purity" of the material and ideological capital they carried with them on their return trips to Brazil.

Sarracino (1988) documents the trans-Atlantic travel and commerce that united Cuba with Lagos during this same period. And it was also during this period—in the late 19th and early 20th centuries—that Cuban Ocha, or Lucumí religion, took its current form and began its rapid expansion across the Cuban landscape (Palmié 2002:162–63, 331n8). Like Candomblé, Ocha was shaped during and after the 19th century by voluntary transnational migration and textual reproduction. The famous African-born Adechina is said to have been enslaved in Cuba but to have returned to Africa for initiation as a *babaláwo* diviner, later returning to Cuba (D. H. Brown 1989, 2:94; Cabrera 1974:27). The oral history also identifies a freeborn African woman named Efunche (also Ẹfúnṣetan or La Funche) who traveled as a free person to Cuba and there reformed Afro-Cuban religion in the 19th century. In the 20th century, Cuban and Cuban-American priests continue to rely on handwritten notebooks, or *libretas*, inherited from such free travelers as sources of authoritative religious and linguistic knowledge.

Two of the most influential of these *libretas* had been written by free persons who had freely emigrated from British colonial West Africa. *Santero* Sixto Samá had come from Sierra Leone (Cabrera 1980:2), whereas Andrés Monzón had "learned to read and write in an English mission in Nigeria" (Cabrera 1986[1957]:16 translation mine). His *libreta* included a version of the Lord's Prayer in a Hispanized version of

WASHINGTON

Figure 14. George Washington in Masonic regalia, 1794. Nineteenth-century lithographic reproduction of a painting held at the National Masonic Monument, USA. Courtesy of the American Antiquarian Society.

the Yorùbá orthography generated by Anglo-Yorùbá missionaries in Freetown, the British West African colony founded as a refuge for freed slaves (see Cabrera 1986[1957]:17; Matory 1999a, and 1999b). Another of Cabrera's priestly informants in the mid–20th century, Miguel Allaí, had lived in Sierra Leone and, according to Cabrera, had learned to speak Yorùbá fluently (Cabrera 1986[1957]:17). Also among the

Figure 15. Freemason-inspired seal of the Reformed Ogbóni Fraternity on the door of Adeniran's shrine room in Ìgbòho, Nigeria. Photograph by the author (1989).

most influential writers of *libretas* is Nicolás Angarica. In the mid–20th century, he employed a West African Yorùbá dictionary and the Yorùbá translation of the Bible (also written by Anglo-Yorùbá missionaries in British colonial West Africa) to verify the orthography of his multiple primers of Afro-Cuban religious language and ritual practice (see Angarica 1955a:4). Today, scores of industrially printed catechisms and other publications circulate among the Cuban, Puerto Rican, and African-American orisha worshipers of the United States.

The trans-Atlantic migration and commerce of the late 19th and early 20th centuries thus closely connect the current form of Cuban Ocha not only to the Lagosian Cultural Renaissance but also to the emergence of Cuban *negrigenista* nationalism. Though Fernando Ortiz's writings about Afro-Cuban religion are initially pessimistic about its retarding effects on Cuba's modernization, even his earliest and most pessimistic writings tie nationalist claims of Cuba's superiority to reports of the superiority of "Yorùbá" religion, whose towering influence in Cuba supposedly makes that island's prospects superior to those of Haiti, the southern United States, and the French Antilles (Ortiz 1973[1906]:51, 52, 59, 60, 101–2, 105, 106). Ortiz also conspicuously embraces the Lagosian Renaissance vocabulary of "purity" (1973[1906]: e.g., 74, 89–90). Thus, like

Figure 16. Şàngó possession priest in the attire of the Aborigine Ogbóni Fraternity of Nigeria chanting before a Şàngó shrine in Ìbàdàn, Nigeria (1986).

Raimundo Nina Rodrigues in Brazil, Ortiz inaugurated a school of nationalist writing that not only endorsed "Yorùbá" religion as an emblem of Cuban national dignity but also thereby facilitated its emergence as the popular, "purist" standard of Afro-Cuban practice.

Collectively, the Alakija family linked extraordinary success in West African British colonial society and culture, including Freemasonry, with the will to reshape indigenous religious institutions in the service of African nationalism. (Yet, as the North American case reveals, Freemasonry could also empower assertions of *independence* from the

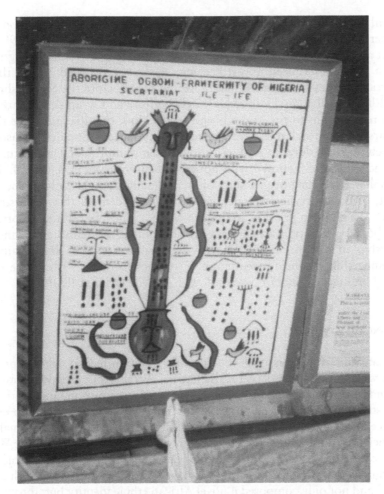

Figure 17. Certificate of membership in the Aborigine Ogbóni
Fraternity of Nigeria belonging to a Şàngó possession priest in
Ìbàdàn, Nigeria. Photograph by the author (1986).

British overlords. Figure 14 depicts George Washington in his Masonic
regalia.) "Traditional religion" (èsìn ibílè) in the 20th-century Yorùbá
hinterland—which is often represented as the primordial origin of Candomblé's and Santería's "survivals" and "memories" of Africa—is
deeply inscribed with the evidence of this ongoing Afro-Atlantic dialogue. The Freemasonry-inspired iconography of the Reformed Ogbóni
Fraternity has penetrated the òrìsà priesthoods of even the "Yorùbá
Proper," the Òyó-Yorùbá. The shrine room door of my Ìgbòho friend,
the Şàngó priest Adeniran, bears various insignia of his membership in a

local Reformed Ogbóni-style fraternity—three unblinking eyes on an inverted V and three vertical shapes within it (see figure 15). In 1986, I met another Ṣàngó priest in Ìbàdàn who chose to pose for a photograph in a Masonic-inspired apron bearing, again, the inverted V and three vertical shapes within it. In figure 16, he poses before the Ṣàngó shrine in Ìdí Arere, Ìbàdàn. Despite his wifely coiffure and wrap skirt (marking his relationship to the god who possesses him), this Ìbàdàn priest thus bears an unmistakable resemblance to his African-Brazilian Masonic forebear, Sir Adeyẹmọ Alakija (known to my informant Dr. Alakija as "Uncle Plasido"), shown with apron and inverted Vs in figure 13. The number 3 has long been a shibboleth in the unreformed Ogbóni Society. But, of course, syncretic relexifications are evident in this "ancestral" garb. The Ìbàdàn priest is no Freemason, nor a member of the Reformed Ogbóni Fraternity. Rather, these Masonic signs authenticate his membership in a later, explicitly nativist organization. His hand-painted membership certificate (see figure 17) comes from the "Aborigine Ogbóni Franternity [sic] of Nigeria," with its secretariat at Ilé-Ifẹ̀.

Conclusion

What came to be classified as "Yorùbá" tradition fed on cultural precedents in the hinterland of Lagos, but its overall name, shape, contents, standards of membership, meaning, means of transmission, and degree of prestige would have been radically different—if they had come into existence at all—were it not for the intervention of a set of transnational financial, professional, and ideological interests that converged on the West African Coast. Returnees from Brazil, Cuba, Jamaica, North America, the Virgin Islands, and Sierra Leone converged on Lagos during the 19th century and not only composed a novel African ethnic identity but, through a literate and politicized struggle, guaranteed that it would be respected in a unique way by generations of students of Africa and its diaspora.

This historical revision is offered, first, for its potency in repealing the model of agentless, collective "memory" and survival that is usual in the representation of African culture in the Americas and at home. What is often called cultural "memory," "survival," or "tradition" in both the African diaspora and at home is, in truth, always a function of power, negotiation, and strategic re-creation. The tract writers of the Lagosian Renaissance and the English professors of Brazil selectively reshaped the meaning and value of shared practices; they reified and redefined African cultures as much as they transmitted them. Thus, at this locus classicus of African-diaspora studies—and of diaspora studies generally—we are well advised to restore agency to our models of cultural reproduction.

None of this is meant to suggest that the agency of literate nationalists has been the lone or sufficient vehicle of black Atlantic cultural reproduction. Powerfully mnemonic ritual song and dance genres are the most conspicuous reminders of the African legacy and emblems of community in the African diaspora. However, these bodily practices possess no intrinsic historical meaning and embody no self-evident program for community building. Rather, they are almost always selectively reproduced according to what is practically allowed and needed by powerful contemporary political actors. These genres of song and dance are regularly mobilized as emblems and as proof for influential *verbal* assertions of meaning and community, the most broadly influential of which have been published. Even in these supposedly "remembered," "preserved," and "traditional" cultures, written texts have become major and transformative vehicles of cultural reproduction and identity formation. It was a long-running, transnational literary movement that differentiated an extraordinarily successful black Atlantic nation, like the Yorùbá, from a merely successful one, like the Congo/Angola nation. The Yorùbá-identified òrìṣà worshipers of Brazil and the United States now also propagate their legacy at academic conferences and on the Internet.

My argument affirms Anderson's 1983 argument that vernacular print media have enabled the citizens of the nation-state to "imagine" themselves as a community, sharing the same communal experience as fellow citizens whom they might never have met. However, I broaden Anderson's concept of "imagined communities" by suggesting, first, that nation-states are not the only communities so united by machine-produced vernacular texts. Second, I call attention to the flexibly "imaginative" power of *multivernacular* communities like the trans-Atlantic nations of Candomblé. Third, as we shall see, the Afro-Atlantic nations remind us that the ritual, musical, and bodily practices shared by dispersed populations (including those practices recommended in texts that are distributed over long distances) also enable powerful "imaginations" of communally shared experience among unacquainted parties. Nationalist writers often rely on and reinterpret these same ritual, musical, and bodily practices to substantiate their own novel and interested visions of community.

Diasporas are often studied as though time had stopped in the homeland. This position inspires the second major critical point of this argument. The diasporas and the homelands of the Atlantic perimeter have engaged in a transnational dialectic of mutual transformation for more than five hundred years, of which the 19th-century "Yorùbá" and their diaspora are but one case in point. The Yorùbá have arguably become *the* preeminent exemplar of sub-Saharan African civilization, and the Candomblé is perhaps *the* paradigm case in ethnohistorical studies of the African diaspora. The foregoing historical revision argues that the

West African Yorùbá and the Quêto/Nagô Candomblé are less like the roots and the tree or like the origin and the outpost than like Siamese twins—the fate of each affecting the other. Of course, the Siamese twin simile, like Deleuze and Guattari's "rhizome" metaphor (1987[1980]), is imperfect, in that it implies that the connected cultures or sites of cultural reproduction are of the same species, the same origin, and the same substance. As I will argue later, the "dialogue" metaphor of translocal cultural reproduction posits no such homogeneity of substance; it simply posits coeval interconnection and continual, mutual transformation. This case, then, is recommended as the paradigm for a *dialectical* approach to the relationship between diasporas and homelands generally (see also Skinner 1982). Here we see how transnational politics have reshaped a diaspora and its homeland through their radically coeval dialogue. A new light is thus shed not only on the perennial question of what is African about African-American cultures but on the question of what diasporas generally—Jewish, Irish, Indian, Chinese, and so forth—have to do with the nations they call home.

The next chapter concerns a nation born in the diaspora and short-lived in the homeland. The short life of the West African Djedji, or Jeje, nation demonstrates not only the antiquity of transnationalism but also the coevalness of the religious, national, and imperial forces that continue to constitute it. To the same degree that the Yorùbá/Nagô case exemplifies the role of literacy in the making of an Afro-Atlantic "nation," the Djedji/Jeje case illustrates the importance of maritime commerce in the making of today's most "purely African" of Afro–Latin American nations.

CHAPTER TWO

The Trans-Atlantic Nation

RETHINKING NATIONS AND TRANSNATIONALISM

> "The 'Nagôts-yorubas' are particularly deceitful and,
> generally, more difficult in their behavior than the Djedj."
> — FRENCH CHIEF ADMINISTRATOR,
> *Cercle* of Porto-Novo, 17 April 1921

IN THE 19TH AND EARLY 20TH CENTURIES, at the same time that white creoles were "imagining" and reifying a nation called Brazil, Africans in Brazil and along the Lower Guinea coast were "imagining" and sustaining a nation of a sort *un*imagined by Benedict Anderson. And it is with no sense of irony—and with no demand for correction—that these Africans and their descendants in the Afro-Brazilian Candomblé religion still speak of their trans-Atlantic communities as nations. I respect this parlance throughout the book not because *I* regard it as the appropriate analytic category—in some ways, "denomination" would be well suited— but because members and neighbors of the Candomblé are fully aware that the citizens and analysts of nation-states use the same term with monopolistic pretensions. Its usage thus teaches a significant lesson to those who believe that nation-states monopolized or dominated all classes' imagination of community. This term illustrates the co-occurrence of, and overlap among, Candomblé adherents' citizenship in multiple imagined communities. Worldwide, people's simultaneous or situational belonging in multiple, overlapping communities is—like the Candomblecista's multiple nationalities—a taken-for-granted at least as old as the nation-state.

This chapter concerns the surprising history of Candomblé's second most prestigious nation. Lest my argument be regarded as a form of "but among the Bongo-Bongo . . ." contrarianism, intent on disproving a general phenomenon by reference to an obscure exception, I forewarn the reader that the case described in this chapter implicates the French, British, and Portuguese empires; the now-independent nation-states of Bénin, Togo, Nigeria, Cuba, Haiti, and Brazil (the last being the largest nation in Latin America and the greatest outcome of Portuguese colonialism); and the kingdom of Dahomey (a major source of the Afro-Atlantic diaspora, itself the largest transoceanic migration in history up until the late 19th century). The Djedji/Jeje nation discussed here is but one among an enormous number of transnational imagined communities that, far from hailing the end of the nation-state (as much of the recent literature would suggest), began before the nation-state, have coexisted with the nation-state, or have subsidized and been subsidized by the nation-state.

Because of their foundational role in nationalism worldwide, the United States, France, and England are often mistaken for the *typical* nation-states, and their long-distance relations with the rest of the world are treated as the typical cases of transnationalism. Predictably, their peculiar historical vicissitudes become the source of a new "universal history," a grand narrative of all nationalism and transnationalism. This chapter presents, instead, a view from the South, which requires us to acknowledge, for example, the central but embarrassing reality of transnational slavery and colonialism behind the piously bounded "national" histories and subsequent transnationalist teleologies now narrated from the first world. The equally important vicissitudes of the Djedji/Jeje nation require us to ask whether the capitalist logic and technology that have best defined British and U.S. imperialism over the past two centuries are truly the only significant engine of transnationalism in every time and place. I argue that religion, the antiegalitarian projects of nationalists, and the low-tech strategies of import marketing are equally important engines of transnationalism. This chapter is, then, an effort to confront two recent and highly provocative literatures—on nations and transnationalism—with the impious realities of the African diaspora.

My claim that print technology has been critical to the genesis of Yorùbá identity and its international prestige benefits from Benedict Anderson's 1983 argument that vernacular print media enable citizens of the territorial nation-state to "imagine" themselves as a face-to-face community. What I doubt is the Andersonian view that the territorially nationalist imagination replaced other kinds of imagination and is now giving way to others still.[1] Theorists of the nation as an "imagined community" (e.g., Anderson 1991[1983]) and of what is represented as its transnational sequel (e.g., Harvey 1989; Appadurai 1996; Sassen 1998; Verdery 1998; also Anderson 1998:66–67) tend to imagine world history in stages driven by the advance of capitalism, particularly in terms of the speedup in transportation, migration, and communication that capitalism requires for profitability and enables through technology. In one stage, empires or religions monopolized or dominated our imagination of the communities to which we belonged. In the next stage, territorial nations monopolized or dominated our sense of community, and these are now giving way to transnational communities. Such "stagism"

(Buck-Morss 2000) and the single "engine-of-history" model are also the stock-in-trade of 19th-century evolutionisms (including the grand historical schemata of Maine, Marx, Durkheim, and Weber), which often render sensitive portraits of the authors' "present" but falter as analyses of the past and of the contemporary peoples mislabeled as "primitives."

Likewise, today's writings about "transnationalism," apocalyptic announcements of a recent historical "break" or "rupture," often preface insightful and surprising accounts of translocal commerce, migration, and communication in our time. These works hail the recent disruption of an old Eurocentric order but simultaneously adopt the same orientalizing narratives about the cultural and temporal Other. The reader is led to assume, rather than being shown any evidence, that the past was devoid of such colossal types and degrees of translocalism and/or that history has been a linear march toward ever-increasing degrees of such interaction among everybody everywhere. The Silk Road, Islamic civilization, and the globalized African civilization that has flourished since the 16th-century takeoff of the Atlantic slave trade disappear into a homogeneous past of cultural localism and social stasis. The nation-state, until its recent disruption, is credited with a degree of autonomy, control over its territory, and completeness of citizen loyalty that was in fact rare even within the most exemplary of nation-states—France, England, and the United States.

The transnationalism theorists' recognition that *not all imagined communities and flows of people, ideas, and money are isomorphic with territories of governance* is, however, indispensable to the argument of this chapter. However, they do a deep disservice to history by implying that such non-isomorphism is radically or qualitatively new.

As Clifford (1997) points out, movement and communication between locales (quantitative changes of degree and speed notwithstanding) have always been normal to the human condition. And as Cooper observes, there have always been long-distance flows, but none of them, including "global" capitalism, has been unrestricted in speed or direction. Moreover, the historical trajectory of translocalism was and continues to be discontinuous, uneven, and, in a word, "lumpy" (Cooper 2001:200). It is a mistake to regard history as a teleology of "everything-was-leading-up-to-this" without regard to the agency, inconsistency, one-purpose-then-another modus operandi of real human actors. As Trocki (1997) suggests, it is appropriate to limit the analytic term "transnational" to those social processes that cut across the boundaries of *territorial nations*. However, we must not then beg the question by assuming that the advent of nationalism diminished or terminated translocal movement and communication, made them into a qualitatively different phenomenon, or caused them to have qualitatively different effects. And we must avoid being seduced by the salesmanship whereby an old phenomenon swaddled in a litany of "New! New! New!" blinds us to a great deal of old form and content.

This chapter is both theoretical reflection and historical revision. I attempt to piece together the historical genesis of the mysterious Djedji/Jeje nation and to demonstrate that, without grasping the breadth of 19th-century transnationalism, we could not understand how this nation came to exist in Brazil and only thereafter came to exist in Africa. Nor could we understand the patterns of death, resurrection, and ritual transformation of the Brazilian Jeje nation from the 1890s to the 1940s, long after the rise of territorial nationalism and the end of the Atlantic slave trade. Candomblé came about and continues to change—despite its usual reputation as either the transplant of a "frozen" Africa or a mechanism of white bourgeois manipulation—primarily because of its place at the ongoing conjuncture of multiple transimperial, transnational, national and subnational imagined communities. Jeje priests and their sponsors have prospered by drawing on the resources of diverse imagined communities in the same historical epoch.

VOODOO NATIONHOOD

If Yorùbá identity owes much of its international fame to the publications of nationalists and scholars, a further cluster of black Atlantic identities has been made famous by voodoo movies. Much like stagisms of 19th-century evolutionist history and of recent theorizing about "transnationalism," white American references to "voodoo" rarely rest on any factual information about the Other; instead, they invoke the a priori conviction that the subject position of the speaker is so positively distinctive that the Other merits no more than a stereotype or dismissal. Since the early 19th century, following the Haitian Revolution, the Haitian Vodou religion has served European and Euro-American political commentators as the perfect trope for the deficient rationality of the *white* politicians they wished to disparage. The American critique of "voodoo economics" is but the latest example.[2]

The term "voodoo" is cognate with the word *vodun*, meaning "god" in the Ewè-Gèn-Ajá-Fòn (EGAF), or "Gbè," dialect cluster of the Gulf of Guinea—to the west of present-day Yorùbáland (see map B).[3] The term *vodun* and its cognates—like the gorgeous rites of music, dance, sacrifice, and spirit possession associated with it—are spoken by worshipers in the Bénin Republic, Togo, Haiti, Brazil, Cuba, and Louisiana.[4] These religions have for centuries shaped the political orders of a civilization as contemporary and rational as any of the Abrahamic traditions and their secular offshoots. Indeed, Vodou's presence is arguably what distinguished the only successful slave revolution in history.

In Bahia, there are three major nations, or denominations of the Candomblé religion, identifying themselves with three African origins—the Yorùbá-linked Nagô nation (of which the Quêto became the eponymous subset), the Angola nation, and the Jeje nation. Demographically, the *orixá*-worshiping Quêto nation's greatest rival is the *inquice*-worshiping Angola nation. In turn, both it and the Quêto nation dwarf Bahia's tiny *vodum*-worshiping Jeje nation. In 1983, a group of 1,211 Candomblé high priests in metropolitan Salvador (the Bahian capital city) was surveyed by the Secretaria da Indústria e Comércio/Instituto do Patrimônio Artístico e Cultural da Bahia (SIC/IPAC). Of this total, 447 (or 36.9 percent) avowed membership in the Quêto nation alone, 384 (or 31.7 percent) in the Angola nation alone, and only 30 (or 2.5 percent) in the Jeje nation alone (Santos 1995:17–19). However, in newspaper coverage, Bahiatursa recommendations to tourists, and state funding of projects for temple preservation, the Jeje nation receives at least as much attention as the far larger Angola nation. The disproportionate public attention to and support of the Jeje nation derive from its reputation for "purity," or the exceptional fidelity of its religious practices to their alleged African prototypes. Yet the origins of this nation's name and the nature of the community it implies pose a special riddle, the answer to which requires us to rethink much that influential scholars have posited in recent decades about the origins of nations and transnationalism generally.

The closely related *names* of this nation—"Djedji" in West Africa and the cognate "Jeje" in Brazil—are unrecognizable to most Africanists. Their origin remains a mystery (see appendix B). The Yorùbá-related "nations" of the Americas are a classic and exemplary case in the Herskovitsian study of African "origins" at least partly because the African sources of their ethnic labels are so easily identified—for example, "Lucumí," "Nagô," "Quêto," and "Ijexá," which New World scholars and worshipers alike have long and easily identified with the eponymous "subgroups" of today's West African Yorùbá ethnic group. The Palo Mayombe, or Regla de Congo, denomination of Cuba, as well as the Angola and Congo nations of Brazil, offer similarly transparent pointers toward their African "origins." Even the *vodun*-worshiping Arará nation of Cuba and the Rada nation of Haiti are easily identified with the eponymous "Allada" town and kingdom of the present-day People's Republic of Bénin.

On the other hand, the term "Djedji" and its numerous orthographic variants appear in the West African historiography only for a brief, seventy four-year period—between 1864 and 1938—that falls long after the effective end of the slave trade. Only for the first half of that period was the term used in West Africa as it is now in Brazil—that is, to refer collectively to all the Gbè-speaking and *vodun*-worshiping ethnic groups between the former slaving fort of Elmina and Yorùbáland.[5] It will become clear that central aspects of "Jeje" religion in Brazil and of "Djedji" ethnic identity in West Africa were also directly shaped by the back-and-forth movement of free Afro-Brazilians across the Atlantic. Indeed, this movement seems to be precisely what *created* the West African Djedji identity on the Coast.

However, the Djedji identity did not achieve the same success in the African interior as did the Yorùbá identity, though the Djedjis' descendants on the Coast came to play founding roles in two nation-states— Dahomey/Bénin and Togo. The Djedji identity invoked a persuasively transnational religious community and drew energy from the rivalry between two empires. Though once persecuted in Brazil, the Jeje and Nagô nations of Candomblé have, since the 1960s, become symbols of Brazilian national identity. Yet much of these nations' symbolic importance in Brazil relies on the conviction that their enduring ties to Africa anchor Brazil against the tsunami of European and Anglo-American cultural imperialism. Though most Brazilian aficionados of Candomblé regard these ties as primordial, what I see in the Jeje nation is an ongoing dialogue of mutual transformation among empires, territorial nations, religions, and transnations at the height of what Anderson (1998) calls the "classical nationalist" project. This chapter will detail both the transnational reasons for the rise and fall of the Djedji nation in West Africa and the related reasons for the extraordinary reputation of the Jeje nation in Brazil.

In the pursuit of the long-neglected history of the Jeje nation, the concrete aims of this chapter are three. The first third of the chapter sets out the extraordinarily complex and hybrid political setting in which the transoceanic Djedji/Jeje nation took shape and explains the central riddle posed by both its name and by the late arrival of the EGAF snake god, Dã, in Bahia. The search for an answer requires us not only to exploit the numerous insights of the recent literatures on nationalism and transnationalism but also to question some of their central postulates about historical change and the collective imagination. The second third of the chapter attempts the difficult task of piecing together the interests and conduct of this transnational religious community based on records left largely by the literate partisans of empires, nation-states, and European religious missions. Answering the central riddle of this case will require some historical reconstruction, but the forms of dialogue and mutual interest manifest across these diverse "imagined communities" will leave us in no doubt that none was simply superseding the others historically.

Finally, I abstract the primary lesson of this case for the literatures on nationalism and transnationalism—that translocalism is the precondition of nationalism. Transnationalism is not only a structural condition of nationalism but also an important idiom of nationalists' political discourses. Far from prefiguring the demise of the Andersonian territorial nation, these and other such transnational phenomena have been critical to its very constitution.

JEJE: A NEW NATION WITH NEWER ROOTS

Both the ethnonym "Djedji"/"Jeje" and the snake god Dã/Dangbe have repeatedly disappeared and reappeared over the past two and a half centuries. And it would be difficult to understand their vicissitudes without grasping the simultaneous imperial, religious, and national projects that provided the motives and the material contexts of their appearance.

In West Africa, the speakers of the Gbè dialects are diverse in language, political allegiance, and ethnic identity. This is what linguist Hounkpati Capo writes of their linguistic diversity: "There is mutual intelligibility between dialects that are contiguous, e.g., Ewè and Gèn and Ajá, Ajá and Fòn, etc.; but the degree of mutual intelligibility is related to 'geographical' distance, e.g., although there is some mutual intelligibility between Ewè and Ajá, it is less than between Ewè and Gèn, and there seems to be none between Ewè and Fòn[,] which are the extreme ends of the dialect cluster" (Capo 1984:168). Despite their distant-past linguistic connections, the EGAF, or Gbè, dialect cluster speakers never shared the same political allegiance or identity. In fact, the constituent political groups often warred against each other in ways quite profitable to the European slave traders on the Coast. It was these European slave traders who first grouped the Gbè speakers together under a sequence of shared "trade marks" (Kubik 1979), such as "Rada" and "Mina." In the 17th and early 18th centuries, the kingdom of Allada dominated the trade with Europeans in this region. To the west of the kingdom was the famous Elmina trade fort, which had long played a prominent role in Afro-European commerce. So, in the 17th and 18th centuries, European travelers and slave traders identified various *vodun*-worshiping peoples collectively as "Arda/Ardra/Arder/Ardres" (from the name of the kingdom of "Allada") and "Minas" (after the Fort of Elmina or, in Portuguese, São Jorge da Mina). Consequently, we thereafter find populations in Haiti called "Rada" and in Cuba called "Arará." In Brazil and French Louisiana, they were called "Mina."

However, in Brazil at some point in the mid–18th century, these same *vodun*-worshiping people came to be known as "Jejes," a term first used there as early as 1739 (Verger 1976[1968]:6, 7, 17, 381, 450, 462, 593ff.; R. N. Rodrigues 1945[1905]:176).[6] The fact that these people worshiped the *vodun* left no doubt that they had come from somewhere between Elmina Castle and the western boundaries of Yorùbáland, where the gods are known by that name. For Brazilianists, however, the origins of the peculiar name "Jeje" have remained highly contested and uncertain (see Matory 1999c: 62–63 and appendix B for a history of the term).

Whatever the origin of this term, large numbers of the people subsequently so described flowed into metropolitan Salvador (Bahia), into the surrounding Recôncavo region, and into the Brazilian states of Maranhão and Pará during the 18th and 19th centuries. As a significant element in the population of Northeastern Brazil, they predated the ancestors of today's Yorùbá-identified Nagôs (Verger 1976[1968]:1, 7, 381, 595ff.).

The origin of the ethnonym "Jeje" is especially mysterious if we suppose that it lies in Africa (see appendix B). Although the Gbè speakers were being exported from West Africa in the greatest numbers before 1800, I cannot find the term "Jeje" anywhere in the accounts of traders, travelers, and missionaries who wrote about the region until 1864—a century after its appearance in Brazil and nearly a decade after the effective end of the slave trade to Brazil in 1851. Thus, the use of the term "Jeje" as an important ethnonym seems to have originated in northeastern Brazil in the mid–18th century. Apparently, even before its spread to other parts of Brazil, it was transplanted to West Africa, where it became the first ethnic designation to unite all of the EGAF, or Gbè, dialect cluster speakers under one class.[7]

The record of how this ethnonym became prominent in *mid-19th-century to early 20th-century West Africa* is clearer than its Brazilian or pre-Brazilian origins. From the second third of the 19th century, *Brazilian* Jejes were returning by the hundreds, and perhaps thousands, to the Gulf of Guinea Coast: to Lagos, Porto Novo, Ouidah, Grand Popo, Petit Popo (Anexɔ), Agoué, and Porto Seguro, the last of which they in fact founded and named after the first Portuguese settlement in Brazil. As traders and slaveholders themselves, some of these returnees regularly traveled back and forth among the Gulf of Guinea, Bahia, and Cuba. It was apparently these sons and daughters of the diaspora who labeled all the West Africans *they* considered their kin with the name "Jeje," even though it is unlikely that those "kin" had ever previously thought of themselves in such terms.

We know of these developments through the writings of priests in the French Society of African Missions (Société des Missions Africaines: [SMA]), who, as the guests of prosperous Brazilian returnees, missionized this region (J. M. Turner 1975[1974]). These clerics were the first West African–based Europeans to call the Gbè dialect cluster and its speakers all "Jeje." Following this lead, a French commercial agent wrote an instructional manual of the Djedji language based on the subdialect of Fòn spoken in Porto Novo (d'Albéca 1889), where both the Brazilian returnees and the SMA were the strongest in the late 19th century (see also Salvaing 1994:47–48). Thus, from around 1864 until around 1889, the Brazilian returnees, the SMA, and those French people whose business interests encouraged them to study the local language from d'Albéca's primer institutionalized the name "Djedji" and made Porto-Novo into the capital of a local cultural imperialism.

Of the five to eight thousand Afro-Brazilians who returned to the Coast between the late 18th and the early 20th century, about 60 percent returned specifically to the Gbè-speaking region. Most of these were Jejes, but more than a few others were Nagô (J. M. Turner 1975[1974]; R.N. Rodrigues 1988[1905]:106). A greater number of the Gbè speakers returned to Porto Novo than to most other towns, and it was there that they most stood out in the trans-Atlantic trade and in the promotion of Djedji/Jeje ethnicity.

This case suggests an important real-world corrective to both the Andersonian and the transnationalists' representation of the 19th-century nation-state and its context. Porto Novo owed its prosperity to and reflected its prosperity in the *multiplicity* of imagined communities that overlapped within it—multiple religious communities (Catholic, Protestant, and *vodun*-worshiping), multiple national identities and *kinds* of national identity (French, Portuguese, and German, Nagô and Jeje), and multiple imperial jurisdictions (British and French). Yet it was not a space where specificities were dissolved or where all other genres of imagined community withered before the growth of one such genre. Rather, the local dialogue sustained multiple genres of imagined community, and the salience of any given symbolic community was both situational and dialectical. The Djedji/Jejes' half century of material and cultural distinction on the Coast, then, relied not on the homogeneity of local imagination about community in the region but on their leadership's deft exploitation of a heterogeneous and changing local imagination. In the real world, these are also the hallmarks of successful nationalist movements and states, religious communities, and dynastic empires as well.

Indeed, the arrival in rapid succession of the Brazilian Jejes, the SMA, and French government officials intent on protecting French commercial interests created a momentary syzygy of interests, which converged most dramatically at Porto Novo. From the middle of the 19th century, the kingdom of Porto Novo established a special relationship with France, the power that would eventually colonize the broader region. Although

the inclusive *colony* of Dahomey would not be established until the conquest of the kingdom of Dahomey in 1894, the Porto Novo kingdom had
for decades sought French protection against the Dahomean kingdom.

By 1902, Porto Novo had become, as one French official put it, "the
richest jewel" of the Dahomean colony.[8] Even before Porto Novo became
the "richest jewel" of the Dahomean colony, it had become the "richest
jewel" of the trans-Atlantic Jeje nation. From 1864 onward, rival religious communities competed for followers in Porto Novo. Both Protestants and Catholics had established churches and schools in Porto Novo,
which remained their largest and most stable outpost on the Mina-Djedji
Coast (Pliya 1970:46, 125; Salvaing 1994:47–48). Therefore, at the turn
of the century, Porto Novo hosted the largest population, the most
schools, and the most students of any city in the Dahomean colony.[9] And
these schools were full of children from Portuguese and Brazilian families, many of whom remained active in the trade with Brazil (see, e.g.,
Pliya 1970:46). By 1921, Porto Novo hosted a population of about
191,000.[10]

As Anderson has argued, print capitalism is a critical and active mechanism in the imagination of new communities, and Porto Novo was
particularly "imaginative" in this regard. Porto Novo hosted the publication of the first African-controlled newspaper in Dahomey—*La Voix
du Dahomey*, founded in 1926. In the 1920s and 1930s, Ouidah,
Abomey-Calavi, and Cotonou each had one newspaper, while Porto
Novo alone had three.[11]

Let us, then, summarize how Porto-Novo became the preeminent city
of both the Dahomean colony and the trans-Atlantic Jeje nation. The
exceptional commercial acumen of Porto Novo's Brazilian returnees, the
stability of its Catholic mission, and the Porto Novo kingship's early
embrace of the future colonial power all guaranteed its preeminence
among the cities of the EGAF region. From the mid–19th century onward, Porto Novo became the focal site of cultural work by bourgeois
Jeje returnees from Brazil, who, for the few decades of their active cooperation with the SMA, managed to impose their own ethnic label upon
numerous peoples who had never before been described as "Jeje," or
"Djedji," and to institutionalize the "standard" language they supposedly shared with all the peoples of the EGAF Coast. This ethnonym and
social category came to guide the policies and actions of multiple powerful social classes and administrative agencies.

From around 1902 to around 1938, however, the terms "Djedji,"
"Djedj," and "Gège" came to be applied specifically and exclusively to
the indigenous, non-"Yorùbá" inhabitants of Porto Novo. While the
term ceased to refer to the totality of Gbè speakers, it remained central
to French colonial efforts to preserve French sovereignty over what was

by then the colony of Dahomey.[12] The French administrators constantly feared not only a British military takeover of the region but also a commercial and cultural takeover.[13] They complained about and imposed tariffs on British imports and penalized the use of British currency. French administrators complained that Yorùbá Protestant missionaries had founded schools where *French* subjects were being taught in *English*.[14] Even the French-based Society of African Missions had chosen *Yorùbá* rather than French or Djedji as the language of choice in religious instruction.[15] This organization did so because the black Yorùbá missionaries of the British Church Missionary Society had already created a substantial literature in this local language. To French colonial administrators, British language, military power, and business—together with Yorùbá language, religious leadership, and business—were tandem threats to French sovereignty.

Though they had been the founding fathers of egalitarian nationalist ideology, the French propagated and subsidized ethnic hierarchies in the administration of overseas subjects and citizens. Even though the Dahomean colonial capital, Porto Novo, was home to many locally born Nagôts, or Yorùbá, the French spared them no invective. In contrast to the Djedji, the Yorùbá were called "deceitful," "difficult," "miserable," and foreign.[16] Though Ọ̀yọ́ and the other proto-Yorùbá peoples had been a strong and old cultural, linguistic, and political influence in this region, the French worked hard to naturalize the distinction between the "Yorùbá" and the "local" people over whom the French could then unambiguously claim sovereignty. In other words, the French campaign against Anglo-Yorùbá imperialism gave a pride of place to the Djedji ethnic identity and fostered its development in the Dahomean colony until at least the early 1930s.

By the late 1930s, however, this ethnic label had virtually died out, gradually replaced as a reference for non-Yorùbá Porto Noviens by the terms "Ègùn" (in Yorùbá) and "Gùn" (see, e.g., Akinṣọwọn 1930[1914]; Kiti 1929). The half century of success enjoyed by the "Djedji" ethnic label had, in sum, rested on a foundation of Porto-Novien/Brazilian returnee privilege in the emergent economic, political, and cultural order of the Mina Coast. Yet the French policy of assimilating especially the Brazilian elites apparently deprived them of the incentive that, for example, the Yorùbá in racist Lagos had to play up the dignity of their African nation (see chapter 1). Thus, by the 1940s, the Brazilian-coined "Djedji"/"Jeje" ethnic denomination had grown obsolete in West Africa.

Nonetheless, the limits of assimilation had also become evident to the West Africans and Antilleans resident in Paris. Some of their responses, like the Négritude movement, received transnational, transimperial, and

transreligious inspiration from the Harlem Renaissance and the Gar-
veyite movement in the Anglophone world. It was not until the 1930s
that those Paris-dwelling *assimilés* embraced another term—"Fòn"—to
identify their unity and their dignity.[17] The Fòn ethnonym and identity
were centered not on the point of French colonial entry and capital of
the colony—Porto Novo—but on the kingdom that had longest *resisted*
French colonial intrusion. The term "Fòn" had referred specifically to
the ethnic identity of the Dahomean kingdom's ruling dynasty. Hence,
after the horrors of French colonialism had overshadowed the horrors
of Dahomean rule, the Dahomean kingship became a convenient symbol
of African protonationalism.

To recapitulate this historical sequence, the term "Djedji" and its cognates were used in writings between 1864 and 1889 to identify all the
EGAF, or Gbè, dialect cluster speakers collectively. In writings from
1902 to the early 1930s, the term was used in a manner loaded with
French Anglophobia to describe the non-"Yorùbá" inhabitants of Porto
Novo. The term died out by the late 1930s and was replaced by "Fòn"
as a seemingly proto-African nationalist reference to all the EGAF dialect cluster speakers in the French colony of Dahomey.

What is perhaps most frustrating and most special about the trans-
Atlantic Djedji/Jeje nation is that its past is known primarily through the
records left by the agents of other institutions—namely, territorial
nations, empires, and missionary organizations. It is therefore difficult to
document, for example, the precise reception of this ethnonym and the
logic of community it implied among its nonliterate referents. However,
the term's very appearance in the documents and in the ethnological conceptions of so many types of imagined community on three continents
suggests not only the transnational nature of the Djedji/Jeje nation but
also the fact that dialogue and interaction across diverse types of imagined community were significant during this era of "classical nationalism," and well into the mid-20th-century history of third world
nationalism as well. The history of the Djedji/Jeje nation suggests the
complementarity of interests and the mutually transformative conflicts of
interest among such imagined communities that shape each such community. The vicissitudes of this trans-Atlantic nation muddy any a priori
teleology of succession and supersession in the history of imperially, religiously, nationally, and transnationally imagined communities. Indeed,
the little that we do know about the "Djedji"/"Jeje" nation demonstrates
that projects of each sort of imagined community are often well served
by the invocation of other sorts of imagined community.

Like the Nagô returnees to Lagos, the Jeje returnees to the Mina
Coast secured an elite role as businesspeople and administrators in
the service of the European colonizers. They and their descendants

continued to dramatize their ethnic and class distinctiveness, calling themselves the Brésiliens, emphasizing their prestigious diasporic origins, celebrating their Afro-Brazilian festivals, and relishing their Afro-Brazilian cuisine. The Brésiliens thought themselves superior to their non-Brésilien compatriots, enjoyed the European empires' role in bolstering their ethnic privilege, and perceived their non-Brésilien compatriots' ambivalence about their elite status. To this day, they also remain fascinated by their transnational link to Brazil (see J. M. Turner 1975[1974]: 345–89, esp. 349, 368, 380, 384, 386–87).

However, the newspapers published by the Brazilians in French West Africa articulated not only their specific ethnic grievances over exclusion from privileges to which they thought themselves entitled in the colonial system but also some of the grievances of their African compatriots generally.[18] In all these ways, the Brazilians resemble the white Creole elites of the New World (not to mention the Saros and the Brazilian Amaros in Lagos). Their position in the colonial hierarchy placed them disproportionately in a position to speak for their African compatriots and to assume leadership in the independent nation-states of Dahomey/Bénin and Togo (not to mention the parallel Nigerian case). The Brésiliens by no means monopolized the leadership of these emergent nation-states. Relative to their small numbers, however, they occupied a disproportionate number of leading roles in these new states. Indeed, the first president of Togo was a Brésilien. From the 19th to the mid–20th century, then, the Jejes and their descendants in West Africa simultaneously participated in projects local, national, imperial, transnational and transimperial, imagining their collective selves in simultaneous and alternating ways.

A SERPENTINE PHOENIX: ON THE DISCONTINUOUS HISTORY
OF BAHIAN JEJE RELIGION

I have so far detailed the late appearance and early disappearance of the West African "Djedji" identity, which, among other lessons, illustrates the nonlinear history of the returnees' imagination of community. In connection with French imperialism, certain 19th- and 20th-century transnational population movements gave an originally Brazilian pan-ethnic identity a brief life in West Africa and a trans-Atlantic fame. The remainder of this chapter addresses two ironies that invite explanation in terms of these transnational population movements. The first is the early disappearance and later resurrection of the Bahian Jeje nation, whose fame in the early to mid–20th century reversed more than a century of cultural dissolution in Bahia. The second is the dominance of the Jeje-Màxí subnation in today's Bahian Candomblé, despite its status as a tiny minority of the Jeje population extant in Bahia at the end of the 19th century. Both of these phenomena relied on a transnationalism that in no way foreshadowed the end of the territorial nation. Rather, it became a class fraction or "fragment," to borrow Chatterjee's (1992) term, of at least three territorial nations—not only Dahomey and Togo but also Brazil.

Dating the moment when Brazil became a territorial nation after the "classical" model is difficult. It became independent from Portugal in 1922 but remained under the governance of the resident Portuguese monarchy until an elitist republican revolution in 1899. Not until after Getúlio Vargas's revolution in 1930 did Brazil commit itself officially to the ideal of secular, egalitarian "brotherhood" among its citizens. However, the gradual movement of that country toward the ideal type of the territorial nation nourished, rather than eclipsed, other ways of imagining community. From the 1890s to the 1940s, both the black merchants and the light-skinned folklorists of Northeastern Brazil increasingly embraced a number of highly religious, hierarchical, royalist, and transnational nations. These nations were not only the structure of several trans-Atlantic communities in their own right but also (as I will detail in chapter 4) symbols of Northeastern dignity in the face of political domination by other regions of Brazil.

Though it constituted a major stream of forced immigrants and a source of soldiers and laborers in the mid–18th century, the Brazilian Jeje nation declined from the late 18th century onward, as first-generation Nagô captives came to outnumber the Jejes in Bahia. Gradually, the kindred religious practices of these two nations were amalgamated into a single Nagô-dominated complex, which Raimundo Nina Rodrigues (1945[1905]: e.g., 365) therefore called "Gêge-Nagô," or, in the current orthography, "Jeje-Nagô." The use of the term *vodum* (plural, *voduns*) is documented in Bahia at least as far back as 1892 (Reis 1986) and in the neighboring state of Minas Gerais at least as far back as 1741 (Peixoto 1943–44[1741]: e.g., 18, 23). Gbè-inspired terms, such as *vodum* (god), *vodunça* (wife, or possession priestess, of the god), *Loucouce* (wife of the tree god Lôcô), *pegi* (altar), and *ogã* (male official), were well integrated into the Nagô-dominant Afro-Bahian religious parlance of the mid–19th century.[19] There are references to the Bogum temple and its African-born founder, Ludovina Pessoa, in the late 1860s.[20] These are undoubtedly ancestral to today's Bogum, the Jeje nation's most famous temple in Bahia. However, that ancestral temple, like the snake god and cognates of the Gbè term *vodun*, disappears from the Brazilian historical record between the mid–19th century and the 1930s. Indeed, the available details of succession in the Bogum temple suggest that the temple itself was shut down for a period of nearly four decades, between 1879 and 1917.[21]

In the 1890s, Rodrigues described the virtual disappearance in Bahia of people and institutions, including work crews (*cantos*), that called themselves "Jeje" (R. N. Rodrigues 1945[1905]:179–80). This is not to say that all the people who descended from Jeje or called themselves Jejes had died. However, they became sufficiently indistinct from their non-Jeje neighbors, sufficiently unlikely to call public attention to their distinctiveness, and sufficiently unlikely to organize distinctly Jeje institutions to go unnoticed even by scholars anxious to find them. Where they appeared, EGAF theological concepts and ritual forms no longer signaled the distinctiveness of the Jeje nation; they had been subsumed under a Nagô ethnic rubric. Despite his sustained search, the only snakelike sacred sign he saw was explicitly related not to a Jeje god but to the Nagô god of iron, Ogum.[22] As late as 1934, Arthur Ramos still found no instances of the term *vodum* in Bahia (R. N. Rodrigues 1945[1905]:366–68; Ramos 1940[1934]:54; Ramos in Pereira 1979:[1947]:12). In sum, from the 1860s until the 1930s, the institutions of the Bahian Jeje nation, the term *vodum* and its derivatives, and the snake god Dã were probably not entirely extinct, but they were in an advanced state of decline and unpopularity. The Africanist researcher A. B. Ellis (cited in R. N. Rodrigues 1945[1905]:366–67) believed that *vodun* and specifically snake worship would be found everywhere in the Americas settled by large numbers of captives from the EGAF region. Several apparent cases of snake veneration appear in Bahia between 1842 and 1870, but the name Dã is absent from the records, and even such cases as these had become invisible by the 1890s.[23] Rodrigues observed that the organized worship of the EGAF snake gods, or Dã, had died out in Bahia by the end of the 19th century.

However, something brought the Jeje nation back to life in Bahia during the 1920s and 1930s, when numerous distinctly Jeje religious houses were found to be thriving. During this period, the Bogum temple was apparently reactivated, as was another temple in Salvador—which, in order to preserve the confidentiality of some of my interviewees, I shall call the Terreiro de Omolú. Moreover, both the EGAF snake god Dã and the term *vodum* suddenly became preeminent symbols of Jeje distinctness. Two and a half decades after Rodrigues's death, the next generation of researchers identified numerous "survivals" and "vestiges" of EGAF religion among the Jeje-Nagô temples. Among these were Jeje gods named "Niçasse," "Oulissá," "Tobossi," "Loco," "Leba," "Sapata," "Anyiewo," and "Hoho," as well as a corruption of the term "Dahomey" itself (Querino 1988[1938]:37; Ramos in Pereira 1979[1947]:12–13; Ramos 1946[1937]:302; see also Ramos 1940[1934]:53–54).

Four decades after Rodrigues's frustrating search, a flood of snake iconography, snake veneration, living snakes, and songs containing the word *vodum* appeared to other researchers (Correia Lopes 1943: 559–65; Correia Lopes 1939, cited in Ferraz 1941:272–273; Ramos 1940[1934]:55–56; 1946[1937]:303; Ramos in Pereira 1979[1947]: 12–13; Carneiro 1948:51; 1986[1948]:72).[24] Among the jewelry of Aninha, the founder of the Ilê Axé Opô Afonjá temple, were found several rings either embossed with or in the shape of snakes. She must have collected these rings before she died in 1938. Mãe Aninha's successor, Mãe Senhora, informed researchers in the early 1940s that she was familiar with "the cult of Idangbé, the sacred snake, who was a *vodunce* [*sic*]" (Ferraz 1941:273–74). "Dangbé" is the name of the preeminent Dã, or EGAF snake god, whose head temple stands in the Gbè-speaking city of Ouidah on the Mina-Djedji Coast.

Between the 1930s and the 1940s, not only snake iconography and Dã worship but distinctly Jeje temples multiplied in Bahia. In 1937, Édison Carneiro counted eight Jeje temples in Salvador, and Correia Lopes identified a further temple in 1943. By 1948, Carneiro reported, likely with some exaggeration, "Dã is present in all the Jeje Candomblés... in Bahia" (Carneiro 1948:27–28, 50–51; 1986[1948]:52–53, 72; Correia Lopes 1943:559).

How does one explain the resurrection (or, possibly, the initial *birth*) of a vibrant and expressly EGAF-inspired snake cult in Bahia long after the end of the slave trade? After a long obsolescence, how and why did Dã worship and the term *vodum* suddenly proliferate in Bahia in the 1930s and 1940s, and then become the distinguishing marks of a resurrected Jeje nation?

It would be difficult to answer this question with certainty. But three bits of evidence suggest that a 19th- century and early 20th-century transnationalism was responsible for this ethnic and religious revival. The first evidence lies in the fact that, from the 1820s to the 1930s, the Brazilian Jejes were not just "returning" to West Africa. They were going back and forth on business and pilgrimage. Secondly, a wealthy Jeje trans-Atlantic merchant—Joaquim Francisco Devodê Branco—was close friends with Mãe Aninha and the godfather of Mãe Senhora in the second and third decades of the 20th century, after which, as we have seen, Aninha was found to be in possession of her collection of snake-related religious paraphernalia and Senhora demonstrated a knowledge of Coastal EGAF religious terminology unprecedented among the priests of the prior three decades (Olinto 1964:213–15). Finally, the Jeje nation resurrected in Bahia was disproportionately of a subethnic group that was poorly represented among the few self-identified Jejes remaining in Bahia at the end of the 19th century—that is, the Marrim, or Màxí, subnation. (Whereas the Brazilian term "Marrim" is pronounced MAH-hee[n], its Fòn cognate, Màxí," is pronounced MAH-HEE.)[25] Following the early 20th-century Bahian renaissance of Jeje identity, the newly over-represented Jeje subgroup in Bahia was precisely the one that made up the majority of the identifiable Jeje *circular* migrants—again, the Marrins (this plural form of "Marrim" is pronounced MAH-hee^{n}ss).

All these factors point to the influence of a class of itinerant African-Brazilian merchants in the resurrection of Bahian Jeje ethnic identity and in the selection of its major religious "diacritica."[26] In turn, the class origins of these innovations seem to have allowed the Bahian "Jeje" nation to achieve a level of dignity and media recognition far out of proportion to its numbers—a level beyond that of the vastly more populous Angola nation and nearly equal to that of the Nagô nation.

CLASS, NATION, AND TRANSNATION IN THE BLACK ATLANTIC

Rather than assuming that history consists of a predictable succession of political forms, I ask why certain rubrics and visions of community appear to prevail over others at a given moment in a given place. For example, if Harding (2000) is correct, 19th-century Candomblé was a pan-African identity and practice of resistance to white domination. Indeed, this interpretation is consistent with the apparent disappearance of distinctly Jeje temples in the late 19th-century Bahia of Raimundo Nina Rodrigues. I explored in chapter 1 why it became so common and consequential in the early 20th century to claim a specifically Yorùbá genealogy for one's temple and sacred practices. So why did claims of Jeje ethnic affiliation suddenly grow in popularity during the 1930s? In fact, Butler observes a proliferation of different Candomblé "nations" (1998: 207–8), even as Euro-Brazilian elites increasingly embraced the logic of Nagô superiority. Why? And why do many priests in the 1990s and the present decade still hold out against the juggernaut force of what they call *anagonização*, or the compulsion to adopt Nagô practices?

I posit that any given imagination of community will favor certain strategic interests over others. The imagination of community is therefore an arena of interclass and interethnic rivalry, and that some classes are more empowered than others to *project* their interested imagination of the community in which diverse classes are pressured to participate. While such pressure cannot guarantee unanimity, people's imaginations are not infinitely free of constraint by their discursive or material context. They are shaped in dialogue with others' imaginations and material priorities. Those classes and ethnic groups whose imaginations of community subdivide or crosscut those of the nation-state cannot be discussed simply in terms of their exclusion from the nation-state, or as though their visions of community had nothing to do with the nation-state. The trans-Atlantic Jeje nation and its cultural projects are a "fragment" of the nation, to borrow Partha Chatterjee's term (1992), but they are not merely that. Nor does such an African "nation" represent a mere alternative to or withdrawal from the nation-state, as Bastide has implied (1978[1960]). Rather, this trans-Atlantic nation represents an important economic function of the nation, mediating its exchange with other territorial entities, generating tax revenues for the state, and providing the symbols of regional diversity. The Jeje nation crosscut the British, French, and German empires as well, and its transoceanic commerce played a role in their economies, too. The most obvious lesson in this phenomenon is that the dialogue among mutually crosscutting forms of imagined community is not only old but normal.

Class diversity is the fundamental structure of this dialogue and cannot be overlooked if we are to understand the very nonteleological history of Afro-Bahian collective self-imagination from the mid–19th century onward. Many authors assume that poor and isolated black populations are the most likely to practice African-inspired cultures (see R. N. Rodrigues 1935[1900/1896], 1988[1905]; Ortiz 1973[1906]; Price-Mars 1990[1928]). The trans-Atlantic Jeje and Nagô nations are two of the numerous cases that undermine this assumption (Matory 2000). Other scholars have tended to read African culture in the Americas as a product of resistance to the fundamentally European culture of the American nation-state's dominant classes (Raboteau 1980[1978]; Creel 1987; Stuckey 1987; Levine 1977; Dayan 1995; Deren 1983[1970]; and especially Harding 2000); as a lament about black exclusion from Euro-American society (Gilroy 1993); or as a product of a clientelism that naturalizes black marginalization and instills a false consciousness in the lower-class membership of the Candomblé (e.g., Motta 1994; Dantas 1982, 1988; D.D. Brown and Bick 1987; Fry 1982; also Henfrey 1981; Bastide 1971[1967]:214).

The most undoubtedly false representation of the role of class in the constitution of Candomblé comes from the most influential theorist of the "African religions of Brazil," Roger Bastide (1971[1961], 1978[1960]), who posits that the African prototypes of these religions and their African social contexts are communalistic and class-free, that these Afro-Brazilian religions are African only insofar as they *escape* the influence of "class society," and that cultural self-whitening and de-Africanization are a precondition of blacks' social ascent in Brazil, not to mention other American societies. Only Verger attributes the strength of African-inspired culture to the class *privilege* of its bearers. In the case of *vodum* worship, he reflected on the example of a Dahomean queen mother, whose presence in early 19th-century São Luís de Maranhão elevated the Dahomean royal gods to preeminence in that Brazilian city (Verger 1953; also 1976[1968]:1). What seems more evident in the postabolition era is that the Jeje and Nagô nations were allied ethno-classes characterized similarly by their commercial links to the West African Coast. They share with many other trade diasporas and social elites a concern with racial and cultural "purity" as an index of membership and rank. The Angola nation of Candomblé allied itself symbolically with a third African territory, Portugal's longtime colony of the same name. That nation was consolidated far less than the Jeje and Nagô nations by trade with Africa but nearly as much by the provision of "African" ethnic services in healing and food vending.

Trade deserves particular attention in this account of the genesis, transformation, and prestige of the Jeje. For international merchants and pilgrims, the trans-Atlantic nations were a means to power and prosperity, and in a multicolored and class-stratified trans-Atlantic society, the prosperity of certain ethnic spokespersons in turn helped to dignify their ethnic identities and spread the religious emblems that had come to be associated with them. Indeed, like many of the Nagô/Yorùbá travelers, some Djedji/Jeje travelers cut an impressive figure in their Western suits and watch fobs (e.g., Manuela Carneiro da Cunha 1985:130), but they also promoted their African ethnic identities and conducted a transoceanic trade in both information and the finest supplies for the practice of African religion.

Manuela Carneiro da Cunha (1985:149–51) cites Abner Cohen's work (1969) on the Hausa trade diaspora in the Yorùbá city of Ìbàdàn to explain the signs of cultural distinction from other Lagosians that Brazilian returnees cultivated in order to consolidate their ethnically based trade network. Cunha is attentive to the strategic shifts of oppositional identity—particularly through the returnees' dramatic display of their Roman Catholicism on the Coast—that helped to create useful political and cultural alliances in Lagos. In Lagos, Roman Catholicism was the returnees' main distinguishing feature (and therefore a useful way of bounding their shared interest group), since Lagos was otherwise populated by practitioners of Protestantism and African religions. As we shall see, mid-19th-century Jeje returnees to the Mina Coast, such as Joaquim d'Almeida, made similar displays.

We should not assume, however, that displays of *Christian* piety, which were most noteworthy to the colonial and missionary officials, were the returnees' only form of religious expresion or the only one that allowed them to invoke useful communities of interest. Particularly for the trans-Atlantic travelers, idioms of *identity* and *sameness* with other West Africans were also necessary. Despite the premise of shared ancestry often invoked by returnees, nothing guaranteed that they and their hosts would consciously share the same values or interests. Yet, because the returnees' lives and livelihood depended on cooperation across culturally diverse networks of erstwhile kin, these international merchants had a strong interest in creating idioms of identity and terms of cooperation across the borders that they traversed.

These itinerant traders were not articulating their identities in the context of the Coast alone. Nor was *opposition* to the other groups on the Coast the sole logic of their assertions of identity. Returnees who continued to trade with and travel to Brazil possessed a distinctive set of interests in promoting the value and sources of their imported products. As we have seen, much that is considered legitimate, original, and ritually powerful in the Candomblé is described as being from the Coast, which in Brazilian parlance refers specifically to the Guinea Coast from Sierra Leone to Lagos. *Palha da Costa* (straw from the Coast), *pimenta da Costa* (pepper from the Coast), *pano da Costa* (cloth from the Coast), and *sabão da Costa* (soap from the Coast), for example, are ritually significant and costly because of their association with Africa and are, in general, considered less valuable insofar as they are produced outside geographical Africa. In Bahia, no group more than the Jejes and the Nagôs depended on the credibility and singular worth of commodities from the Coast.

From the mid–19th century to the mid–20th, a number of well-publicized fortunes were made through the Bahian trade with West Africa and such as those of Lourenço Cardoso and Joaquim Devodê Branco. In Bahia, such transnational "African" merchants sold strip-loomed "cloth from the Coast," or *aṣọ òkè*, "African" cowry shells, West African peppers, black soap, shea butter, alligator pepper, kola nuts, herbs, and raffia (see Herskovits 1966[1958]:249). During some of his trips, Martiniano do Bomfim took coral and wool to sell in Lagos and returned with "cloth from the Coast" to sell in Bahia (Costa Lima in Oliveira and Costa Lima 1987:52). Felisberto Sowzer exported tobacco to West Africa and imported *sabão da Costa* into Bahia (Irene Topázio Sowzer dos Santos, personal communication, 3 January 1996). The value of their sacred merchandise in Bahia depended on its "legitimacy," which, in local terms, meant its *Africanness*. So highly valued was African merchandise that many Candomblé gods refused any "non-African imitations" (Herskovits 1966[1958]:249). Not every participant in the import trade grew wealthy, but the link between Africanness and pecuniary value relied on and amplified an aesthetic value increasingly pervasive in the local religious and secular milieu—"purity."

These transoceanic black merchants traversed multiple empires and nations and, through their movements, created novel, hybrid communities. As in the Yorùbá/Nagô case, their international networking appears to have vastly amplified the degree of influence that their small numbers would otherwise lead us to predict.

THE RISE OF THE MARRINS IN BAHIA

Although almost all of the Bahian Jejes at the end of the 19th century had originated from the West African seaside towns of Ouidah, Grand-Popo, Petit-Popo, Abomey-Calavi, and Cotonou (R. N. Rodrigues 1945 [1905]:180), most of the Jejes in present-day Salvador identify themselves with an altogether different Jeje subgroup, whose exceptional mobility may hold the key to its preeminence in the Jeje renaissance of the 1930s in Bahia. Most Bahian Jejes today belong specifically to Jeje subnations, or *terras* (lands), called "Savalú" (named after the foremost city in Màxí country, Savalou), "Marrim," "Marrino," or "Marrinje" (all Brazilian derivatives of the West African word "Màxí," though the diverse European-language orthographies conceal the similarity of pronunciation). Here we find further evidence of a major shift in Bahia Jeje identity between the 1890s and the 1930s. Not only did it virtually rise from the dead, but a subethnic minority of its constituents became a majority. Why?

The poverty of the historical documents makes any conclusive answer to this question impossible. However, the changing historical conditions on the Coast suggest one possible explanation. The Marrim/Màxí *terra* of the Bahian Jeje nation had originated from a West African interior region to the north of the Dahomean kingdom. However, rather than returning there, Brazilian Marrins, like most other Jeje returnees, concentrated on the littoral. Moreover, the Marrins concentrated themselves especially in the western seaside town of Agoué, far away from the capitals of colonial commerce, administration, and assimilation. The Marrins thus remained not only mobile in the trans-Atlantic trade but also, in all likelihood, more committed than other groups to a Jeje (as opposed to a Brésilien and French) identity. They lacked the opportunities for assimilation enjoyed by the Jejes of Porto Novo and Cotonou. To put it another way, theirs was the strongest remaining fragment of an otherwise moribund identity. As we have seen, the Djedji ethnonym was becoming obsolete in the commercial and political capitals of the Dahomean colony by the 1930s. Thus, at the late date of the resurrection of Bahian Jeje identity, it is not surprising that the touchstones of Djedji identity in West Africa would be among the Marrins of the Coast. Amid the Bahian renaissance, their leadership and prosperity in trade made their ethnic identity all the more appealing for other Djedjis to adopt, just as the Aku/Yorùbá ethnicity and cultural practices attracted a wide swath of less prosperous groups in Freetown and Lagos.

Both the Brazilians and the Cubans identified as Màxí (and its cognates) returned disproportionately to the Coastal port of Agoué, where, from the first half of the 19th century, they and Nagô returnees formed important communities.[27] At midcentury, the leading figure in the economy and politics of Agoué was a Màxí slave trader who had himself returned from captivity in Brazil: Joaquim d'Almeida had been born to the Azata family in the village of Hoko, near Savalou. He was captured by Fòn raiders and sent to Bahia as a slave, but he had probably purchased his freedom in Brazil before 1815, since he had made at least one return trip to the Coast before that year. He first settled permanently in Agoué in 1835. There, he is said to have lived like a Bahian planter. He had a Roman Catholic chapel built, the first of its kind on the Coast, and named it after the Church of Senhor Bom Jesus da Redempção in Bahia, which had long been identified with an all-Jeje Roman Catholic brotherhood in Bahia.[28] Following the pattern of other similar brotherhoods in Cuba, Bahia, and elsewhere in Brazil, these brotherhoods were foci of not only Roman Catholic practice but *vodum* worship as well.

Not only did d'Almeida return repeatedly to Brazil, but, as the owner of slaves in Cuba, he must have sustained repeated contact with that Caribbean island as well. According to his will, he owned, at the time of his death, thirty six slaves in Havana and others in the Brazilian province of Pernambuco, not to mention the passel of "Nagô," "Jeje," and "Mina" slaves whom he kept in Agoué (Verger 1976[1968]: 475–76, 476n16; J. M. Turner 1975[1974]:102–5).

What makes this businessman and traveler most interesting is that he links the emergent Màxí politico-cultural traditions of Cuba, Brazil, and the West African Coast. The 19th-century voyages of Joaquim d'Almeida, his slaves, and/or his agents thus seem a likely medium of ongoing ideological communication among bearers of the emergent Màxí ethnicity on three continents.[29] Though he may be the most famous, he is not the only free traveler shaping the ongoing communication among these locations.[30]

Their travels clearly transformed and integrated multiple local cultures, helping us to understand a number of remarkable similarities between Bahian and Cuban religious practices and ethnic identities that distinguish both from those of the inland and precolonial African societies usually identified as their origins. First, both Afro-Brazilians and Afro-Cubans collectivize the ethnic identities associated with the EGAF dialect cluster: what Brazilians call "Jeje" is in Cuba called "Arará," after the dynastic city of Allada. Second, both Cubans and Bahians have coined compound terms for subgroups within the Jeje, or Arará, nation. Whereas Cubans speak of the "Arará Sabalú" (from the town of Savalou), "Arará Cuevano" (from the Màxí-speaking town of Covè), and "Arará Magino" (from Màxí)[31] subnations (Ortiz 1921:24; Sogbossi 1996), Bahians speak of the parallel "Jeje-Savalú," "Jeje-Caviano," and "Jeje-Marrim" subnations.

On the mid-19th-century Coast, the wealthy Jeje-Marrim and Savalou-born merchant Joaquim d'Almeida commanded influence from

Agoué to the towns of Petit-Popo, Comè, and Grand-Popo and owned a residence in Porto Novo. By the 20th century, his descendants and those of his slaves constituted an elite dynasty on the Coast and still identified themselves as Coastal Marrins, or Màxí, regarding Agoué as their ancestral home (J. M. Turner 1975[1974]:102–6). The Coastal Marrins' geographical distance from Porto Novo, the French administrative capital, and from the sinecures it made available, placed them among the least assimilated Coastal Djedjis and the most motivated to speak for the Djedji/Jeje collectivity in its trans-Atlantic manifestations. Their descent from Joaquim d'Almeida, furthermore, imbued them with the pride to maintain and advertise their distinctness from other peoples, including the descendants of other Djedji/Jeje subgroups.

THE PREEMINENT BLACK ATLANTIC MERCHANT:
JOAQUIM FRANCISCO DEVODÊ BRANCO

The most famous distinguishing mark of Ewè-Gèn-Ajá-Fòn religion and the new tutelary god of the Bahian Jeje nation in the 1930s, Dã is not only a snake but a god of dynamism and wealth. He could have found no greater human embodiment than Joaquim Francisco Devodê Branco. From the early 19th century until the 1930s, various Coastal Djedjis were traveling back and forth to Brazil, and, for Bahians, Branco was the most renowned among them. The contrasts between him and Joaquim d'Almeida are instructive. Whereas d'Almeida had been a slave trader, Branco was active only after the end of the slave trade; he and many other Coastal Màxís of the period traded in products critical to Afro-Bahian cuisine and religion, such as shea butter, kola nuts, and bitter kola (see Mulira 1984:145; Pliya 1970:46).[32] It seems no accident, then, that d'Almeida is most famous for his Roman Catholic piety, while Branco is most famous for his association with the priestesses of the Bahian Candomblé. Thus, the public religious affiliations of these merchants mirrored those of their critical trading partners—that is, for d'Almeida, the Portuguese, Brazilian, and French slave traders and, for Branco, both the Bahian Candomblé priesthood and the nationalist bourgeoisie of Lagos. Like the cultural diacritica of all ethnic groups, Djedji/Jeje religious affiliations were historically dynamic, situational, and strategic. Branco's commercial interests and the changing conditions of the Atlantic trade do a great deal to explain why his generation, rather than that of d'Almeida, presided over the rebirth of the Jeje nation and the snake god in Bahia.

Branco's tombstone describes him thus: "born in Asante (Mahié), resident in Lagos, businessman in Pôrto Novo" (Olinto 1964:213–15).[33]

Born in 1856, he had been taken as a slave to Bahia in 1864 (Lindsay 1994:31). Once freed by his Bahian master, he went to Lagos and became one of the richest men there. His enterprises supplied the goods that became emblems of Brazilian identity on the Coast and of African identity in Brazil. Branco took innumerable trips to Brazil, to which he supplied kola nuts and bitter kola. In Brazil, he bought dried meat, rolled tobacco, and *cachaça* (Brazilian rum) to sell in West Africa.

He also traded in goods *among* the cities of the Coast, as far away as the Gold Coast (Ghana) and perhaps other areas of West Africa as well. He owned more than twenty one houses in Lagos and Porto Novo. By the time he died in 1924, he had grown so wealthy that, according to his 1919 will, he bequeathed £1,220 to the survivors of his then-deceased ex-master (Olinto 1964:210–16, 265, 267; Laotan 1943:16; Marianno Carneiro da Cunha 1985: "Casas, Um Investimento"; Manuela Carneiro da Cunha 1985:122–31; Lindsay 1994:31).[34] Joaquim Francisco Devodê Branco had apparently taken this *nome de branco*—that is, his Western name—from his ex-master, João Francisco Branco. The insertion of "Devodê" seems to have taken place amid the Yorùbá-centered African cultural nationalism of Lagos at the turn of the century. While his change of name reflects a Lagosian nationalist idiom for the expression of political and cultural difference, the name is Fòn. "Devodê" is the Lusitanized transcription of "Devodie," a Fòn praise name for a person of extraordinary accomplishment. It means "This-Is-a-New-Fashion" or "Something-Unheard-Of."[35]

Branco's travels along the Coast surely exposed him to the snake worship of Ouidah and Porto Novo as well, and the Lagosian Renaissance of the 1890s established the practice among the Westernized Coastal bourgeoisie of invoking African religions as emblems of local cultural distinctiveness and dignity. A friend to several highly placed Candomblé priests, Branco undoubtedly influenced not only trans-Atlantic ethnic identities but religious ideologies as well.

Branco's back-and-forth travels across the Atlantic made him one of several Màxís whose personal renown reflected well upon their *terra* (land), or subnation. His personal friendship with Opô Afonjá's chief priestesses Aninha and Senhora allied his name and his nation with the most prestigious of Bahia's Candomblé nations and the most famous of its priestly genealogies. However, Branco would not be the last African-Brazilian to travel back and forth between French colonial Dahomey and Bahia during the period of the Bahian Jeje Renaissance, in which priests of his own Màxí/Marrim *terra* took the lead.[36]

Along with Martiniano do Bonfim, Felisberto Sowzer, and Lourenço Cardoso, Branco represents the zenith of an early 20th-century generation of high-profile merchants and pilgrims who deeply influenced

Candomblé practice, iconography, and parlance. The influences that Branco bore from the Coast cannot be known with the same certainty as those of Martiniano, but a comparison of their circumstances and the moments of change in their respective nations lends more than a little credibility to the case that developments on the Djedji, or Mina, Coast wrought effects in Bahia similar to those wrought in Bahia by the Lagosian Cultural Renaissance—but with a difference. This difference enabled the Bahian Jeje nation to demonstrate its distinction from, and therefore nonsubordination to, the mighty Yorùbá/Nagôs—that is, the famous snake god Dã.

Like these other travelers, Branco spent much of his West African time in the British West African colony of Lagos and experienced the influence of the Lagosian Cultural Renaissance of the 1890s, during which Yorùbá culture, religion, and language (and a defense of their "purity") became political weapons against British racism. Branco, who had commercial interests stretching across the French colonies of the Gulf of Guinea Coast, appears to have embraced the French-endorsed Djedji identity and the Agoué-centered Marrim/Màxí identity that he shared with his trade partners there and to have expressed it in a style typical of the Lagosian Cultural Renaissance.

The Brazilian Nagô travelers knew and discussed Lagosian Renaissance era publications about Yorùbá society with leading figures in the Bahian cultural world (see chapter 1). They also carried into Bahia the Lagosian Renaissance–inspired discourses of Yorùbá superiority and of the virtue of African "purity." It would be no surprise if Branco became an equally active advocate of Jeje identity—and perhaps even of its superiority—in Bahia. The fact that two of the priestesses with whom he associated in Bahia later turned up with icons and knowledge of the Coastal snake cult, previously unknown for decades in Bahia, suggests that Branco trafficked in information and supplies for the practice of West African EGAF religion as well. A trader in one set of commodities essential to the practice of Candomblé—kola nut and bitter kola—he is very likely the same person who, before his death in 1924, supplied Mãe Aninha (head priestess of the Opô Afonjá temple) with her collection of snake rings and Mãe Senhora (her successor) with her remarkably Coastal-sounding knowledge of "Idangbe" (Ferraz 1941:273–74).

The veneration of the snake gods Òṣùnmàrè and Ògún (yes, Ògún was, in some regions, a snake god—Peel 1995) had existed among the ancestors of the Yorùbá, but these gods have never rivaled Ṣàngó, Ifá, or Odùduà as divine emblems of Yorùbá pan-ethnicity. Snake gods known as Dã were worshiped in many towns of the Gbè-speaking region, but nowhere with the political centrality and ritual elaboration that their worship has enjoyed since the early 18th century in the Dahomean

Coastal city of Ouidah. The foremost temple of Dã was established then by Agadja, king of Dahomey from 1708 to 1740, during which time he first extended the kingdom's sovereignty to the seashore. He is said to have founded the Dangbe temple in the Coastal town of Ouidah to "ensure the continued prosperity of the profitable slave trade" (Herskovits and Herskovits 1976[1933]:57–58). Much in Afro-Bahian religious practice recommends that the god of pestilence Omolú, the goddess of death Nanã Burukú, and their straw-and-cowry-shell iconography as unifying emblems of pan-Jeje identity in Bahia. However, the sacred snake is the god most prominently displayed by the ranking Jeje temples in today's Bahia—Bogum and the temple of Pai Francisco.

It was by no means inevitable that Dã would become an emblem of shared identity among the Gbè-speaking peoples in West Africa or Bahia. The pride of place given to the snake god on what would become the Mina-Djedji Coast resulted from a Dahomean imperial strategy and from the slave-trading wealth that allowed Agadja and his successors to sponsor the cult so lavishly in the region to which so many Brazilian Marrins would return. The Màxí ancestors of these Brazilian returnees had not worshiped Dã. The European exoticism that made powerful white traders and missionaries particularly attentive to snake veneration, above all the other African religious practices on the EGAF littoral, probably also enhanced its value as a "cultural diacritic" (Barth 1969) of general Djedji/Jeje national identity, much as Western fretting about the veil has enhanced the veil's value as a symbol of Arab and Muslim distinction. Indeed, generations of European travelers to the EGAF Coast had, in their writings, given priority and publicity to ophiolatry as the defining feature of the Gbè culture.

Though his wealth and friendship with the great Nagô priestesses of Bahia must have made Joaquim Devodê Branco especially influential in Bahia, Branco is not the only Jeje traveler to communicate with the Candomblé priestesses in the 1920s and 1930s. Nor are the Nagô priestesses Aninha and Senhora the only priests with whom Branco communicated. Trans-Atlantic Marrim merchants like Branco are probably responsible for making Marrim the foremost *terra*, or subnation, and for establishing the snake god Dã as the preeminent public emblem of the Jeje nation in today's Bahia.

The Bahian Jeje Renaissance followed the late 19th-century and early 20th-century canonization of Jeje ethnic identity in West Africa and coincided with its dignification under French West African colonialism. Intercontinental Jeje business travelers had made these African intellectual and political developments available in person to the Jejes of Bahia. At the Coastal crux of late 19th-century interaction among Dahomey, the Lagosian Yorùbá, European visitors, and Brazilian returnees, the snake god had clearly emerged as a diacritical feature of EGAF religion and, by extension, ethnicity. The lapse of the slave trade and its replacement by a "legitimate" trade in Candomblé ritual supplies make it unsurprising that the strongly Roman Catholic cultural diacritica employed in public identity assertions by early Jeje merchants gave way to diacritica more compatible with the Candomblé-based markets that the later merchants would seek to cultivate. Gifts of information and religious adornments related to the Coastal snake god Dã seem to have lubricated new Jeje networks and communities in Bahia beginning in the 1920s. These Jeje-Marrim-dominated transnational networks fertilized the rebirth and iconographic transformation of the Bahian Jeje nation in the 1930s.

WHAT MADE THE TRANSNATIONAL MERCHANTS PERSUASIVE IN BAHIA?

One can understand the interest of trans-Atlantic merchants like Branco in cultivating a Jeje identity in West Africa. From the 1860s to the 1930s, a syzygy of French and Brazilian returnee interests made it an increasingly useful idiom of African cooperation in French West Africa. But why would so many priests, worshipers, and patrons in Bahia follow them? Why would people living in Bahia find Jeje identity and its religious emblems attractive in the 1930s? First, as wholesalers, the trans-Atlantic merchants enjoyed a wide network of contacts useful to Bahians. Second, the money, literacy, and cosmopolitan education of these extraterritorial merchants lent special credibility to their interested claims about the authenticity and singular worth of their merchandise, while their access to print media amplified those claims. Third, their public personae, their travels, their wares, and the profitability of those wares to local retailers allowed them to chart new and persuasive symbolic geographies. The Jeje "nation" and the Marrim "land" were so vivid and alive with networks of mutual dependency that the symbolic geographies of territorial nations and empires were hardly more credible.

What a territorial nation, an empire, and a diasporic nation all have in common is that they clothe the interests of particular classes in such vivid symbolic geographies and collateral networks of profit that more than one class can affectively embrace any of them when it is useful to do so or costly not to. Indeed, imagined communities typically rely for their success on their frequent persuasiveness to people of classes other than the founding class (Max Weber in Gerth and Mills 1946:267–301).

The interests and enterprises of transoceanic Jeje and Nagô wholesalers linked them socially and ideologically to groups beyond their immediate class and clothed that alliance in the Anglo-Atlantic language of "race." And these factors seem to account for a large measure of these nations' success in Bahia. Four generations after the arrival of the last captive from Africa, a whole range of sedentary local retailers and priests in Bahia were still described as "Africans," as much because of their typical occupations as because of their complexions. In defiance of Brazilian racism and with the support of the Lagosian Cultural Renaissance, they voiced pride in the "purity" of their ancestry (see, e.g., Pierson 1942[1939]:284, 292, 302).[37] However, their "purity" became equally useful as a promotional logic: their professed ancestry gave proof to the quality of their merchandise.

It was not only the trans-Atlantic traders but also a range of retailers and craftspersons who made their livelihood based on African ethnic trade monopolies and on an aesthetic of the distinctiveness and value of the ethnically African. The commercial motives and ritual consequences of such leaders' embrace of the Coastal merchants' "purism" are detailed in the next chapter, but a few examples underscore the point here. Chief priestess Mãe Aninha, who also grew wealthy on the Bahian retail end of the trans-Atlantic trade, was a foremost advocate of African religious "purity" (Pierson 1942[1939]:293–94; Santos 1962:20). The founder of the Jeje Casa da Ventura temple in the Bahian Recôncavo town of Cachoeira, Tia Júlia, also founded a major commercial house, called Casa da Estrela, in which she and her daughters marketed African products sent by her brothers, who resided on the West African Coast. From the late 19th or early 20th century until the mid–20th century, Tia Júlia and her daughters profitably supplied imported African religious goods for markets in Salvador and the Recôncavo region (Nascimento and Isidoro 1988:21–23; Wimberly 1998:86, 89n41).

Candomblé's strength as a "diacritic" of ethnic identity, as a structure of commercial cooperation, and as a creator of consumer needs has relied on a diffuse affective and spiritual power—linked not only to purposes but also to the rich array of feelings that the Candomblé ritual orders and integrates for the community that imagines itself through its symbols. Candomblé's service to political and economic projects relies on its power to resolve human suffering in believable and palpable ways. In the 1930s, when many victims of slavery were still alive, no small part of Candomblé's affective and spiritual power arose from its content as a vivid and self-confident response to the cruelty, indignity, and social disability suffered by generations of Brazilian blacks. The "Africans" of Bahia were somewhat culturally and phenotypically different from many of their neighbors, and that difference had been used to oppress them. Yet Candomblé priests found the psychological and literary ammunition to valorize that difference. It may be that only a small minority of "Africans" grew wealthy or made competitive political gains by emphasizing those differences. But they did so with such a vivid sartorial, culinary, gestural, and ritual iconography that millions of people now imagine their community, either periodically or regularly, in terms of this iconography and the symbolic geography it entails.

Next to the symbolic geography of the Yorùbá/Nagô nation, that of the Jeje was, from the 1930s onward, the most specific, linguistically marked, literarily verifiable, and for those reasons commercially profitable emblem of "African purity" available in Bahia. Not only its cultural icons and its identification with a known place in Africa but also the evidence of its bearers' dignity were available in the persons of Joaquim Devodê Branco and Lourenço Cardoso on the eve of this Jeje Renaissance. Like the Yorùbá/Nagôs, the Jeje merchants brought their own variety of transformative textual "proof" from West Africa. Dignified textual representations of EGAF religion, replete with endorsements of the snake god Dã, made their way into Bahia between the 1890s and 1930s (e.g., Herskovits and Herskovits 1976[1933]; Herskovits 1938; also Carneiro 1948: 51, 106; Oliveira and Costa Lima 1987:185–86; Ramos 1940[1934]:55–61; 1946[1937]:303). Ellis had written of the "Ewe-speaking peoples" of the Gulf of Guinea and declared that they possessed nearly the cultural sophistication of the "Yorùbá-speaking peoples"—a message that he impressed upon Brazilian psychiatrist Raimundo Nina Rodrigues, who in turn conversed with numerous Candomblé priests at the end of the 19th century.

THE TEMPLES OF THE JEJE RENAISSANCE

In Bahia, the momentum of the Jeje Renaissance reached a peak in the 1930s and 1940s. Bahian Jejes had suddenly come to think of themselves as a nation rather than as fragments submerged in a Yorùbá-Nagô-dominated Jeje-Nagô amalgam. And, like Branco, most of the Jeje priests embraced the Marrim subnation. As though reporting a "survival" from the past rather than the 20th-century innovation that it was, Carneiro cited three temples for having "valiantly defended the purity of the Jeje religion"—the Bogum temple of Mãe Emiliana, the Pôço Bètá temple of Pai Manuel Falefá, and the temple of Pai Manuel Menêz in the São Caetano neighborhood (Carneiro 1948:27–28, 50–51, 86; see also Correia Lopes 1943:559; Carvalho 1984:123). The first of these—and, by Carneiro's report, the most important house—displayed a painting of a snake on the wall of its public dance space, a novelty linked not to the inland home of the Màxí people but to the West African "Coast," where Jeje-Marrim traveling merchants conducted their business.

Though this Jeje snake iconography was new, it stuck. Since the time of my earliest observations in Bahia—the late 1980s—Bogum's inner walls have featured an elaborate mural depicting *multiple* snakes, and the priestesses of that temple have called their numerous snake gods by the Fòn name "Dã" (also Carvalho 1984:52, 54). In the early 1990s, my friend Pai Francisco mounted multiple bas-relief sculptures of snakes on the walls of his temple. Beyond the Bogum temple, the term "Dã" is now widely known but infrequently used, while the once-dead term *vodum* has become common parlance for the Jeje gods, used self-consciously to distinguish them from the Nagô *orixás* and the Roman Catholic saints.[38]

RETHINKING NATIONS AND TRANSNATIONALISM

This tale speaks to the literatures on nations and transnationalism in a number of ways. First, it embraces Anderson's conviction that the territorial nation is a relatively recent phenomenon and that particular economic classes of people tend to spearhead both national independence movements and the idioms of community that normally accompany them. Second, it challenges the conviction widespread in both of these literatures that transnationalism is new or structurally inimical to the territorial nation. A quasi-evolutionary logic of supersession underlies both literatures and distracts us from the variable circumstances and class agency that invariably keep multiple imaginations of community in play in the same time and space.

Whereas Anderson identifies printers and locally born administrators as the avant-garde of the territorially imagined community, in this chapter I have identified international merchants as the foremost agents of the trans-Atlantic nations of the African diaspora. The case of the diasporic Yorùbá nation suggests the further importance of musicians, literati, translators, and priests, including alienated Christian missionaries and leading Freemasons of the oppressed group (Matory 1999b). The better-studied case of the Anglophone black Atlantic recommends that we include black politicians (see esp. Stuckey 1987; Kasinitz 1992), sailors (Linebaugh and Rediker 2000), and record producers (Gilroy 1993) in the vanguard of transnational imagined communities. This case bears comparative significance for the study of the Jewish, Chinese, Irish, and South Asian diasporas, which do not simply exist but, on the contrary, must first be profitably "imagined" by interested classes in order for *populations*, periodically, to become *communities*.

Thus, I stop short of affirming that the territorial nation superseded or replaced the dynastic imperialisms, transnational religions, races, and other transnational imagined communities, only some of which preceded it. These diverse units of collective self-construction—not to mention what is nowadays called the "global community"—are interpretive frames that require *nothing but* imagination to make them real. It is just that, when they are materially profitable or our participation in them is militarily enforced, they become more real for more people. It is true that improved transportation and communications technology have increased the persuasiveness of transnational and "global" imaginations to wider audiences or publics, but these imaginations have hardly overridden or displaced other logics of community. In fact, the technological improvements that brought the transnational and the global to broader publics have also allowed empires, nations, and regions to render more vivid portraits of themselves, to broadcast their imaginations farther afield, and to find new, defensive motives to reinforce themselves.

These diverse imaginations of community have in fact been coeval and overlapping for the past two centuries, and each has been real to the degree and under the circumstances that it suits powerful class interests to dramatize, enforce, or act in terms of them. For example, in the 19th century, the "classical" territorial nations of northwestern Europe were also massive international empires based on the premise of their citizens' shared neo-aristocratic right to rule over racial Others. The imperial dominance of one's nation was a psychological equalizer that enabled all citizens of the metropolitan nation to pretend at times that their national brotherhood mattered more than the nation's internal hereditary class hierarchy.

And, of course, many an Old World territorial nation retained its antecedent monarchy, restored a defunct one, or invented one to go along with its national aspirations. Consider, for example, the United Kingdom of the Hanovers, Napolean's France, Juan Carlos II's Spain, Bokassa's Central African Empire, Faisal's Saudi Arabia, Thailand, and, according to at least one U.S. proposal in the wake of September 11th, Afghanistan. In the Americas, Haiti under Emperor Dessalines and King Christophe is not exceptional at all. The "classical" American nation-states typically retained another form of aristocracy. In the Americas, it was typically slavery and racial oppression that enabled even "white" peasants and sharecroppers to pretend that they were "kings" in their own homes, that they were racial aristocrats, and that their national brotherhood with other creole "whites" mattered more than their poverty or class endogamy. Citizenship in most of the Americas meant (and in many ways still means) "not-slave," "not-black," or "not-Indian" and depends for its reality on the real or symbolic presence of the disfranchised black or Indian.

Seen from the perspective of the slave and her descendant, the so-called American Revolution, for example, was a palace coup, in which the revolt against literal monarchy was less a move into some new historical stage of imagined community than the "white" creole elite's self-interested vacillation among the multiple idioms of community equally available in their time. Consistency of democratic principle had little to do with the motives of the white "revolutionaries" or with the outcome of their palace coup. Moreover, argues Susan Buck-Morss (2000), even the metropolitan French bourgeoisie at the time of the French Revolution, the mother of all Old World nation-states, was so symbolically and economically dependent upon Haitian slavery that the nonexistence of Haitian slavery would have made democratic nationalism unthinkable in the Old World. Furthermore, she argues, it is because the most influential philosophers and historians of "modernity" contrived to relegate a coeval and massive slavery to some falsely antecedent historical period that they have both (1) misunderstood the role of Africa and of slavery in "modernity" and (2) projected a false "stagism" on world history. Stagism, one might add, is the European nation-state's and the European settler colony's contrivance to silence the ongoing evils that they have jointly created.

It is true that, since the late 1990s, more and more analysts have acknowledged the major antecedents of late 20th-century transnationalism (Foner 1997; Robotham 1998; Szanton-Blanc 1997; Duara 1997; Abu-Lughod 1997), as had David Harvey in certain widely ignored portions of his widely cited book, *The Condition of Postmodernity* (1989). However, individually and collectively, these works have continued to emphasize not the vast qualitative similarities and quantitative differences between prior and present phenomena but, instead, a *typological* difference between stages.

Harvey is best known in the transnationalism literature for his concept of "time-space compression"—or the speeding up of change and the increasing ease of movement between places—whereby capitalism is said to have brought about transnationalism. His time horizons, however, are much longer than those stipulated by most of the anthropologists and sociologists who cite him. Whereas they tend to see "recent decades" as sufficient to cover the beginnings of transnationalism, Harvey sees the European explorations of the 16th century as the starting point. Harvey makes no effort to hide the fact that he is interested exclusively in the history of the West. It is vastly more puzzling that the anthropologists who propose universal histories intended to encompass both Western and non-Western cultures overlook the vastly translocal world knitted together by Islam, long before the 16th century.

Following Harvey, the most influential book-length studies of transnationalism (e.g., Appadurai 1996; Sassen 1998; Hannerz 1996; also Pries 2001), continue to represent the world as a manifestation of a single, capitalism-driven trajectory, in which regional diversity and local ebbs and flows in the degree of transnational involvement simply disappear or are seen as merely temporary interruptions. Because Western capitalism is treated as history's sole motor and rudder, history is divided into periods according to some moment or moments identified as "ruptures" in Western economic history—the European maritime revolution of five hundred years ago, the implosion of European colonial empires and the revision of U.S. immigration laws in the 1960s, or, most often, technological improvements and the crisis of capitalism in the early 1970s. Before one or another such "rupture," it is said that almost all social life everyplace in the world was lived and thought about in terms of bounded, autonomous social units.

Thus, transnationalist studies of the recent acceleration in the movement of people, ideas, media images, and investments across national boundaries tend to rest on the unnecessary premise that the nation-state and every other antecedent sociopolitical unit was once a closed container of political, economic, social, and imaginative life (see esp. Pries 2001). Before this recent acceleration, the nation-state is understood to have enjoyed a monopoly over its territory and over the loyalty of its citizens; therefore, transnationalism cannot but threaten the integrity or even the survival of the territorial nation in unprecedented ways (e.g., Appadurai 1990, 1996; Sassen 1998; Anderson 1998).

Ong (1999) has presented the most detailed ethnographic challenge to the view that transnational communities are superseding sovereign territorial nations but, like Sassen, she regards transnationalism as new and as requiring the nation-state to sacrifice bits of its sovereignty in unprecedented ways. However, to my mind, the historical standard of comparison is a suspicious one. Was there really a time in the history of most of today's nation-states when their territorial sovereignty was absolute, and not continually renegotiated with the agents of stronger nations, stronger international coalitions, and stronger commercial entities? It is unclear why we should overlook the ways in which the third world nations that became independent between the mid–19th century and the mid–20th century never gained the sort of cultural, economic, demographic, or political autonomy and territorial sovereignty that is idealistically attributed to the pre-transnational nation-state.[39] Throughout the existence of nation-states, many have suffered continual interventions in their internal affairs by imperial powers like the U.S., France, and the Soviet Union, not to mention interventions by regional powers like Cuba, Tanzania, and Vietnam. Recent theoreticians' silence about these matters seems curiously complicit with the similar silences in the national narratives of such imperial and regional powers.

The brilliance of Ong's ethnography and of theoretical overviews by Sassen and Appadurai is easier to appreciate when their work is read as a study of the contrast between the current transnational *reality* of nation-state and its border-bound *ideals* than as an account of how a national reality suddenly turned into a transnational one.

Where precedents are recognized in this literature, they tend to be described as the diminutive roots of today's *real* thing or classified as irrelevant because they happened before territorial nations had existed. Massive and ongoing forms of migration, communication, and exchange that predated and/or co-occurred with the nation-state are ignored without explanation (e.g., Ong 1999; Ong and Nonini 1997; Hannerz 1996; Pries 2001; Sassen 1998). Inattention to Islam is apparently justified by the teleological premise that only Western capitalism is *the* engine and rudder of history, or by the equally false premise that territorial nationalism so overwhelmed other forms of collective self-imagination that today's transterritorial communities only began once the territorial nation began to die.

Noncapitalist transnationalisms occurring at the height of "modern" nationalism are sometimes treated as some dead-end offshoot of history's *real* trajectory. For example, Verdery's iconoclastic account of the transnational forces that are now shaping nationalism in the former Soviet bloc minimizes the mightily transnational character of socialism itself (Verdery 1998:291). Socialist networks, alliances, commercial exchanges, and publications had long undergirded communities and created governments far beyond Eastern Europe—in Africa, Asia, the Americas, and Western Europe. Thus, Verdery's description of the pre-1989 "Communist Party states" as "partly insulated from transnational processes beyond those with other socialist countries" acknowledges the transnationalness of socialism but privileges Western capitalism and republicanism as somehow uninsulated and unrestricted in their flow across the world of social spaces, *except for those of parochial spaces occupied by socialism*. The ultimate triumph of Western-style capitalism, free trade, and republicanism over other systems' parochial insulation and restrictedness seems the foretold and inevitable overall trajectory of history. As first world steel, grain, and dairy industries inflict death by a thousand cuts on free trade, as "outsourcing" threatens the voting middle class, as third world dictatorship proliferates with subsidies from the West, and as Islam spreads through Western Europe, Mexico, and Rwanda, how long can we hold on to the seemingly foretold historical trajectories of the present moment? And as transnationalists heralded the global decentralization of power and deterritorialization of sovereignty in the 1980s and 1990s, who could have predicted the enduring efficacy of U.S. unilaterialism and territorial conquest in Iraq? Excessive confidence in teleologies and foretold trajectories blinds us to such realities.

Yet the transnationalism literature is not without its virtues. Perhaps its greatest insight is that we cannot study social units—even the ones that profess their own impermeability—as though they were bounded and autonomous from broader, global forces (see esp. Pries 2001; Hannerz 1996; Sassen 1998). Of course, the suggestion is often phrased as "we can no longer" rather than "we cannot" terms, as though the bounded autonomy of local cultures in a previous age was once a tenable premise (e.g., Pries 2001) and as though no scholars had previously considered the permeability and long-distance context of local sociocultural experience. Methodological precedents ranging from Malinowski's *Argonauts of the Western Pacific* (1922) to Sidney Mintz's *Sweetness and Power* (1985) and Gilroy's *Black Atlantic* (1993) are generally ignored.

The conventional vocabulary of "transnationalism" is shockingly amnesiac. Hayden White (1981), Deleuze and Guattari (1987[1970]), and much of the subaltern studies literature argue that the very structure of historical writing and of its sources tends to reproduce the interests and dramatize the legitimacy and inevitability of the state. They marginalize or render "unthinkable" (Trouillot 1995) those human projects that are of no interest to or contradict those of the state. Of course, the theorists of transnationalism are eager to capture a present moment and a future in which the interests and projects of the state are themselves marginalized. This history substitutes one unthinkability for another.

For example, I have yet to encounter a scholarly discussion of the slave trade as an instance of transnationalism, though it endured into the mid–19th century—at the height of "classical nationalism"—and was structurally central both to 19th-century capitalism and to the economies of a dozen nation-states. And theorists of transnationalism have only lately addressed the massive 19th-century immigration of Europeans to the Americas. Late 19th-century and early 20th-century Italian and Swedish immigrants to the United States, for example, sustained significant ties with their homelands (see Foner 1997; R. Smith 2001; see also Thomas and Znaniecki 1974[1918–20] on Polish immigrants). But even these cases are presented as finite in duration and, implicitly like earlier European migrations to the United States, doomed to be superseded by the assimilative powers of the territorial nation. I submit, on the contrary, that these transnational communities were superseded not by the national imagination but by imagined ties to a homeland more and more broadly reconceived—first as Protestant or Catholic Europe, then as "Europe" generally, and then as "the West," all of which professed homelands guaranteed potentially "white" immigrants' right to "equal" citizenship in the U.S. nation-state (see Ignatiev 1995; Roediger 1991; also Alba 1990). Euro-Americans' continued identification and communication with the European homeland have been as important in New World history as has the celebration of homogeneous citizenship within the bounded territory of the nation-state. Indeed, one has been the logical complement to the other.

Thus, while the imagination of indigenous and shared "brotherhood" in the nation has mattered greatly in these nation-states, so has the imagery of *foreign and diasporic racial origins*. On the one hand, the citizens of the territorial nation regularly imagine themselves as an indigenous, homogeneous, and egalitarian "brotherhood" within the territory of the nation-state. The imagery of egalitarian brotherhood, which many before me have called "indigenism," is associated with appeals for cross-class cooperation and for solidarity against foreign— and usually continental European colonialist—enemies. Various Latin American movements advocating *indigenismo, mestizaje/mestiçagem, negrigenismo*, and *mulatez* vividly illustrate the style of nationalist mythology that dramatizes such egalitarian "brotherhood." So do the novels of James Fenimore Cooper, the mythic origins that U.S. Americans ascribe to Thanksgiving, the Pocahontas and Sacagawea legends, the names we give to our sports teams, and the myth of shared Indianness that we *estadounidenses* celebrate at summer camp and in the Boy Scouts. Thus, American nations regularly rehearse national myths about the glorious friendship, miscegenation, and cultural convergence that have united whites with Indians or blacks (Anderson 1991[1983]; Sommer 1991; Hoetink 1973:131–38).

On the other hand, the imagery of distant origins and an ever-available discourse of shared community with them—which I call "diasporism"—is associated with appeals to hierarchy *within* the nation. In other words, "It is because we come from (and continually re-vitalize our ties with) some other, superior place," the strategic argument goes, "that we have the right to rule here (or to join the race that rules here)." Consider, for example, the Ku Klux Klan's appeals to white Americans' Anglo-Saxon-ness and the "classicism" that defines the dispersion of people and knowledge from ancient Greece and Rome, or at least from France, England, and Spain, as the source of all that is worthy within the New World nation. The imagined communities uniting New World elites with Europe are lashed together not only by enormous amounts of commerce but also by ceaseless streams of back-and-forth travelers bent on study and pilgrimage and, particularly in the 20th century, hundreds of thousands of New World soldiers willing to give their lives to save their European "brothers" from conquest by the Germans, the "Huns," the Spanish Fascists, or the "eastern bloc" Soviets (see, e.g., Levine 1977). By this logic, blacks and Indians can be represented, when it is useful to do so, as negative figures of the citizen, standing in contrast to a white citizenry that is represented as a transnational outpost of "Western civilization."

This case contrasts in detail but not in substance with a number of famous traits of Old World nationalism, in which long-resident transnational communities, such as Jews and Gypsies, have repeatedly been imagined and used as symbolic antitheses of the indigenist territorial nation. These diasporas have at once threatened and dialectically constituted the self-image of the European territorial nation. In both Old and New World scenarios, the territorial nation has been imagined in terms of both hierarchies and boundaries, of which the symbolic contrast between the indigenous and the diasporic has been a central metonym.

The French even employed this contrast in their construction of the hierarchies and boundaries of France's Dahomean colony. They imagined the Djedjis as indigenous, in contrast to the "foreign" and unwelcome Nagôts/Yorubas, despite the fact that both ethnic categories were in fact transnational and the ancestors of both had had an ancient presence in that colony. This is one of many cases in which the African and American colonies became testing grounds for the atrocious logics or genocidal deeds of 20th-century nationalists within Europe.

The New World nationalist logic of diasporic superiority and social hierarchy is quite standard within the Afro-Atlantic nations as well. These first two chapters illustrate how transnationalism established certain black ethno-classes in elite niches of Brazilian religion and commerce. Indeed, the notion of the Nagô/Yorùbá and Jeje nations' distant and superior origins is precisely what has attracted the light-skinned elites of Northeastern Brazil—not to mention the Cuban-American exile community—to these particular Afro-Atlantic nations.

In sum, the territorial nations emerging over these last two centuries are characterized less by the constant imagination of a territorially bounded and brotherly lot of equal citizens, as observed by Benedict Anderson, than by the strategic alternation between discourses of *indigenism* and *diasporism*—each involving its own selective constructions of community through memory, ritual, and text.[40] Hence, the same skepticism that Anderson advises toward indigenist territorial nationalisms is properly directed toward diasporisms as well. The myths told in their service often naturalize the social boundaries, cultural features, and political privileges of diaspora communities but recently invented.

Twentieth-century maps might have convinced some scholars that the identities associated with territorial nations are the all-defining juggernaut of the past two centuries of political history. And the latter-day proliferation of green cards has convinced other scholars that transnational identities really just got going thirty years ago. On the contrary, Afro-Atlantic diasporic nations like the Jejes illustrate a massive phenomenon that started a long time ago. National identities and national cultures have always been formed in a *transnational* context and have always crosscut alternative symbolic geographies. In general, I have argued that territorial nations, transnational communities, dynastic empires, religions, and races represent not evolutionary phases of world history in which one universal or even predominant form supersedes the others in a predictable sequence. Rather, the territorial nation added a way of imagining community to an already broad and still-viable range of others. These diverse ways of imagining community, as the vicissitudes of the Djedji/Jeje case suggest, are coeval, strategic alternatives invoked in the service of class interest. What makes one class's interested image of community more or less useful to other classes is less a matter of historical stage than of historical circumstance and strategic human imagination.

Many of the theorists predicting the imminent, transnational demise of the nation-state are themselves recent transmigrants into the upper-middle class of academic communities in the United States. Moreover, they are among the most talented, skilled, and sought after of immigrants, and, unlike their countrymen in the computer and atomic energy industries, they have tenure. They have experienced the United States at its most permeable. They might therefore take for granted the state that guarantees their exceptional privileges and fail to note how few people share those privileges. On the other hand, blacks, Native Americans, dark Latinos, and, in the wake of September 11, people of Middle Eastern origin or Palestinian sympathies in the United States cannot take for granted the inclusiveness of the nation or the lightness of the state's touch. Nor are they likely to imagine that the nation-state is melting away.[41] Hence, in one way, the transnationalist narrative reproduces the U.S. nation-state's official narrative of "freedom" and its white population's habit of taking for granted the privileges that the state confers upon their race. Those classes and races whose movements, commerce, and safety are *restricted* by the racial logic of U.S. nationhood are "silenced" (Trouillot 1995) in the conventional transnationalism narrative, just as they are in the U.S. nationalist narrative itself.

Also "silenced" in the transnationalism narrative are those in the past who were marginalized by the nation-state during its "classical" period and who therefore imagined geography and history in terms not dictated by the territorial nation. Some of their imaginations were decidedly transnational, such as early 20th-century pan-Africanism and ancient religious communities ranging from Islam and Judaism to the Yorùbá and Jeje diasporas. Sometimes those translocal ethnic groups marginalized in one nation-state drew symbolic or material strength from the leadership of that nation's subregions, from other nation-states, from foreign empires, or from empires previously overthrown in the host nation. Hence, the conventional transnationalism narrative not only silences a range of early translocal communities but also keeps its silence on much that the contemporary nation-state's official narratives also silence, such as those who are too poor or language-bound to seek a job abroad and those who still look to the nation-state to defend their jobs and property at home.

It might be fair to say that current transnationalism narratives are subsidiary to the nationalist narratives of the major metropolitan host countries. And while the Andersonian narrative restores a logic of historical change and agency to the genesis of nation-states, the transnationalism narrative simply appears to naturalize and extend the teleological history of capitalism. Transnations are represented as the "late capitalist" product of the juggernaut of history rather than one of several contemporaneous imaginative geographies by which intelligent actors with diverse interests and capacities have always vied to invoke the most advantageous among the multiple imagined communities available at any given time.

In my view, we will misunderstand the nature of today's transnationalisms if we imagine, following Harvey (1989), that all classes and races (not to mention genders, which are beyond the scope of this chapter) experience the material conditions of any given historical period in the same way and therefore imagine time and space in the same way. Such historical stagism is not sufficient to explain the sorts of communities that people can and do choose to imagine. As Clifford shows, any given space can host multiple readings of its social geography and history (1997:299–347).

Conclusion: On the Coevalness of Empire, Nation-State, and Transnation

The Djedji/Jeje transnational community came about before the "classical" age of nationalism and has endured well beyond it. Indeed, it shared its Atlantic space with multiple empires, nation-states, regions, diasporas, and religious communities (including some units that were several of these kinds of community at once, depending on how their members and neighbors experienced them at the moment).[42] The Djedji/Jeje nation cannot be understood outside the context of the multiple genres of imagined community that subsidized it *and* were crosscut by it, even at the very height of the period when nationalism ostensibly monopolized the imagination of communities. For example, if it was British-influenced priorities, language, orthography, and ethnological traditions that informed Nagô purism, then it was French imperialism that informed the brief Coastal rise and long-lived Bahian success of the Djedji/Jeje nation.

The "Congo" and "Angola" nations, too, were products of the Afro-Atlantic dialogue. They emerged from ongoing historical changes in West-Central African social identities, which accelerated amid the warfare, Roman Catholic missionary activities, and Portuguese settlement that accompanied the slave trade (see Miller 2000; Thornton 2000). According to Heywood (2000b), an Afro-Lusitanian Creole culture flourished around the ports of Luanda and Benguela in the eighteenth century; spread, through the agency of the enslaved, to Brazil, South Carolina, and Haiti; and facilitated the captives' adaptation to the Euro-American-dominated cultures of the Americas. Yet, without a living African language, literary representations of its African creole sources, or personal contact with West-Central Africa, the Congo-Angola ethno-class of Brazil embraced 19th-century imagery of the Brazilian nation-state itself—the Brazilian Indian and the Euro-Indian mixed-race *caboclo*, as well as the post-1930s nationalist imagery of *mestiçagem*, or cultural and racial hybridity (see Wafer and Santana 1990).

Like all imagined communities, the Djedji/Jeje nation has been a somewhat protean idiom of survival, prosperity, and healing for people who, at other times, claim membership in (or are claimed as members by) other imagined communities. The transnational Djedji/Jeje nation is a fine example of how forcefully a class of international merchants was able to reshape and indeed invent national identities, along with their cultural signifiers, across multiple territorially imagined communities. Moreover, this particular class of merchants and travelers did so *in dialogue with* (and not in contradiction with) several emergent territorial empires and nation-states. That is, the discourses and activities by which the itinerant Djedji/Jejes were shaping a trans-Atlantic nation were useful in the French imperial project of defining the cultural boundaries of the proto-state of Dahomey/Bénin Republic, and the Brazilian returnees themselves played a central role in the founding of Dahomey and Togo as independent nation-states just over a half century later (see J. M. Turner 1975[1974]). They constituted the first "indigenous" bourgeoisies, articulated the first cultural nationalisms, and contributed disproportionately to the postindependence ruling and administrative classes of Dahomey, Togo, and Nigeria.

As we shall see in chapter 4, the discourses and activities of the transnational Jeje and Nagô nations also facilitated the efforts of Regionalists in the Brazilian Northeast—such as Raimundo Nina Rodrigues, Édison Carneiro, Arthur Ramos, and Gilberto Freyre—to dignify the African roots of Brazil's *mestiço* Northeast. The Yorùbá played the same role in the Cuba reimagined by *negrigenistas* Fernando

Ortiz and Nicolás Guillén. Thus, over the two centuries since the territorial nation-state's emergence in the Americas, transnational imaginations of community have coexisted with territorially national ones. These seemingly contradictory imaginations of community not only *coexisted* within the cradle of the nation-state but also have been *vital to the nation-state's very empirical and logical constitution.*

The destruction of the World Trade Center on September 11, 2001, and Osama bin Laden's highly effective invocation of Islamic unity in its wake stand as an exclamation point on this complaint about the inadequacies of the literatures on nationalism and transnationalism to date. As an enormously popular imagination of community, Islam preceded and has coexisted with the nation-state throughout the nation-state's existence in the world they both share with Christianity and Judaism. Yet U.S. support for transnational settler colonialism in Palestine and for Islamic nation-states that tolerate U.S. military policy in Iraq and Afghanistan has left the populations of the Middle East and central Asia divided along class lines about the relative importance of the nation-state, the Ummah, and the global neo-empire of the United States. Even if we conclude that the Islamic revival is a reaction against the recent injustices and inadequacies of the nation-state, we must concede that Islamic imaginations of community have long been absolutely transnational. They were as transnational in the 19th and early 20th centuries as they are now. Whether and how its forms of translocal action have changed over the past millennium and a half, the past half millenium, the past two hundred years, or the past thirty years is an open question that is neither asked nor sufficiently answered by a capitalism-based teleology. In this chapter, the same question has been asked of the two-century-old Jeje nation.

In every age, scholars in the West are a bit too anxious to declare the dawning of new ages and, based on technological or ideological change, to declare the imminent end of all that has come before. It is high time for us to set aside teleological narratives and pay attention to the ever-hybrid present in our midst. The form of community imagined by newspaper publishers, the form imagined by transnational capitalists, and the form imagined by the adherents of widespread religions might be very different, but, under the right circumstances, each can mobilize enormously powerful movements out of the populations of other imagined communities.

An imagined community can outlive its immediate advantage, and it seldom equally serves the interests of all the people invoked in its ideal membership. Consequently, the availability of a new idiom of imagined community hardly predicts the extinction of others. Old and deeply sedimented idioms of community, particularly those committed to paper or embodied in ritual, can forevermore be resurrected in the assertion of new interests. Therefore, my point is not that things are now just the same as they always were, but that things are never just the same as they always were and that the present is not unique in this regard. Human beings—even under the dire circumstances of New World chattel slavery— have the power to alternate among and combine a wide range of imaginations of community within the same epoch. Evolutionary teleology seldom places past imaginations of community far beyond real people's reach.

Lest changing ethnonyms, ethnic boundaries, and cultural diacritica be mistaken for the *only* consequences of these circum-Atlantic identity politics, the next chapter concerns how deeply the class interests of an early 20th-century trade diaspora changed Candomblé ritual from its 19th-century Bahian antecedents and from what priests and scholars normally regard as its West African prototypes.

CHAPTER THREE

Purity and Transnationalism

ON THE TRANSFORMATION OF RITUAL
IN THE YORUBA-ATLANTIC DIASPORA

> My temple is pure Nagô, like Engenho Velho. . . . But I
> have revived much of the African tradition which even
> Engenho Velho has forgotten. Do they have a ceremony
> for the twelve ministers of Xango? No! But I have.
>
> —MÃE ANINHA, founder and chief priestess of the
> Ilê Axe Opô Afonjá temple[1]

> Dona Aninha. . . . really tried to study our ancient religion
> and reestablish it in its African purity. I taught her a lot, and
> she even visited Nigeria.
>
> —MARTINIANO DO BONFIM, diviner, merchant,
> trans-Atlantic traveler, itinerant religious
> consultant, and mentor to Mãe Aninha[2]

AMID THE PROLIFERATION of self-consciously hybrid postcolonial identities, why do leading adherents of Candomblé embrace and, as we shall
see, embody the principle that their material and spiritual well-being depends on purity? And there is a further puzzle posed by my own two
decades of ethnographic inquiry among Nigeria, Brazil, and the
Caribbean Latino diaspora: Why is it that Brazilian Nagô Candomblé
and the Cuban Lucumí Regla de Ocha pursue ritual objectives of "purity" and "cleansing" that are virtually absent from the cognate Nigerian òrìṣà religions that are typically regarded as their origins?

PURITY AND DISPERSION

Mary Douglas (1984[1966]:92) argued that, in "primitive" societies, the
logic of purity and pollution furnishes vivid images of mystical punishment for antisocial behaviors that these societies have no forcible means
to punish. I would argue, pace Douglas, that such logics are not lost to
so-called modern social groups—and particularly not to such diasporic

groups as Jews, Chinese in Southeast Asia, South Asians in East Africa, and the Nagôs and Jejes of Bahia (see also Kotkin 1993:36–38; Gungwu 1992[1991]:181–216; Bharati 1972; pace Douglas 1984[1966]:92).[3] For example, Jewish dietary laws and the *miqvah* (both based on a logic of ritual purity) dramatize the hereditary boundaries, the solidarity, and the cooperation of geographically dispersed religious/ethnic communities that often endeavor to unite around shared economic and political goals.

I will not be arguing that idioms of ritual purity exist or come about *only* as a function of the need to guarantee the solidarity of so-called primitive or dispersed non-state-based social groups. Nor will I argue that such idioms exist only in these social groups. Idioms of ritual purity have been elaborated actively by some soi-disant "modern" and state-based groups as well, such as Nazi Germany and the Anglo-Saxon nationalists of the 19th-century U.S. South. Moreover, the discourse of purity is not the *only* ritual idiom found in so-called primitive or diasporic groups. As we shall see, the religion of the ancient Ọ̀yọ́ kingdom, from which the founder of the greatest Nagô Candomblé lineage originated, celebrated ethnic hybridity and, as far as experts are able to discern, gave little emphasis to ritual purity. Moreover, the diasporic Angola and *caboclo* nations of Candomblé endorse their own brands of ethnic and cultural hybridity.

What I *will* argue is that, amid a range of ritual idioms and precedents available to the priests of the Yorùbá-affiliated diaspora, the idiom of purity has enjoyed the greatest success—a success out of line with its scarcity in both Ọ̀yọ́ religion and the reigning *mestiçagem* ideology of the Brazilian host state, even at the height of Candomblé's elaboration of ritual purity. I *will* argue that the disproportionate and anomalous success of the purist ritual idiom in the Brazilian Candomblé bears a Weberian "elective affinity" to two factors: the condition of diaspora and the role of merchant interests in making diasporas endure as communities.

Not all ritual idioms are equally suited to the consolidation of dispersed communities that cut across multiple ethnically plural societies. Such communities' transterritorial dispersion and their frequent role in long-distance commerce create, even in the "modern" world, special incentives to selectively highlight, preserve, and elaborate on discourses and ritual metaphors of "purity." Indeed, since the territorial nation-states that govern diasporas often throw their considerable material power behind efforts to homogenize their citizenries, religions of purity become the *safest* (albeit only partially successful) means of compelling the solidarity of the dispersed groups that crosscut nation-states.

On these matters, the Jeje and Nagô nations of the Brazilian–West African axis are cases in point. Among all the nations of Candomblé, it is these two that have both elaborated the idiom of purity to the greatest degree and conducted the greatest commerce in goods from the African "homeland." Indeed, I will argue that a class of transnational merchants and the local priests who retailed their merchandise gained the greatest benefit from the "purity" idiom that the transnational merchants introduced.

Hence, this is not so much a case study in the economic determinism of religious ideology or in the inevitability of "purity" as an idiom of social control in stateless and diasporic groups. Rather, it illustrates the convergence of social conditions and class interests that have made religious purism hegemonic, in the Gramscian sense, in a range of diasporic societies. It illustrates the forces that made "purity" (as opposed to a range of other available idioms of racial, religious, and personal identity) a useful bargaining chip for an influential class of Afro-Bahians in the early 20th century. This case also illustrates the forms of leadership behind the naturalization of the "purity" idiom as the institutional foundation of an entire family of religions.

I have argued that the foremost historical source of the discourse of African racial and cultural "purity" in early 20th-century Candomblé was, at one level, a late 19th-century movement of racial and cultural nationalism in colonial Lagos, British West Africa. The people who introduced this discourse into 20th-century Bahia were not Euro-Brazilian elites but the hundreds of Afro-Brazilian elites who, as merchants and pilgrims, repeatedly traveled between the West African Coast and Bahia from the 19th century until at least the 1930s.

Why the Euro-Brazilian elites of the Northeast went along with the countercultural assertions of the Candomblé priests and merchants is the question explored in chapter 4. The purpose of this chapter is to explore the convergence of black Bahian class interests that converted the Coastal literary and commercial discourse of "purity" into Candomblé's preeminent logic of ritual practice and embodied "imagination" of community. Whereas chapter 1 documents the transnational role of the Coastal literary discourse, and chapter 2 documents the role of transnational commerce, the present chapter explores the influence of a community's transnational dispersion on its choice of "purity" as its preeminent ritual idiom. The current chapter explores how this idiom is lived through bodily practice in an exemplary diasporic site. This case challenges the premise of Connerton (1995[1989]) that traditions of bodily practice exempt the "social memories" they embody from controversy and from a role in the tendentious reinterpretation of the past. Indeed, this case shows that the same leaders who articulate tendentious verbal reinterpretations of the past can redesign the bodily practices of enormous groups of people.

Though itinerant Afro-Bahian elites introduced the discourse of racial and cultural "purity" into Bahia from the 1890s onward, it became persuasive to multiple local occupational groups who made up the class known as the "Africans" in 20th-century Bahia. Moreover, this discourse validated the worth of the African merchandise and ethnically "African" services of these travelers and their allies in the Brazilian marketplace. Hence, with the support of numerous local retailers of "African" goods and priestly purveyors of "African" services, "purity" grew into the normative standard of identity and practice in the Bahian Candomblé as well. The discourse of purity also dignified the black merchants themselves and, by extension, all people of African descent in Brazil, thus facilitating the governance and solidarity of a dispersed community that crosscut an ambivalent nation-state. Thus, the selective privileging of "purity"—amid a range of principles and precedents available to Candomblé priests—took place not because of but *despite* the preferences of Euro-Brazilian elites, who, from the 1930s onward, came to advocate *mestiçagem*, or "hybridity," as the preeminent standard of Brazilian national identity and culture.

Merchant-priest Martiniano do Bonfim and merchant-priestess Mãe Aninha canonized the highly selected range of West African and Brazilian precedents that dramatized "purity" and "purification" in the religion's ritual technology of healing and community-building. Mãe Aninha's promotion of the Waters of Oxalá ceremony, which I will detail later, is perhaps the best example of these interested reforms.

PURITY AND COMMERCE AMONG THE "AFRICANS" OF BAHIA

In the 1920s and 1930s, a number of well-publicized fortunes had been made through the Bahian trade with West Africa and continued to be sustained by it, such as those of Lourenço Cardoso, Maximiliano Alakija, and Joaquim Francisco Devodê Branco. The "Africans" of Bahia—including these transnational merchants and their local retail partners—relied on the import trade for their supplies of strip-loomed "cloth from the Coast," or *aṣọ òkè*, "African" cowry shells, West African peppers, soap, kola nuts, herbs, and raffia (see Herskovits 1966[1958]: 249). Mãe Aninha grew wealthy on the retail end of that trade. She owned a shop in Bahia's historic Pelourinho neighborhood and another on the temple grounds, where she did a thriving business in Brazilian goods and African religious supplies well reputed for their "legitimacy" (Pierson 1942:293–94; Santos 1962:20). Ruth Landes describes numerous other "African" women in Bahia during this period with "stands selling spices, soaps, beads, and other specialties imported from the west coast of Africa" (1947:16–17). Among them was the *mãe-de-santo* of my friend Pai Francisco. She sold medicinal and sacred herbs. During some of his trips, Martiniano took coral and wool to sell in Lagos and returned with "cloth from the Coast" to sell in Bahia (V. Lima 1987:52). Felisberto Sowzer exported tobacco to West Africa and imported *sabão da Costa* (black "soap from the Coast") into Bahia (Irene Topázio Sowzer dos Santos, personal communication, 3 January 1996). The value of their sacred merchandise in Bahia depended on its "legitimacy," which, in local terms, meant its *Africanness*. So highly valued was African merchandise that many Candomblé gods refused any "non-African imitations" (Herskovits 1966[1958]:249). Not every participant in the import trade grew wealthy, but the link between "legitimate," or unadulterated, Africanness and pecuniary value relied on and amplified an aesthetic value increasingly pervasive in the local religious and secular milieu—"purity."

In Bahia, the "Africans" figured prominently not only in the overseas commerce with Africa but in butchering, dressmaking, laundering, herbal pharmacy, and market vending generally, while they monopolized certain other spheres of trade, such as the sale of cow heads and inner organs and the vending of food on the street. Indeed, it was often the *orixás*, or Candomblé gods, who ordered female initiates to take up the latter profession, and the gods were said to confer special protection on street vendors generally. The foods they sold were regarded as "typically" or "purely African," and vendors usually wore dresses and beads that are still associated with the Candomblé priesthood, including a shawl called a *pano da Costa*, because it is ideally made of strip-woven cloth from the West African Coast (Herskovits 1966[1958]:250–51; Landes 1947:92–94; Querino 1988[1938]:138–41). Thus, a range of retailers and craftspeople also made their livelihood based on African ethnic trade monopolies and on an aesthetic of the "purely African."

The "Africans'" monopoly on the supply of these goods and services enhanced both their wealth and their influence. The Africans used many means to demonstrate the pure Africanness of their wares and services, the most dramatic of which was the "Africans'" own daily sartorial, verbal, literary, genealogical, and ritual assertion of where *they themselves* had come from. Proudly, they proclaimed themselves and their ancestry "purely African" (e.g., Pierson 1942[1939]:284, 292, 302).

Substantiated by the literature and ideologies arriving in Bahia from Lagos, Ouidah, Agoué, and Porto Novo, these claims were not only commercially profitable but psychologically empowering during the transition from slavery to semi-freedom in early 20th-century Bahia. Hence, as it moved around the Atlantic, the discourse of Yorùbá superiority and "African purity" did not remain simply a diatribe against

British racism or a product endorsement. It became a balm to Bahians and other African-American peoples under psychological and political assault. In the wake of Martiniano and Aninha's religious revolution in the 1930s, this idiom of ethnic and class distinction not only contested the Brazilian premise of black inferiority but also overwhelmed rival idioms in the ritual management of personhood in the Afro-Brazilian nations of Candomblé.

The collaboration between "Professor Martiniano" (as he was often called) and Mãe Aninha to rescue the dignity of the Black race and "reestablish" the purity of what was, to them, its exemplary religion had probably begun decades before the 1930s, but only the groundwork had been laid in the 1890s, when, after his return from colonial Lagos, Martiniano was instructing forensic psychiatrist and ethnologist Raimundo Nina Rodrigues. At that point, the "Africans" had been recognized as an ethnic category in Bahia, while the emergent, literarily informed "Yorùbá" identity was only beginning to assume its paradigmatic status within the "African" ethnic group.[4] Casa Branca and a range of other Candomblé temples had already existed, but they were quite comfortable with the simultaneous or hybrid adoration of Roman Catholic and African divinities. And, though Nina Rodrigues had been persuaded to accord some value to the purity of African practice, his analysis largely concerned blacks' *inability* to understand any accurate form of Catholicism (R. N. Rodrigues 1935[1900/1896]:20). For example, he casually linked the term *bossal* (for "new arrival from Africa"—literally, a person "of the forest") with the term *ignorante* (ignorant).

By the 1930s, the socioeconomic status of numerous "Africans" and their rapport with the Regionalist intellectuals (such as Gilberto Freyre, Édison Carneiro, and Arthur Ramos) had overshadowed such anti-African reasoning and given way to literary and social scientific discourses that were remarkably like the political and religious discourses of Professor Martiniano and Mãe Aninha. Before 1911, Mãe Aninha had not even founded her own temple and lacked both the authority and the platform to articulate any purist apologetics. Her collaboration with Martiniano occasioned her embrace of this ideology. By the late 1930s, their teamwork exercised a transformative influence on the broader Candomblé priesthood, from Rio to Pernambuco, on the local and international press, on the academy, and on the governing class (e.g., V. Lima 1987:61).

Professor Martiniano and Mãe Aninha knitted together the Ọ̀yọ́ royalist ritual idioms of the Casa Branca temple, the political priorities of the Lagosian Cultural Renaissance, and the commercial devices of the trans-Atlantic Jeje and Nagô nations. They galvanized Afro-Bahian society and charted a new course for Candomblé discourse and practice by the late 1930s. These two priests appear to have been a major inspiration behind a growing emphasis on ritual idioms of purity, cleansing, and the radical extirpation of exogenous "influences" from the sacred "nation" and its subjects. They led a significant minority of Afro-Bahians—including trans-Atlantic merchants, local retailers of African religious supplies, and itinerant vendors of African foods—who found commercial advantage in claims concerning the "legitimacy" and "African purity" of their merchandise, further dramatizing these with claims that they themselves were of purely African ancestry. From their position of relative wealth, power, and prestige, they waged a battle for the dignity of blackness that many others found persuasive.

For an ethnic group organized around the importation of distinctive goods and the provision of distinctive services, "purism" might seem to be a structurally conditioned form of taste, but it took Martiniano and Aninha's leadership to integrate their nation around it. Hence, with the material and ideological support of ethnic merchants and tradespeople and with the moral support of many others, the Quêto/Nagô and Jeje priests of Bahia elaborated a pervasive ritual system of purification and recruitment unprecedented among their forebears in Africa.

In the wake of Martiniano and Aninha's public leadership and ritual innovations, the Quêto nation grew from a distant third in numerical rank among Bahian Candomblé temples in 1937 to the first-ranked in 1983 (Carneiro 1986[1948]:52; Santos 1995:17–19). Once ranked behind the Angola and *caboclo* nations, the Quêto/Nagô nation is now significantly ahead of both. Yet, proportionally speaking, much of the Quêto/Nagô nation's gain during this period was at the expense of the other purist nation, the Jejes, rather than of the self-consciously syncretic Angola nation. Since the 1980s, however, the Jeje and Nagô nations of Candomblé have also gained tremendously at the expense of Umbanda and Kardecism in the metropolises of Center-South Brazil—that is, Rio and São Paulo.

THE IMPURE ROOTS OF PURITY

The results of Martiniano and Aninha's revolution are illuminated by a comparison between Candomblé ritual practices and the historically related ritual practices of the West African Ọ̀yọ́ people. Amid the vast formal similarities between the two, what is remarkable is the unimportance of "purity" (and even of "cleansing") as a rationale of ritual practice among the Ọ̀yọ́ ethnics I have studied and, by contrast, the *hypertrophy* of these rationales in the Ọ̀yọ́-inspired Candomblé temples of Bahia.

On the Brazilian side, the focus of this article is the ritual practices of the most influential family of temples in Candomblé—those that grew from the root of the Casa Branca do Engenho Velho temple (also known as Ilê Iyá Nassô), which was founded around 1830 by "three black women from the Coast (i.e., West Africa) . . . Adêtá, Iyá Kalá, and Iyá Nassô" (Carneiro 1986[1948]:56). This last figure clearly ranked the highest, as the temple still bears her name, the House of Iyá Nassô. Her personal name is not known, but the priestly title indicates either (1) that before her initial arrival in or return to Brazil, she had been the highest-ranking priestess of the thunder god Ṣàngó in the Ọ̀yọ́ Empire (which would make her the highest-ranking of all òrìṣà [Nigerian Yorùbá, "god"] priests in the empire besides the emperor himself), or (2) that she and her Bahian followers held Ọ̀yọ́ religious models in such high regard that they named her after Ọ̀yọ́'s highest priest of the empire's tutelary god (see also Bastide 1961:65–66; V. Lima 1977:24–25).[5] Initiates of the ancient and prestigious Casa Branca do Engenho Velho temple, or Ilê Iyá Nassô, founded the now equally famous Gantois and Ilê Axé Opô Afonjá temples.

The non–Ọ̀yọ́ Ìjèṣà, Ànàgó, and Kétu peoples, as well as the Jeje-affiliated Màxí people, have left distinctive verbal marks on the oral history and nomenclature of Candomblé. But an Ọ̀yọ́ genealogy, language, and set of religious protocols left a mark like no other on the Iyá Nassô's Casa Branca temple and its scions. Even Mãe Aninha, the woman responsible for the vast expansion of these temples' prestige and influence in the 1930s, had been born of "Grunci" parents (from an entirely non-Yorùbá region in what is now Ghana and Burkina Faso) but was initiated at Casa Branca as a possession priestess of Ọ̀yọ́'s tutelary god Ṣàngó (Santos 1962:17). Xangô (in the Brazilian Portuguese orthography) remains the best-known orixá, not only in Bahia but throughout Northeastern Brazil. It is for this reason that the religion of the Ọ̀yọ́-ethnic town of Ìgbòho serves as an appropriate West African point of comparison in the following discussion. In the early 17th century, Ìgbòho was the capital of the Ọ̀yọ́ Empire. As the capital shifted southward, Ìgbòho remained to the north, more isolated from developments in Lagos and on the Coast generally.

PURITY AND THE MAKING OF THE *AXÉ*: THE INSTITUTIONAL
  LOGIC OF CANDOMBLÉ

Some symbolic themes in Brazilian Candomblé have been elaborated in ways far more reminiscent of the ideologies of the turn-of-the-century Lagosian Cultural Renaissance, and far more characteristic of the discourses of merchant diasporas and their religions, than they are of òrìṣà religion in Ọ̀yọ́ and Ọ̀yọ́ North. Purity is only one of them.

If there is one key term that encapsulates the religious and social differences between Brazilian Candomblé and the religion of the contemporary Ọ̀yọ́-Yorùbá, it is àṣẹ (Yorùbá, pronounced "AH-SHEH"). Àse (Yorùbá)/axé (Portuguese—pronounced "ah-SHEH") is, first, the power and authority embodied in human agents, animals, and inanimate objects. Hence, it denotes the "life force" that is inherent in all things and creatures. It is concentrated and transferable in the form of animal blood. That blood is used to amplify the "life force" and power of West African òrìṣà and Brazilian orixá altars, including the possession priests themselves, who, at pivotal moments, embody the gods through possession-trance. West African òrìṣà priests and Brazilian orixá priests share the related view that initiatic families (as opposed to birth families) can be constructed through the applications of sacrificial animal blood that link a person undergoing initiation (the initiand) to an initiating priest and, in turn, to the priest who initiated the initiator. West African and Brazilian worshipers share the sense that people, communities, and their residences are like *vessels*, which are rendered functional, orderly, and effective by their regulated containment of the proper àṣẹ/axé (see Matory 1986, 1994; Apter 1992).

As in West Africa, so in Brazil, the term àṣẹ/axé can be used to denote "power" and "authority," and can also mean "amen," or "let it be so." Virtually the same word, aché, is widely used and understood among Cuban and Cuban-influenced oricha worshipers in the United States and around the Caribbean basin. Uniquely in Brazil, though, the term axé is shouted by young people and sung by hip musicians as an exclamation meaning "Right on!" or "Wow!" or "Cool!" The expression has become a commonplace in daily Brazilian conversation and advertising.

In this, and several other ways to be detailed here, the concept of àṣẹ has been weighted differently in West Africa and Latin America. In colonial and postcolonial Nigeria, where Alfred Lugard's strategy of indirect rule fortified indigenous systems of top-down royal government, the term àṣẹ is most commonly used as a reification of governmental command. It refers to ritually constructed hereditary authority and posits the consistency of Yorùbá and proto-Yorùbá royalism with the entire God- and god-centered universe. In Latin America, on the other hand, aché (Cuban Spanish) and axé (Brazilian Portuguese) reify solidarity within geographically dispersed groups that often crosscut nation-states—groups that may include black and *mulato* Brazilians, the Black race as a whole, or specific transnational religious nations and families. Nigerian àṣẹ, Cuban aché, and Brazilian axé are all accumulable; people can lose àṣẹ/aché/axé, gain more of it, and possess more or less of it than others do. Thus, all three correspond quantitatively to rank within a community. However, far more than West African àṣẹ, Latin American *aché* and *axé* are detachable from states or governments. They link their bearers to relatively decentralized, geographically dispersed, and noncoercive communities.

In Brazil, *axé* can be a *countable* thing. An *axé* is the membership of a temple (or of a family of temples), united by the same continuous ritual transmission of *axé* from a founding priest or priestess to the "children-in-saint" (*filhos-de-santo*) whom he or she has initiated and, in turn, to all the "children-in-saint" whom *they* have initiated and so forth. Thus, one might ask a worshiper which *axé* he or she belongs to. In reply, he or she might identify the initiatic lineage by the name of its founder, its temple of origin, or its currently ranking temple. An *axé* (temple community or family of temples) is held together by shared *axé* (ritually constituted life force) and, ideally, by a shared set of ritual conventions that are supposed to have remained unchanged since the founding of that *axé* (family of temples). Members of the same *axé* are more likely to trust each other than to trust outsiders to have contact with their young and vulnerable novices, to witness and participate in their important and secret rites, and to manage a possessed member of that *axé*, because the ritual techniques from one *axé* might foul up or harm the gods, altars, and lives of people in another *axé*. The *purity* of an *axé*'s rituals is an important idiom of social boundedness and distinction.

In the Ọ̀yọ́ kingdom, the structure of the initiatic family does appear to have changed over the past two centuries. In the precolonial and colonial period, initiation seems to have bound possession priests to the monarch or other hereditary authority, thus replacing royal kinsmen who might usurp the monarch's power if allowed to serve as important delegates and counselors for the ruler (Matory 1994b:8–13). In the late colonial and postcolonial periods, the possession priesthoods acquired the additional role of dramatizing the continued rights of female agnates and their children to the increasingly exclusive patrilineal households of both commoners and hereditary rulers (Matory 1994b:141–69).

In Bahia, by contrast, black royals did not rule states, and patriliny neither really nor ideally defines membership in the corporate groups that have dominated most Afro-Bahians' lives. Thus, ironically, while the West African Ọ̀yọ́ possession priesthoods tend to be initiatic *kindreds* centered on the living ruler of the host kingdom or chieftaincy, Afro-Bahian *axés* (like their Cuban counterparts) are *lineages*. Far away from the forms of state control that shaped the priesthood in precolonial Ọ̀yọ́ and other proto-Yorùbá kingdoms, Candomblé is largely made up of multiple, autonomous lineages of priests and temples. Moreover, unlike the small corps of priests and priestlike royal delegates that made up the Ọ̀yọ́ possession priesthoods, the Candomblé priesthood has become a mass organization, with tens if not hundreds of thousands of possession priests.

Yet from the viewpoint of the individual initiate, merely *being initiated* is not a sufficient guarantee of personal well-being, respect, and power. The diagnosis and cure of spiritually caused afflictions are thought to rely on the legitimacy of the curing priest's links to higher powers. His or her credible association with a well-established and respected *axé* is similar in importance to a U.S. healer's display of a medical degree from a reputable university. That is, despite the existence of other medical colleges and other healing systems, with their less widely understood protocols of training and authorization, the prestigious medical degree inspires a qualitatively and quantitatively different sort of confidence.

Within any given temple, an initiate's authority and degree of respectability depend in large part on his or her initiatic age—that is, how long he or she has been initiated. In the broader community of initiates and among knowledgeable laypersons, each initiate's authority, degree of respectability, and very legitimacy depend greatly on the legitimacy of his or her lineage of temples and on the clarity of the initiate's links to it. This legitimacy relies, in turn, on the time depth of its traceably uninterrupted priestly genealogy and (in theory) its inherited, unadulterated liturgical practice. The age of a temple is frequently used as a metonym of its respectability and authority. Thus, various *axé*-founding Jeje-Nagô temples in Brazil—such as Casa Branca and Alaquêto (or Ilê Maroialaje)—are described by their affiliates, with some *symbolic* truth, as three hundred or four hundred years old. According to these indices, clearly, some *axés* rank higher than others.

There is no single priest with authority over the entire Candomblé religious community, even within any given Brazilian state, but some *axés* have the institutional power and some priests achieve degrees of projection in the media that give their public pronouncements and ritual protocols overwhelming authority, inspiring widespread deference, quotation, and imitation. These are called the "great houses" (*grandes casas*), and the "greatest" of these are linked to the *axé* of Iyá Nassô. (I employ this term in scare quotes because, as my friend Pai Francisco often reminds me, size and power are not necessarily preconditions of greatness.)

Of course, assertions of the uninterrupted and pure transmission of ancient *axé* and of specific ritual practices are seldom literally demonstrable. Sublineages within an *axé*—even that of Iyá Nassô—do diverge in their practices. Any given priest and his or her temple might abandon their original *axé* and seek affiliation with an older and more prestigious *axé*. Some temples are founded and run by uninitiated practitioners, who claim inspiration by other human or spiritual means. However, rising in the world of the Candomblé almost always depends on the public *recognition* of one's claims of descent from publicly *recognized and allegedly pure* genealogies of *axé*. The broadest and most comprehensive *axé* in Candomblé is the African "nation," which ideally shares not only a language and musical standard but certain broad commonalities of ritual practice.

Yet back-and-forth travelers have posed a challenge to the lineal transmission of tradition, authority, and sacred life force, or *axé*—and to the genealogical standard of purity that would otherwise seem inherent in this religion. If back-and-forth travelers, or even newly arrived Africans, bring credible information about supposedly original African ritual practices in the imagined homeland of the diasporic nation, does a priestess reproduce the practices of the previous Brazilian generation of priests in her family of temples or follow the recommendation of those who claim direct contact with the African roots of the more encompassing African "nation"? The founder of the Ilê Axé Opô Afonjá temple, Mãe Aninha, set a disruptive precedent and a prestigious example with her highly publicized answer. She placed the "African purity" of information and models derived through the Lagosian Cultural Renaissance squarely above the "purity" of the lineally inherited Brazilian standard. She thus jumped to the head of the queue of Candomblé leadership and powerfully reoriented the priorities of her Quêto/Nagô nation.

Between 1911 and 1938, Mãe Aninha enlisted the trans-Atlantic travelers—such as Martiniano do Bonfim and Joaquim Devodê Branco—in "restoring" to her then-new temple the supposedly primordial African practices that she claimed had been "forgotten" in the mother temple of her *axé*. Trans-Atlantic traveler Martiniano do Bonfim said, "Dona Aninha. . . . really tried to study our ancient religion and reestablish it in its African purity. I taught her a lot, and she even visited Nigeria" (Landes 1947:28). Aninha herself said, "My sect is pure Nagô, like Engenho Velho [the founding temple of her Brazilian *axé*]. . . . But I have revived much of the African tradition which even Engenho Velho had forgotten. Do they have a ceremony for the twelve ministers of Xangô? No! But I have" (Pierson 1942:293). Professor Martiniano was precisely the person who "revived" this ceremony and the practice of appointing *obás*, or "ministers," of Xangô from the ranks of the temple's economically or politically elite protectors and sponsors. However, this "revival" appears to be the product of Martiniano's Lagosian literary imagination, rather than the mere "reestablishment" of any Òyó practice.[6] Whatever its source, Martiniano's purist reform enhanced his own fame and influence as much as it did Aninha's.

Today, many priests and scholars of Candomblé remember Martiniano's name. While far more people remember Aninha and rightly credit her success to her personal brilliance and charisma, few contemporary priests or scholars recognize Martiniano's role in making her—that is, in creating the terms of distinction that enabled her to become more famous, more influential, and wealthier than any of her sisters in the *axé* of Iyá Nassô. Far from restoring Aninha's Candomblé to the state of its Ọ̀yọ́ antecedents, what Martinano actually did was to establish—for the first time in a syncretic religion with syncretic Ọ̀yọ́, Fòn, and Bahian antecedents—the organizing discourse and ritual standard of "purity." On the grounds of his transnational expertise, Martiniano came to be known and consulted by Afro-Brazilian priests in Bahia, Rio, Pernambuco, and probably other Brazilian states as well.[7] However, in the decades since his collaboration with Mãe Aninha, no priest has been able to ignore the prestige of this standard, even as the objective possibility and virtue of "African purity" are debated by scholars, journalists, artists, and priests alike. Little aware of the history of its genesis as an ideology of Candomblé practice, subsequent generations of priests have reproduced its logic not only in the patterns of the struggle for prestige but also in the ritual logic of Candomblé healing, recruitment, and initiation.

THE CHANGING CONDITIONS OF "PURITY"

World War II interrupted the trade between Lagos and Bahia that had initially favored this discourse in Bahia (Herskovits 1966[1958]:249). However, from the 1950s onward, the prospect of privileged diplomatic and commercial relations with Nigeria and Dahomey/Bénin persuaded the Brazilian government to subsidize a series of visits by Brazilian-descended Nigerians to Brazil and by Afro-Brazilians to West Africa (Verger 1976[1968]:553–56; Olinto 1980[1964]:295; Dzidzienyo 1985:135–53). In the 1980s, Nigerian travelers, traders, diviners, and a publisher flooded into Brazil with a well-precedented combination of religious zeal and profit in mind. Hence, like the 1930s, the 1980s witnessed a trade-empowered drive and a literature-backed putsch to "Africanize" the Afro-Brazilian religions and "purify" them of their allegedly non-African elements. In the 1980s, it was Opô Afonjá's Mãe Stella and a new cast of West African Yorùbá scholars—from the Yorùbá-centric University of Ifẹ̀/Ọbafẹmi Awolọwọ University in Nigeria— who led the way.

Indeed, there is a new Martiniano—Professor Wande Abimbọla— Ọ̀yọ́ man, former professor and vice-chancellor of the University of Ifẹ̀, itinerant *babaláwo* diviner, official spokesperson (*àwíṣẹ́*) of the Ifá divination priesthood, polyglot, and published author many times over. Yet as his authority threatens to displace that of his sometime Afro-Bahian,

Afro-Cuban, and Afro–Puerto Rican collaborators, the more established priests among them have cooled to talk of religious "purity" and to the idea of restoring an "African" standard of ritual practice.[8] Nonetheless, the mark of "African purity" on the religions of Brazil and of the Cuban diaspora has been inscribed indelibly on Candomblé, while "re-Africanization" has acquired powerful allies at the superwealthy margins of òrìṣà religion—for example, in São Paulo and the United States (see also Prandi 1991).[9]

Hence, dispersion and commerce do not make the ritual idiom of "purity" inevitable; they just create the conditions of certain powerful and persuasive assertions by those who stand to profit by them. Leaders who wield this discourse well have shaped revolutionary changes in a range of Afro-Atlantic religions. Hence, in the Brazilian Candomblé, the organizing discourse of "purity" is subject to contestation, and resistance is often coded in terms endorsing "syncretism." Nonetheless, hierarchy and membership in the most prestigious axés, as well as upward mobility among axés, are still coded in purist terms. Moreover, the ritual pursuit of "purity" remains deeply embedded and pervasive.

THE CONTEMPORARY RITUAL MANAGEMENT OF PURITY AND IMPURITY

The contemporary Candomblé depends on chief priests' expertise in managing the contents of the heads and bodies of a geographically dispersed group of temple clients, followers, and initiates. Priests conduct rites intent on removing improper "influences" from people's heads, from the vessels that represent their heads, and from social spaces like cars, houses, temples, and temple grounds. In turn, expert priests engineer the *deposition* of *proper* forces and entities into people's heads (as well as structurally similar social spaces), all intent on guaranteeing the health of their wards and the orderliness of the communities that these experts rule. A logic of cleansing and purification rests at the center of this series of removals and depositions. In fact, it is no exaggeration to say that most Candomblé rites now include "cleansing" as an indispensable element or as their main purpose. Since the 1930s, certain rites of *collective* cleansing, expressly intent on "purifying" the temple community, have become central in the temples most influenced by Ilê Axé Opô Afonjá.

It seems no coincidence that a wide range of these rites requires the use of imported African materials, as well as foods, clothing, and other products the provision of which was the specialty or monopoly of the "African" ethnics of the 1930s. These include camwood, indigo, kola nut, bitter kola, red palm oil, "soap from the Coast," "unguent from the

Coast" (*limo da Costa*, or "shea butter"), cowry shells, African wild buffalo horns, cow heads, animal innards, wild herbs, handwoven cloth, and "African"-ethnic foods. In the following section I will try only to illustrate these ritual themes and their commercial entailments, as well as the logic of personhood and community that they dramatize.

Purity and Impurity

Since the time of Martiniano and Aninha's collaboration, Oxalá and his rites have enshrined "purity" as the first and highest principle in the political discourse and ritual practice in an ever-widening circle of Jeje-Nagô temples. However, as Douglas observes cross-culturally, purist religions regularly restore what has been rejected as filthy or anomalous as sources of power (1984[1966]:159–79). In Candomblé, these anomalies and boundary straddlers are personified in Exú, the *eguns*, and the *caboclos*. Exú in Brazil is the gender-blending, promiscuous, omnivorous master of communication. His domain is the street (*a rua*) and particularly the crossroads (*encruzilhada*). Exú is also the opportunity-opening, alliance-making, fertilizing, and dangerous intermediary between the outside world and the purity-centered hierarchies of the Candomblé, without which those hierarchies cannot survive. Exú can facilitate or disrupt the purity that is necessary for the personal well-being of the Candomblé member and for the collective integrity of the temple.

The *eguns* are the spirits of the dead—usually personal ancestors whose efforts to help the living actually harm them because their world is upside down relative to ours. The *eguns* suggest both the structural similarity between biological and ritual kinship and the need to keep them apart. The *caboclo* Indian spirits are ethnically hybrid, hard-drinking, and often animal-like beings who typically wear copious feathers and live in holes in the ground (*buracos*). They empower those who seek independence from the purist and hierarchical structures of the Quêto/Nagô nation. Such sources of disruption and insubordination are, if Douglas and I are correct, almost predictable adjuncts to the cult of purity in Bahian Candomblé.

Cleansings

People with all sorts of troubles and from all social classes seek the help of Candomblé priests, who might recommend recourse to other curative systems or prescribe rituals of personal fortification, cure, or retaliation. People whose lives are full of conflict and confusion are often believed to be afflicted by an Exú, and people with the appearance of mental problems are often believed to be afflicted by *eguns*. However, the problems that Candomblé priests endeavor to solve have all sorts of symptoms, and proper diagnosis usually depends on cowry-shell divination.

The typical rite of personal fortification is the *bori*, which, in West African Yorùbá, means "to worship the head." In Candomblé, it is understood to invoke the sea goddess Iemanjá to remove the heat from or "cool down the head" (*esfriar a cabeça*) and to feed the head, thus strengthening its *axé*. Another common sort of personal fortification is "to close the body" (*fechar o corpo*), that is, to seal out malevolent forces. It goes along with the preparation of sacred bead necklaces for the client to wear for self-protection. Because the beads gain their protective powers from the imported "soap from the Coast" in which they are washed and from the herbal infusion in which they are steeped, this comprehensive procedure of closing the body is known *pars pro toto* as the "washing of the beads" (*lavagem das contas*).

One of the foremost and perennial themes in Candomblé healing, initiatic, and funerary rites is cleansing the body of inappropriate contents and sealing it against their future penetration. This theme, much akin to the logic of "purity" in the apologetics of Martiniano and Aninha in the 1930s, has been elaborated to an extreme in Candomblé ritual, as contrasted with contemporary Ọ̀yọ́ ritual practices, which, due to Ọ̀yọ́'s distance from the Atlantic coast, are probably less directly influenced than Candomblé by the Lagosian Cultural Renaissance and certainly less so by the Brazilian trade diaspora.

For example, on the one hand, present-day Candomblé healing rites emphasize the removal and exclusion of the exogenous influences that are the source of affliction. On the other hand, present-day Ọ̀yọ́ Yorùbá medicine emphasizes balance among the fluids and worms that normally occupy the body, their restoration to their proper locations in the body, and the *pacification* (rather than the exclusion) of the malign forces that might impinge on the body. For Yorùbá traditional healers (*oníṣègùn*), these factors are the bases of personal health (Buckley 1997 [1985]). This logic of pacification is explicit in the term for a commonplace element of Yorùbá medicinal practice—*omi ẹ̀rọ̀*, literally "the water of softening." Of the malign influences that afflict, Nigerian Yorùbá diviner Abimbọla explains, "No, you don't drive things off. They would just come back. You soften or placate them so you can coexist" (Wande Abimbọla, personal communications 10 August 1999).

By contrast, almost all Candomblé curative rites, which work by removing and sending away the evil "influences" (*influências*), are also described as forms of "cleansing" (*limpeza*). For example, when insanity results from attacks by the dead, it is cured through "cleansing (away) the spirit of the dead" (*limpeza de egum*). Seen from the viewpoint of an outsider anthropologist, this type of cleansing seems to dramatize the threat to ritual kinship posed by the implicitly rival and parallel principle of social obligation and authority—blood kinship. In this "cleansing," white foods are passed over the body of the client/patient (white being the color associated with the dead, with protection from the dead, and with mourning) and then deposited into a bowl. A cow's head—incidentally, another paradigmatic item of "African" commerce in the 1930s—is similarly passed over the body and deposited in an earthenware bowl (*alguidar*). These items are intended to feed the spirit of the dead and thus lure it away from the human patient's head.[10] The bowl of sacrificial items is then wrapped up and rushed off to a place where the evil influences it now contains can no longer touch the patient. However, once deposited elsewhere, these influences might escape and affect passersby, making "the street" a permanently dangerous place of contagion.

Diagnosed even more often than attacks by spirits of the dead are attacks by the messenger god Exú, who can be bribed into doing anyone's dirty work. The victim of Exú's attack must also be "cleansed" in such a way that draws Exú away from the victim, partly with a bribe of food.

In Candomblé, as in many other religions, restrictions on what one may eat and with whom one may eat it define the boundaries of important social groups. Moreover, in Candomblé, the placement of shrines maps out a series of forces on whose exclusion the purity and survival of the temple depend. Exú is both the foremost guardian of those boundaries and the bearer of the filth that would penetrate those boundaries if his vigilance lapsed. Hence, Exú is enshrined near the exits of a temple and its grounds, and is characterized by his willingness to eat any kind of food that any kind of Brazilian will eat (i.e., *tudo que a boca come*, say Candomblé priests). On the other hand, the gods who typically possess Bahian Candomblé initiates in trance sit in the central, interior shrines of the temple and are highly selective in their eating habits. They typically eat "African" ethnic foods—such as bean fritters (*acarajé*), cornstarch custard (*acaçá*), hominy (*ebô*), mashed yam (*ipetê*), puréed okra (*amalá* [*sic*]),[11] and beans with eggs (*molocum*), as well as the heads and inner parts of animal victims (i.e., the same parts whose sale was monopolized by "African" butchers of Bahia in the 1930s). Moreover, each god and his or her devotees obey certain stringent food prohibitions. And each such commensal group performs a specific role in the temple division of labor and authority. Some food prohibitions are shared by the whole community of initiates, such as eggplant (*beringela*), which is reserved for the spirits of the dead (*eguns*).

Exú's vivid sexuality dramatizes another boundary and logical structure of Candomblé's sacred kinship. Unlike most *orixá* altars, Exú's are often anthropomorphic, featuring exaggerated, erect phalluses or prominent vulvas and breasts. One of the main metaphors of sexual excitement and promiscuity in Brazil is "heat," which also describes the main forms of pollution that an improperly managed Exú can dangerously convey into the sacred community. Exú represents an irony at the core of Candomblé ritual conventions, one similar to that posed by the *eguns*. Prohibitions on sexual activity (*resguardos*) are a defining element of ritual activity and of the purity that ritual activity requires and endeavors to increase. Thus, priests and participants are required to abstain from sex for a specified period before and after any sacred rite.[12] Moreover, priests and their novices are forbidden to have sex with each other, and a person cannot be initiated by a lover or a spouse.[13] It is partly for these reasons that some contemporary priests say that sex has nothing to do with (*tem nada haver com*) and has no place in their religion.

However, in a way beyond the conscious recognition of most priests, this religion (like most others) has a great deal to do with sex. In Candomblé, sex is a recurrent metaphor and metonym in ritual efforts to control what does and what does not penetrate the body of the temple, the religious community, and its members. For example, possession by and other forms of engagement with the god prohibit sexual activity during the surrounding period *precisely because* the devotee's and the temple's relationship with the god is, at its height, structured symbolically like an *exclusive* sexual relationship. Hence, for example, the initiand is called an *iaô*, which many worshipers know means "bride" in Yorùbá, and possession priests are said to be "mounted" (*montados*) by the god. In the 1930s, among the various ways of ending an episode of possession by the god, reported Ruth Landes, was the washing of the priestess's "mouth and sexual organs" (1947:146).[14] In a further example of this implicit metaphor, one purist Bahian priest of the Quêto/Nagô nation told me in 1987 that anyone who serves as a possession priest of more than one god is "promiscuous" (*promiscuo*).

Thus, Exú's eating habits and sexual character identify the nemesis of Candomblé's purity and institutional solidarity, as well as the logic of its membership, internal hierarchy, and ritual protocol. An *axé* is a geographically dispersed community structured by the gods' symbolically sexual penetration of the priests, by ritual parenthood, and by an ethnically coded commensality. Worldly sexuality and indiscriminate eating are therefore generally kept at bay as a condition of the maintenance of a bounded diasporic community and a condition of social ascent within or among such communities. Moreover, "legitimate" African goods and African ethnic merchandise—such as kola nuts, black soap, and cow heads—are among the least dispensable elements in these procedures to cleanse away the "impurity" of sex and blood kinship.

Initiation

Initiation links people permanently to the *axé*. Initiands, like priests, can be male or female. In this case, however, I shall employ the feminine pronoun for ease of exposition and because most initiates are female. The initiand, or novice, is stripped of her old clothes, shaved, bathed repeatedly with "soap from the Coast" and herbal infusions, and subjected to other ritual "cleansings" (*limpezas*). The head of the novice is surgically implanted with the substance of the god to which she is consecrated. Just as the novice's head is prepared, so are vessel-altars containing the emblems of the gods that empower and protect her. Ultimately her primary god will possess her.

Both her head and the vessel-altars will become instruments through which her relationship and the temple community's relationship to the gods will be regulated. These instruments must be handled with great respect and with special attention to rules concerning the "cleanliness" of the body. The initiate must obey rules concerning her menstrual cycle, sexual activity, and bathing. Her well-being and that of the temple community depend on it.

Though the novice is called an *iaô*, cognate with the Yorùbá term *iyàwó* (bride [of the god]), many Brazilians emphasize the metaphor of "birth," also present in Ọ̀yọ́ initiations, to understand the nature of the novice's experience. The novice is *born* into the *axé* and, for some time, must behave like a small child—eating only with a spoon, eating only cool foods, sitting on the floor or on a bench rather than at a table, and so forth. During this period, the novice is regarded as highly vulnerable, her body "open" to dangerous influences, such as "heat" and other dangerous forces "from the street" and from the dead. She must avoid persons just arriving from the street and must wear armbands (*contra-eguns*) and necklaces (*mocãs*) to keep the spirits of the dead at bay. Much of the conduct she learns while in seclusion in the temple is designed to protect her from the penetration of bad "influences" (*influências*) and to maintain proper and exclusive relations with the forces and entities that legitimately penetrate her body.

While the initiation and its purist principles are surely not reducible to their historical relationship to the commerce between Brazil and West Africa, the initiation too relies extensively on that commerce. Initiation subsidizes and prepares personnel for a transnational industry in the production and exchange of drums, ceramics, wooden bowls, beadwork, metalwork, domestic animals, wild animal products, Brazilian and

African plant products, African black soap, African chalk, cloth from every continent, and, most important, stones. Among these products, whatever comes from Africa is the most expensive, the most highly prized, and the most symbolically charged. The initiation itself requires the purchase of hundreds of dollars' worth of these goods, and ritual "obligations" (obrigações) entailing similar expenses and purchases will continue throughout the initiate's life. The demand for "purely African" merchandise does not result from the preferences of Euro-Brazilian bourgeois sponsors. Rather, such sponsors are called in to satisfy a preexisting demand that would not be fulfilled as easily without their resources.

The initiation suggests a further shift in emphasis between the Ọ̀yọ́ and Bahian logics of sacred community. Far more than the Ọ̀yọ́ Ṣàngó priesthood, Candomblé has become antihereditary. Even biological parenthood is anathema to "parenthood-in-saint," or initiatic parenthood. For example, a person cannot be initiated or taken care of spiritually by his or her biological parent. The initiation has become a form of forcible adoption in which the threat of continued illness and misfortune is the overwhelming motive to enter into a new and demanding adoptive family. Not only are dead kin represented as sources of pollution, but the initiation also emphasizes the *removal* of the novice from her family, such that she must be bought back at the end of the ceremony (Herskovits 1966[1953]). Prohibitions on initiating one's children guarantee that initiatic lineages will not be biological lineages.[15] By contrast, most Ọ̀yọ́-Yorùbá people (and especially òrìṣà worshipers) seek to remain so close to the familial dead that they bury them in or near the home and often lounge on the graves. Many children are happily regarded as the reincarnations of deceased grandparents. And, in today's Ìgbòho, replacements for a dead or retiring Ṣàngó priest are compulsorily recruited from his or her kindred. The Ṣàngó possession priesthood in Ìgbòho does appear to dramatize the removal of the initiand from her family (Matory 1994b:193–98), but not nearly to the degree that Jeje-Nagô Candomblé initiations do. Bahian Jeje-Nagô initiations appear not only nonhereditary but antihereditary.

These differences, however, reflect as much change in Ọ̀yọ́ North religious practice as in Jeje-Nagô religious practice. That is, the Ṣàngó worship in Ọ̀yọ́ North has, over the past half century, ceased to be a means of displacing royal kin and recruiting non-kin into the bureaucracy of the Ọ̀yọ́ king's palace. By contrast, the Candomblé priesthood has, over time, become more and more detached from any particular hereditary kin group, nationality, or race. Priestly authority is increasingly in contrast with and often antagonistic to familial authority. Initiation is generally seen as a calling and a matter of inexorable individual destiny, rather than as a matter of family obligation or hereditary ascription.

A Sacred Geography

During a period of seclusion in the temple ranging from seventeen days to six months, the new initiate of the Jeje-Nagô Candomblé will observe and assist in numerous rituals from which he or she might previously have been excluded. These rituals will take her to parts of the temple compound previously unseen. Overall, these perambulations diagram a normative structure of personhood and community. The resident novice will internalize a distinctive, sacred geography.

Whereas Bastide has argued that the compounds of the "great houses" plot out a miniaturized geography of Africa (1961:82–83; 1978[1960]:247–48), it is evident to me that they also plot out a logic of boundaries, hierarchies, and antitheses reproduced in ritual protocols of cleansing and purification as well. In wealthy houses, each *orixá* or group of related *orixás* is given its own detached building, or "house," on temple grounds. But, in more modest temples, the first god that is given quarters apart from the main building is Exú. On account of his role in expelling and keeping out bad, exogenous influences, he is generally kept as far away as possible from the other *orixás* and as close as possible to the exit of the temple compound. The next priority for a separate dwelling is given to the goddess who governs the *eguns*, or the dead (Iansã Igbalé); the lords of sickness and death (Omolú and Nanã Burucú); the *caboclo* Indian spirits; and the spirits of deceased high-ranking priests of the house. Like Exú, these spirits personify the forces that a purist temple must keep at a distance to maintain its internal collective health. These spirits variously personify, it should be noted, the linked phenomena of worldly sexuality, ancestors and biological kin, ethnic hybridity, and loyalty to persons outside the temple community (i.e., the "street").

The street is the most dangerous site in this sacred geography. Proximity to those who have just arrived from the street and have not yet bathed is thought to convey heat, which is a particularly dangerous influence for initiands and recent initiates. They are taught to run away from such people, just as they are taught to avoid the heat of the sun and not to turn their backs to flames (Cossard-Binon 1981:134, 137, 143; also 1970). Heat, the street, and Exú appear to be interlocking representations of the complex forms of interaction that disrupt the purity and focused sociospiritual order of which the new initiate is being made a part.

In that regard, Exú and the spirits of the dead are not the only threatening spirits. Occasionally, a *caboclo* Indian spirit "mounts" an initiand. Unlike new *orixás*, *caboclos* (see figure 11) are ill disposed to following orders, and they have been known to defy head priests and to object to the initiand's choice of a temple to join. One priestess told me that her father-in-saint (i.e., her *pai-de-santo*, or the priest who initiated her) therefore "suspended" (*suspendeu*) her *caboclo*—that is, he sent it away. That same father-in-saint has for decades resisted this priestess's efforts to establish her own independent house. After some years, it was the *caboclo* who came back and inspired the priestess to set out on her own, and it was the *caboclo* who taught her the necessary information that her father-in-saint had kept away. *Caboclos* embody little concern for cultural and racial purity, and even less for hierarchy. Indeed, the term *caboclo* normally refers to a person with Indian and European ancestry combined. *Caboclo* spirits are both a source of impurity and an emblem of individual self-assertion. Their shrines, too, stand apart from the main temple building.

In sum, virtually all Jeje and Nagô/Quêto Candomblé rituals engineer the removal of bad influences (associated with Exú, with the *eguns*, with the street, and with nonmembership) from the bodies of individual clients, friends, and members. Most Jeje or Nagô/Quêto Candomblé rituals also deposit and nourish in the heads of members the substance of the gods (associated with the "house" and with "Africa"), thus linking those members to the sacred temple lineage, or *axé*. The purist logic of hierarchy and institutional integrity is personified in Oxalá and given the highest priority. Candomblé's main public rituals display the wealth, antiquity, efficacy, and overpowering reality of the gods' presence in the community through spirit possession at the lavish festivals, where the priests don the attire of prosperous wives and royals while the gods don the bodies of the priests themselves. A range of symbolic themes unites the Jeje and Nagô Candomblé with its contemporary Ọ̀yọ́ counterpart—particularly those related to vessel symbolism and the displacement of biological reproduction. However, contemporary Jeje-Nagô Candomblé and Ọ̀yọ́ religion diverge strikingly around the theme of purity—a theme virtually absent from Ọ̀yọ́ ritual rationales but pervasive in Candomblé talk and practice.

THE STRANGENESS OF "PURITY": THE Ọ̀YỌ́ COMPARISON

In Bahia, major ritual elaborations on the cleansing/purification theme and concentration on the worship of Oxalá, or Ọbàtálá, do not appear to come naturally—or by way of mere Herskovitsian "survival"—to such an Ọ̀yọ́ dominant priesthood as the *axé* of Casa Branca, or Ilê Iyá Nassô. So we are left with much the same riddle that sent me searching for the origins of the organizing discourse of purity in Bahia since the 1930s. The ethnographic literature on Yorùbáland in the middle to late 20th century documents several socially central rites of collective ritual self-cleansing, chiefly in locales distant from Ọ̀yọ́ cultural and political influence—such as Ilé-Ifẹ̀ and Ìjẹbú (Awolalu 1979:152–55, 179).

For example, during Ilé-Ifè's annual Èdì, or Ọlọ́jọ́, Festival, citizens of the kingdom sweep out their houses and surrounds, loading the filth into a basket that will be borne to the river on someone's head, cleansing the city of the misfortunes that it has accumulated over the prior year. Thereafter, citizens of the town observe an early-evening curfew to avoid recontamination by these misfortunes. Town-wide rituals of this sort have been identified in other Yorùbá cities as well (see also Awolalu 1979:179–80), but few have been documented with the specificity that I heard from one long-term resident of Ilé-Ifè. Having observed the rite alongside close friends in Ifè's royal family, she tells me that the person chosen to carry the load is a Hausa residing in the neighboring and enemy Ọ̀yọ́-ethnic town of Mọdákẹ́kẹ́. This Hausa has allegedly been lured under the pretense that he is to be crowned king of Ifè, whereupon he is mesmerized by chanting Ifè priests and liberated only once he has dumped his headload into the river (Ọlabisi Ọdẹrinde, personal communication, 23 August 1999).

I cannot verify my informant's details as empirical observations of the actual conduct of this ceremony, and I doubt that she observed all the phases of the ceremony that she reports. What I do not doubt is that the account she related to me is based firmly on what her princely friends in the Ifè think is *supposed* to happen, revealing a culturally and politically central *ideology* of cleansing and ethnic identity in Ilé-Ifè. According to this social imaginary, the person who bears away the filth purged from the in-group is a member of a major, rival political and commercial group in Nigeria (the Hausa) and is said to have been recruited from a residence in the headquarters of Ifè's local Yorùbá archenemy—the Ọ̀yọ́ refugee town of Mọdákẹ́kẹ́. For Ifè royals, ritual cleansing appears to have an ethnic-boundary referent of the sort that ritual purification also displayed in 1930s Candomblé.

Ifè's elaborate and central development of this sort of ceremony or image of ritual cleansing is symbolically consistent with the importance the city gives to the worship of Ọbàtálá, or Òrìṣà-Ńlá (the Nigerian cognate of "Oxalá"), a god for whom cleanliness is a central iconographic and ritual leitmotif. Researchers in this non-Ọ̀yọ́ cultural sphere regularly describe Ọbàtálá as the highest-ranking of all òrìṣà (Ellis 1964[1894]:38–39; Fabunmi 1969:6; Awolalu 1979:21; Idowu 1963:71). Yet it is only relatively recently that West African commentators have taken to describing Ọbàtálá as the god of "purity" and describing these forms of collective cleansing as "purification" (Idowu 1963:73; Fabunmi 1969:6). Even these evolving references in West Africa, and their attachment to the defilement of the ethnic Other, have an explicable political history.

As Andrew Apter points out, Ọ̀yọ́ expansionism inspired the development of two rival "ritual fields"—one "Ọ̀yọ́-centric," of which Ṣàngó is the ranking divine symbol, and the other "Ifè-centric," of which Ọbàtálá and the òrìṣà funfun (white òrìṣà) generally are emblematic. Particularly in polities conquered by Ọ̀yọ́, these "white gods" appear to be identified with the conquered kingships and to dignify the virtue of patience, humility, cool forbearance, and peaceful acquiescence. Thus, even in the wake of military and political domination by Ọ̀yọ́, the conquered claim continuing ritual powers under the divine imprimatur of Ilé-Ifẹ̀ (Apter 1992:27–31, 231n29).

As Mary Douglas has argued, ideologies of purity and the mystical consequences of pollution often bolster social norms and claims that are difficult to enforce by any means other than the fear of those mystical consequences. Ọbàtálá's proto-purist principles seem to acknowledge that the conquered had few other means to protect the small hereditary privileges with which their defeat had left them.

Neither Ọbàtálá/Oxalá nor the proto-purist principles of his worship in Ifẹ̀'s sphere of influence enjoy nearly the same reverence in the Ọ̀yọ́ "ritual sphere." As an incorporative and continually growing military empire, Ọ̀yọ́ did not dearly value purity, perhaps because mixing was so profitable, just as it would become for the expansionist Dahomean kingdom (Yai 1992; Blier 1995c:78). Even the priests of Ọ̀yọ́'s tutelary god Ṣàngó celebrate his and other Ọ̀yọ́ gods' foreign and Islamic roots (Matory 1994b:129–30, 253).

The Ọ̀yọ́-Yorùbá religions that I researched in the late 1980s showed no signs of purism or any suggestions of even the mild forms of ethnic "cleansing" reported in Ifẹ̀. Any given òrìṣà vessel-altar in Ìgbòho, for example, tended to incorporate multiple gods of diverse ethnic origins, depending on the multiple lines of inheritance that had converged on the worshiper-owner of that altar. That is, the calabash representing Yemọja in a particular compound might include small calabashes and dishes for Ọ̀sun, Èṣù, Ṣàngó, Ọbàtálá, Òrìṣà Oko, Ifá, and/or Ògún, and any of those small containers might have come from a range of the owner's lineal and nonlineal kin (Matory 1994:155–61). One Yemọja calabash even contained a cowry-shell neclace that the owner said was brought from Mecca after a relative's return from the hajj. In the panegyrics (òríkì) recited in Ìgbòho, Ṣàngó is described as a non-Yorùbá, Nupe Muslim; Ògún as a native of the distant, non-Ọ̀yọ́ town of Irè; and Òrìṣà Oko as a native of the distant Ọ̀yọ́-ethnic town of Ìràwọ̀. The historical documentation of Ọ̀yọ́ religious hybridity is old (Frobenius 1968[1913]:177, 187, 205, 210; Johnson 1921[1897]:164; Peel 2000: 123, 125; see also Kramer 1993[1987]).

Thus, the Ọ̀yọ́ òrìṣà priesthoods that I have researched develop themes of multiethnic and interreligious bricolage (Lévi-Strauss 1966 [1962]) to a far higher degree than do their current Candomblé counterparts, for whom the idea of a Muslim òrìṣà or an altar for one god that includes multiple others within its vessel would be difficult to understand (see also Frobenius 1968[1913]:187–205). The otherwise largely Ọ̀yọ́-inspired possession priesthoods of both Candomblé and Cuban "Santería," or Ocha, have elaborated rites of purification alien to the practices of today's Ọ̀yọ́ priesthoods (Matory 1994b:226–28).

It is no coincidence, then, that the first West African writers to introduce the discourse of "purity" and "purification" into the exegesis of òrìṣà religion—that is, Fabunmi (1969) and Idowu (1963)—were non-Ọ̀yọ́ and part of a metropolitan bourgeoisie in the shadow of politician Ọbafẹmi Awolọwọ's mid-20th-century Yorùbá ethnic nationalism, in which the Yorùbá people were seen as actual or potential victims of Hausa and/or Igbo domination, and in which Yorùbá individuals or subgroups who allied with non-Yorùbá groups were persecuted. It is also the advocates of this ethnic nationalism who most privilege Ilé-Ifẹ̀ as the Yorùbá cultural capital. On the other hand, across multiple generations of civilian politics in 20th-century Nigeria, no Yorùbá subgroup has been more willing to ally itself with the Hausa than have the Ọ̀yọ́-Yorùbá (Matory 1994b:70–73).[16]

And, while Abimbọla tells me that the Lagos-centered Ijọ Ọ̀rúnmìlà, or Church of Ọ̀rúnmìlà, has of late endeavored to reduce its reliance on Christian models (Wande Abimbọla, personal communication, 10 August 1999), I know of no movements in the Ọ̀yọ́ region to expurgate non-Ọ̀yọ́, non-Yorùbá, or non-African influences from their òrìṣà religions. The Brazilian Candomblé, on the other hand, is in the midst of a growing movement, led by Ilê Axé Opô Afonjá's Mãe Stella, to eliminate religious syncretism altogether.[17]

DOCUMENTING THE CHANGE IN BAHIA

The themes of purity and purification were not entirely absent form 18th- and 19th-century West African practice, but their appearance in the 20th-century Candomblé is no mere Herskovitsian "survival." They are themes selectively reproduced and amplified under specifiable political and economic circumstances. Professor Martiniano and Mãe Aninha shared the motives and the means to make them the preeminent themes of Candomblé ritual practice.

The preexistence of some of these themes in Bahia gave credibility to their project. The Washing of [the Church of] Our Lord of Bonfim (Lavagem do Nosso Senhor do Bonfim) was a major annual event for Afro-Bahians in the 1890s. That a rite of *cleansing*—the actual scrubbing down of the church—became the focus of late 19th-century Bahian Catholicism suggests the influence of motives that included but went beyond Catholicism, such as the theme of washing in Oxalá's mythology. It was no secret that many Afro-Bahians regarded Jesus Christ, to whom the church was consecrated, as the Christian counterpart of Oxalá/Ọbàtálá (R. N. Rodrigues 1935[1900/1896]:176, 180). This fact might be one reason for which Oxalá eventually came to be regarded by all as the highest of all the Brazilian Nagô gods. However, this consensus did not come quickly or automatically. Even in the late 1930s, after centuries of Roman Catholic influence, there was still great diversity of opinion among the temples: "In some *seitas* [or temples] Oxalá is considered the most important of the deities; in others, Xangô; and, in still others, Omolú" (Pierson 1942[1939]:280).

Oxalá's meteoric rise within the Candomblé and his symbolic association with purity appear uniquely indebted to the efforts of Professor Martiniano and Mãe Aninha, which are in turn rooted in the Lagosian Cultural Renaissance. We find evidence in the fact that the period of Martiniano and Aninha's greatest leadership in the Candomblé and of their most highly publicized endorsements of racial and cultic purity—the 1930s—immediately preceded a proliferation in ethnographers' descriptions of rites of "purification" in Candomblé. Over the subsequent decades Oxalá came to be called the most "important," "first," or "highest" of the Candomblé deities, and only then did he come to be associated expressly with "purity."

The Waters of Oxalá

In Brazil, the Candomblé temple most responsible for propagating the ideology of African purity is Mãe Aninha's Ilê Axé Opô Afonjá. There and at many of temples of the Yorùbá-affiliated Quêto/Nagô nation, the annual ritual calendar now begins with a rite called the "Waters of Oxalá" (As Àguas de Oxalá). It is an in-house rite whereby all the members of the temple community help to wash the "god of purity and peace"—Oxalá—and all members of the community cleanse themselves of the impurities that have entered them individually and collectively over the previous year.

The rite commemorates the mythical time when Oxalá insisted on taking a trip outside his kingdom, against the advice of his *babalaô*, or Ifá diviner. The diviner allowed Oxalá to go and visit his son Xangô in the son's distant kingdom, but only on the condition that Oxalá never refuse to help anyone or complain about anything during his voyage. Thus, when Exú the trickster asked Oxalá to assist him in raising his headload of red palm oil, Oxalá could not refuse. Exú then maliciously spilled the oil all over Oxalá, who always wore immaculately white clothes. Without complaint, Oxalá washed his clothes in the river and continued on his journey. But when Exú stopped him again to ask for help in raising his headload of charcoal, Oxalá could not refuse, and Exú dumped the charcoal on Oxalá's clothes. Without complaint, Oxalá again washed his clothes in the river and proceeded on his journey. By the same ruse, Exú then soiled Oxalá with a burnt black palm oil pomade called *àdín*. Oxalá patiently washed his clothes clean a third time and finally reached Xangô's kingdom.

However, on his path, he encountered Xangô's escaped steed, which he secured with the intention of returning it to the owner, but Xangô's soldiers immediately apprehended Oxalá, mistaking him for a thief. During the seven years that they kept him in prison without Xangô's knowledge, the land and the women of the kingdom fell barren. Only when Xangô's own *babalaô* divined to determine the cause of the kingdom's misfortunes was his father Oxalá released, whereupon Xangô begged Oxalá's pardon and ordered his subjects to retrieve water and wash Oxalá three times (see also Verger 1981:260–62; note that Verger himself was an affiliate of Opô Afonjá). It is these three washings that are now reenacted annually in most Quêto/Nagô temples.

In the 1930s, when Mãe Aninha of the Opô Afonjá temple first became famous, not every Quêto/Nagô temple held the Waters of Oxalá ceremony. Moreover, Mãe Aninha's choice to place this ceremony at the very beginning of the festival season diverged from the sequence established by the founders of her priestly line (in the Casa Branca do Engenho Velho, or Ilê Iyá Nassô, temple) in a direction upgrading the ceremony's importance in the life of the religious community she led. This ceremony and the priority given to it in her Opô Afonjá temple appear to ritualize and dramatize the very same "purity" that traveler Martiniano do Bonfim and Aninha established in the 1930s as the core of Candomblé's social and ritual ideology.

The Ilê Axé Opô Afonjá temple had been founded under the authority of Brazil's version of the Ọ̀yọ́ imperial god Ṣàngó, known as Xangô; indeed, he is still the tutelary god of the house. But, by Aninha's hand, Oxalá became the object of the first and most elaborate series of festivals in the annual ritual calendar. Opô Afonjá appears to have reinterpreted the year-opening New Yam ceremony (o Inhame Novo) of the older temples in Iyá Nassô's *axé* in terms emphasizing the role of *Oxalá*, his struggle to keep clean while away from his homeland, and, for the first time, his emblematic status as a god of purity and purification. Opô Afonjá's older-sister temple, Gantois, had called the year-opening ceremony the "New Yam," highlighting Oxalá's agricultural significance, and Raimundo Nina Rodrigues, writing in the late 19th century, described Oxalá as governing the "reproductive potency of nature" (R. N. Rodrigues 1935[1900/1896]:38–39 emphasis), an understanding of the god that is close to what I have heard expressed in Ìgbòho—that is, that Òrìṣà-Ńlá is a god of uterine gestation. In 1916, Manuel Querino described Gantois's New Yam as a right of penitence, without ever describing its purpose in terms of "purification" (Querino 1988[1938]:38–39—this passage had originally been published in 1916 [see p. 17]).

In 1948, following Mãe Aninha and Professor Martiniano's extensive influence on Candomblé and lengthy collaboration with Édison Carneiro, Carneiro became the first scholar to document a ceremony called the Water of Oxalá (Agua de Oxalá) among the temples of Casa Branca's *axé*. He did so with explicit reference to Aninha's Opô Afonjá temple and described the ceremony repeatedly as a "rite of purification" (*rito de purificação*), noting that not all temples performed such a rite in the name of Oxalá (Carneiro 1986[1948]:64, 89–90).[18] Nor does it appear to have been the first rite of the ritual calendar in all of the temples that performed it. In his 1958 publication, based primarily on research in the Opô Afonjá temple, Roger Bastide gives nearly exclusive emphasis to the function of the Waters of Oxalá ceremony to "purify" Oxalá's altar and the temple community (Bastide 1961:112–13).

Like Carneiro, Bastide infers that such "purification" is an expiation of "sin" (*pecado*). If perhaps some Candomblé members offered this explanation between the second decade of the 20th century and the 1950s, hardly any do so now. "Sin" is an idea with almost no force in Candomblé discourses about proper ritual and social conduct. In the 1980s and 1990s, Candomblé members almost uniformly describe Oxalá as the god "of peace and purity" (*da paz e da pureza*). Precisely *what* the community is being metaphorically purified of during the current Waters of Oxalá ceremony remains open to priestly and scholarly interpretation. However, the simultaneous crescendo of ideas about ritual, ethnic, and racial "purity" in the 1930s suggests a gestalt whose unity scholars have persistently overlooked and priests have apparently come to take for granted.

Today at Mãe Stella's Opô Afonjá temple, the weekly Amalá de Xangô (where the tutelary god of the temple is feted with his favorite dish, okra) is the most frequent occasion of community-wide gatherings, but no member of the community is expected to miss the annual Waters of Oxalá ceremony, where the sacred stones of Oxalá are washed. Moreover, it is the most participatory of all the rituals in the temple, with every officer, initiate, and affiliate of the temple ideally carrying sacred water for use in the ritual washing. Every participant has prepared for the rite, through personal self-cleansing with water, herbs, "soap from the Coast," and balls of moistened cassava meal passed over the body before being tossed out the front gate of the temple grounds. To the participant's head is bound one of the few commodities for which there exists no Brazilian-grown substitute—the kola nut, or *obí*. At Opô Afonjá, this rite has been performed since at least 1934, when novel purist discourses about "African" ethnic identity had come into full evidence, with associates of Opô Afonjá taking the lead in articulating them (Santos 1962:96). Since then, it has become a dominant practice in the Jeje-Nagô Candomblé.

Three changes have grown increasingly clear in the "great" *axé* of Casa Branca, Gantois, and Opô Afonjá and in the broader Quêto/Nagô nation of which this *axé* is the virtually undisputed and clearly most imitated leader. First, the Waters of Oxalá ceremony is now almost universally performed in that nation. Second, the name "New Yam" has fully given way to the name apparently established at the Opô Afonjá temple, suggesting the leadership of Opô Afonjá in this gradual transformation. Third, with Opô Afonjá in the lead, virtually the entire Candomblé community has reached a new consensus about the centrality of purification among the functions of the Waters of Oxalá ceremony. The ceremony's former associations with agricultural and sexual fecundation are long forgotten. These changes clearly reflect the ideological lead provided by Opô Afonjá and its merchant allies in the 1930s.

CONCLUSION

This ethnographically inspired historical revision is directed against not only Melville J. Herskovits's and others' vision of African diasporic religion as a "survival" of the past but also against a new, antiessentialist ethnography (e.g., Motta 1994; Dantas 1982) that effaces the black interests and strategies behind "African purism." Leaders of the Quêto/Nagô diaspora selected among and magnified Ọ̀yọ́ and Ifẹ̀ ritual precedents and Lagosian political discourses in a way that captured the imaginations of the "African" ethnic group in the Bahia of the 1930s. Their political and commercial discourses of "purity" appear to have selectively amplified and recast the meaning of certain antecedent ritual themes, which, once recast, came to dominate an ever-widening circle of Candomblé temples throughout Brazil.

Mãe Aninha and Professor Martiniano might not have been the first in this trans-Atlantic Yorùbá ritual complex to conduct rites of cleansing, but their tremendous influence on Candomblé rested on their novel use of the concept of "African purity" as the linchpin uniting the Ọ̀yọ́ cult of possession, a non-Ọ̀yọ́ cult of cleansing, the nationalist discourses of the Lagosian Cultural Renaissance, the social conditions of dispersion, and the ethnic priorites of a diasporic commercial class in

Bahia. Personified in Oxalá, the logic of purity has been increasingly central and increasingly ritualized in the Candomblé since the 1930s. The prestige, media coverage, and protection from police persecution that Aninha, Martiniano, and their friends secured for the purist temples have, since then, guaranteed the expanding prestige of the "purity" principle.

Nonetheless, there is disagreement over the precise meaning and indices of "African purity" (see also Dantas 1988). For example, do the imported African tie-dyes worn by Italian-ethnic Candomblé priests in São Paulo outrank the time-honored *baiana* skirts that remain de rigueur even in Bahia's most "African purist" temples? "African purity," in the end, is less a consensual norm of belief and cooperative action than a shared symbolic meta-logic available for use in cooperative *or* competitive projects. It is used as easily to marginalize or demean fellow worshipers and temples as to include or uplift them. Such meta-logics are in the nature of diasporas and of culture generally.

At the same time, and for the same reasons, the opposite principle holds power in Candomblé. The impurity of sexuality is ritualized and personified in Exú, that of blood kinship in the spirits of the dead (*eguns*), and that of equality and independence in *caboclo* Indian spirits. It is not so much that these forces are categorically condemned as morally bad. Rather, the integrity and hierarchical functioning of the Quêto/Nagô temple, the Jeje temple, and their *axé* depend on continual efforts to keep those forces at the margins of the ritual space and time.

Logics of purity and purification nourished by mercantile diasporas can long outlast the importance of overseas trade. For example, the earlier decline of trade between West Africa and *Cuba* has required the Lucumí nation to endorse a whole range of substitutions for "authentic" African imports—such as coconuts (called *obi* in Cuban Lucumí) for kola nuts (called *obì* in Nigerian Yorùbá) and ground eggshell (*cascarilla* in Spanish) for African chalk (*efun* in Yorùbá). Nonetheless, Lucumí ritual is rich in cleansing procedures, called *limpiezas* (cleansings), *despojos* (acts of stripping away), and washing (as in *santo lavado*, or "washed saint"). Since the 1960s, the participation of Black North Americans and West African immigrants in Lucumí religion has been fertile ground for new movements advocating African racial, religious, and cultural purism (Matory, forthcoming a).

Purism stands among numerous other qualities of Candomblé that reflect the dispersion and commercial dispositions of Bahia's free "Africans." For example, the extraordinary elaborations of Candomblé's sacred cuisine and priestly attire clearly reflect the importance of commercial food preparation and dressmaking among the chief professions of the "African" ethnic community of the early 20th century. Moreover, the

extensive use of sacred languages, literacy, books, notebooks, newspapers, libraries, and museums not only is highly visible in both Brazilian
Candomblé and Cuban Ocha but also is explicable in terms of the way
that Brazilian and Cuban worshipers imagine their relationship to a distant place of origin.[19] This quality of African diasporic religion presents
an obvious parallel to the role of commemoration, sacred language, and
literacy in the Jewish diaspora.

Though all of the U.S.- and Brazilian-affiliated diasporas I have discussed have a powerful religious dimension, they do not all emphasize to
the same degree the exclusivist and hierarchical logic of purity. What I
have hypothesized is that idioms of purity are most likely to be emphasized where exclusivity is most profitable. Cases that involve a monopoly
over the local provision of highly valued foreign goods are probably the
best example, but such cases bear telling similarities to the idiom of purity in South Asia, which privileges the descendants of foreign "Aryan"
invaders, as well as those who successfully embrace the cultural markers
of that descent—Sanskrit literacy, vegetarianism, and so forth. Upwardly
mobile castes thus "purify" themselves. In the Indian diaspora, such
castes might strengthen their case by displaying in their abodes the finest
artifacts available from the South Asian homeland (e.g., Bharati
1972:29–31). The promise of a privileged position for ethnically pure exiles upon their return to the homeland has provided a similar incentive
for purism among Hutu refugees in Tanzania (Malkki 1995).

I do not mean to suggest that the Lagosian Cultural Renaissance and
the social configuration of this West African trade diaspora are the only
possible origins of the rites and vocabulary of purification. Both Roman
Catholicism and Brazilian Kardecism are rich in raw materials for purist
rites and vocabulary. Rather, I am suggesting that the history and sociology of the "African" ethnic community in Bahia, as well as the strategic
priorities of Professor Martiniano and Mãe Aninha (not to mention
their trans-Atlantic merchant contemporaries Lourenço Cardoso, Maximiliano Alakija, Joaquim Devodê Branco, Felisberto Sowzer, and Aninha's
distant but nonetheless purist priestly successor Mãe Stella), do a great
deal to explain *which* of the locally available ritual concepts and terms
were integrated into and most elaborated in post-1930s Candomblé and
which have been neglected or suppressed.

Candomblé ritual and its logic of purity are not merely instrumental
means to the efficiency and profitability of a trans-Atlantic trade network. The profits of this trade are but part of a larger struggle among
classes, by which I mean sets of people bound by at least an implicit
consciousness of their shared interests. The "Africans" of Bahia in the
1930s, with the Jeje and Nagô temples at their symbolic core, were one
such class, or ethno-class, the membership of which was structured not

only by ties to a particular African locale but also by race, gender, economic status, profession, and particular types of access to the power brokers of the Euro-Atlantic world—journalists, anthropologists, generals, and politicians. These interests are not reducible to earnings or type of relationship to the means of production.

As Pierre Bourdieu points out, economic *and* cultural capital "are simultaneously instruments of power and stakes in the struggle for power" (Bourdieu 1984[1979]:315–16). The stakes are not only earnings but esteem and control over the means of self-definition among various broader collectivities, including the Candomblé community, the Brazilian nation, the Yorùbá-Atlantic world, and the African diaspora as a whole. Candomblé ritual practice and politics as we know them today are products of such a struggle, and not products of mere "survival."

I argued in chapter 2 that such transnational communities and the nation-state have evolved in dialogue with each other. For example, Brazil and Cuba have helped constitute and subsidize the trans-Atlantic nations to recruit and organize 19th-century slave labor, to advance their 20th-century international commercial and diplomatic interests, and to assert, as we shall see in chapter 4, new and interested folkloric visions of the territorial nation's character. More recently, Nigeria and Bénin have sponsored and publicized the Yorùbá and Jeje/Fòn diasporas as means of extending their influence abroad and encouraging tourism.

Pace Gellner (1983) and Benedict Anderson (1991[1983]), the nation-state does not rely on the real homogenization of its citizenry, or on the imaginary homogenization of that citizenry's experience. Rather, as Wallerstein has observed (1990), ethnic groups within the nation-state play a central role in producing and reproducing the diverse classes and occupational groups that the nation-state is assumed to comprise, inevitably and integrally. The nation-state of nation-states, the United States is self-consciously made up of multiple, living diasporas—Irish, Italian, Jewish, African, Anglo-Saxon, Cuban, Puerto Rican, Mexican, Salvadorean, Indo-Pakistani and so forth, not to mention "the West" itself.

Such ethno-classes have long been training grounds for diverse labor niches in the U.S. economy. They are also key units of collective bargaining, as diverse groups compete for their share of the profits from the productivity of the nation-state. Despite periodic eruptions of popular xenophobia, numerous official institutions of the nation-state officially document, commemorate, celebrate, and congratulate these diasporas, making them constantly and officially available as alternative identities. Many of them have their own newspapers, television stations, religious holidays and parades, often facilitated by state-supplied infrastructure and policing. Ethno-classes constitute major voting blocs in U.S.

electoral politics. As interest groups and conduits of influence, they are indeed the cornerstone of U.S. foreign policy, particularly vis-à-vis the choice of international allies, clients, and satellites. Hence, the diasporas that constitute ethno-classes typically become inimical to nation-states not when they articulate too much with their homelands but when their interests contradict those of more powerful diasporas within the nation-state.

Those who stand to profit symbolically, electorally, and materially from a united ethno-class or its consumption patterns regularly sponsor cultural activities that give the appearance of primordiality to that ethno-class. It comes as no surprise, under conditions of dispersion, that *religious* activities are foremost among these cultural activities—in synagogues, in ethnically based churches, on Thanksgiving, in Saint Patrick's Day parades and neighborhood-based saint day processions, at weddings, funerals, circumcisions, miqvahs, and naming ceremonies, on the Mexican-American Día de los Muertos, during Caribbean Latino Spiritist sessions and Ocha *tambores*, and amid the recent Yorùbá Revivalism of African Americans. And contrary to Connerton's argument (1995 [1989]), these embodied rituals of commemoration and their public interpretation are manipulable and can be tendentious.

The main error in Gellner's, Anderson's, and Appadurai's assessments of the relationship among the nation-states, religions, and transnational communities lies in their shared teleological convictions. Their analyses rely on predictions about the *direction* of change in the relationship rather than on a hard-nosed examination of the *ongoing relationship* among them, however moribund that relationship might appear in the long term. I, for one, do not believe that we will see any end to this mutually constituting relationship for some time to come.

Herskovits (1958[1941]) suggested that the reason for the preeminence of religion and music among the "survivals" of African culture in the New World is that African cultures are inherently more focused on these areas, producing, reproducing, and elaborating on them with particular care. On the other hand, Sidney Mintz and Richard Price (1992[1976]) suggested that such dimensions of African culture tended to endure insofar as they presented no threat and were therefore allowed to survive amid white-controlled regimes of politics and production. The diasporas I have studied suggest a related hypothesis, one that is as applicable to the African diaspora as to any other: in functioning diasporic communities (i.e., those whose social order is actively shaped by communication with the "homeland"), religion is the *typical* social glue, and it can be elaborated in a way enhancing forms of solidarity, hierarchy, and economic interest that would otherwise—under the reign of the multiple, ethnically plural nation-states across which these diasporas are dispersed—be unenforceable. Perhaps religion is the least common denominator among the ties that bind and guarantee the endurance of transnational communities.

Moreover, in secular or constitutionally pluri-religious states, religions are not just a space of allowance that the authorities do not care about controlling; they are the typical symbolic representation of the private organization and reproduction of the diverse classes that necessarily make up the capitalist nation-state and of their profitable connections to the world beyond the nation-state. Yet religiously coded diasporas do not serve the interests of the nation-state alone, and they are often shaped by priorities at odds with those of the nation-states, such as the value that the Jejes and Nagôs of Bahia gave to "African purity" amid elite Euro-Brazilians' embrace of *mestiçagem*, or "hybridity." Such contradictions reveal that diasporas often possess a relative autonomy of interests, interests that are economically, politically, and socially conditioned.

Even those Brazilians who do not participate in the often-secret rituals of Candomblé are aware of this religion's existence—through the presence of sacrificial offerings at urban crossroads, through television minidramas and news reports, through the national literary canon, and through newspapers. Particularly in the newspapers, the Brazilian bourgeoisie has continuously been shown the diversity of contemporaneous "evolutionary stages," subnational groups, and African transnations that occupy the imagined Brazilian present and, increasingly, make claims on the Brazilian state. The long-running role of these sacred transnations in the self-imagination of the Brazilian nation-state is the subject of the next chapter.

CHAPTER FOUR

Candomblé's Newest Nation: Brazil

Bahia is the black Rome.
—MÃE ANINHA, founder of the Ilê Axé
Opô Afonjá temple[1]

ANY GIVEN "IMAGINATION" OF COMMUNITY is less an evolutionary stage or a consensus than one position in an ongoing "struggle for the possession of the sign," to borrow Hebdige's phrase (1979). Candomblé is one of the signs most struggled over in the imagining of Brazil, of the Northeastern region, and of the transnational Black community. This chapter charts Brazilian nationalists' construction of Candomblé as a "folk" emblem of a "racially democratic" nation-state. It also charts the forms of risk and advantage this literary and journalistic representation has entailed for the cosmopolitan and often highly educated priests of the Jeje and Nagô nations. Through this interclass and interracial dialogue, the Brazilian nation-state has been changed as much as Candomblé has. Indeed, the Brazilian state has, since the 1970s, inadvertently resurrected forms of Black transnationalism that had been in decline in Bahia since the 1940s.

The information presented in this chapter derives from not only newspaper articles but also popular nationalist texts, magazine articles, government publications, commemorative flyers, tourist orientation materials, and books housed in the archives of key national institutions. Most of the materials used here come from the archives of newspapers in Salvador and in São Paulo—the *Estado* of São Paulo, as well at the *Tribuna* and the *Tarde* of Salvador. Each of these newspapers' archives includes clippings from a nationwide array of publications and assorted informational materials, which are considered important enough for reporters on future events to consult.

The archives of these particular newspapers were selected for investigation because they represent the local viewpoints of what I sense are opposite ends of the Brazilian national polemic—historically Europhile São Paulo, the economic capital, and historically Afrophile Salvador. Other materials cited in this chapter come from the archives of the Bahian state tourist agency, the Salvador mayor's office, and the Bahian Federation of the Afro-Brazilian Cult. These are analyzed alongside other canonical texts of 20th-century Brazilian national identity. Together, they are taken as both illustrations and sources of the Brazilian bourgeoisie's sense of the nature of its *national community* and both the symbolic and the material role played in it by *an "African" cultural Other that occupies the same territory.*

A NONLINEAR HISTORY OF BRAZILIAN NATIONALITY

It is difficult to say when Brazil became a territorial nation after the "classical" model outlined by Anderson (1991[1983]; see also chapter 2 of this volume). Brazil gained independence from Portugal in 1822 but remained under the sovereignty of a resident Portuguese monarchy until 1889, a year after the last monarch completed the abolition of slavery. The last monarch so angered white elites that they overthrew her and founded the extremely nonegalitarian First Republic. Only after Getúlio Vargas's overthrow of the First Republic in 1930 did the nationalist fiction of egalitarian brotherhood achieve any official expression or any semblance of widespread acceptance. Yet, contrary to the "stagism" implicit in much Andersonian and transnationalist theory, the 1930s hosted the endurance, the resurrection, and indeed the proliferation of transnational identities (see, e.g., Butler 1998 on the *multiplication* of African "nations" in Bahia during this period). The most prestigious of these transnational identities, far from undermining Brazilian national identity, substantiated Regionalist efforts to represent Brazil as an integral, egalitarian community.

Moreover, despite their residence in a nation-state ever stronger in its commitment to the ideals of the Andersonian territorial nation, the transoceanic black merchants, as well as their "African" class allies, and their followers in Brazil have regularly "imagined" themselves as members of crosscutting, transnational communities. The communities they "imagined," and lived their lives under the authority of, remained both royalist and dynastic, serving gods recognized as ancient kings and queens of their African nations. When these gods possessed their human mediums, they donned crowns combining European and African iconographies of royal authority. Dynastic symbolism and the logic of the communities it constitutes were alive not only in numerous black Atlantic religions of spirit possession but also in the Sebastianist messianism of the early-20th-century Bahian interior (E. Cunha 1944 [1902]), in many 20th-century Brazilian samba clubs, and at the heart of Roman Catholicism itself. For Afro-Bahians since the 1930s, religion, dynastic empire, nation-state, and transnation have all been simultaneously available and highly persuasive idioms of community. As we shall see, the imagined community of the transnational Portuguese empire remains powerfully present, from the early 1930s onward, in the work of Brazil's preeminent nationalist writer—Gilberto Freyre.

GEOGRAPHY AND IMAGINATION

Nigeria, Brazil, and the United States are the three cultural superpowers of the black Atlantic. No other nation but the Democratic Republic of Congo boasts as large a population of African ancestry, and they have been the foremost clearinghouses of the enormous influence on black Atlantic culture wrought by England, Jamaica, and Cuba. Jamaica, Cuba, and Congo have exercised an enormous influence on Afro-Atlantic music, but these three are dwarfed by the Big Three in commerce, literary and media production, and ongoing extraterritorial population movement. The triumvirate of black cultural superpowers has been bound together not only by forced and free migrations but also by an intensely transnational dialogue over race, gender, the environment, and the fate of nations. In one corner, this dialogue has fertilized and been fertilized since the 1930s by Brazilian Regionalism—the revolutionary cultural movement whereby the Euro-Brazilian bourgeoisie rethought the role of race and culture in the Brazilian nation. That movement has fueled monumental cosmetic efforts by the Brazilian state and general population to deny the existence of racism in their society. Almost as consequentially, Brazilians have nearly recovered from thinking of themselves as inferior facsimiles of northern Europe or the United States and have largely accepted the image of Brazil as a profoundly new and laudably *mestiço* or *mulato* country.

Since the beginning of the Northeastern literary movement called Regionalism, cultural emblems once associated exclusively with the "Africans" and blacks have become available to vastly larger Brazilian populations, both as symbols of sociopolitical opposition and as symbols of the entire national culture.[2] Like the audiences of Caribbean Latino music all over the Hispanophone Americas, Brazilian audiences of popular music all know a smattering of African-inspired gods. Numerous artists sing about Oxum (goddess of wealth and sweet water), Xangô (god of thunder and justice), Iemanjá (goddess of motherhood and the sea), Ogum (god of war and iron), and Oxalá (god of peace and purity). Brazilians who watch Carnaval parades hear the rhythms of the gods and see facsimiles of their sacred attire paraded every year. Afro-Brazilian religions are also the source of elaborate gay argots in many of the cities where they are practiced, especially the Brazilian Center-South—that is, Rio de Janeiro and São Paulo.

Since the 1930s, as more and more people have invoked images from Afro-Brazilian religion to imagine their own citizenship in the Brazilian nation-state, the initiates of the Northeastern religion continue to rehearse and invest in the reality of a different sort of "nation"—a form of *transnational* peoplehood that links them not only to a specific place in Africa but to its sons and daughters in Cuba and the United States as well. This chapter explores how such extraterritorially oriented imaginations of community have empowered and been empowered by a nationalist literary movement.

For the advocates of regionalisms, some regions are more regional than others. The term "regionalism" conjures up images of marginality and resistance to the national metropolis. But, technically speaking, culturally and politically central capitals are also regions. In the Brazilian case, one region, more than all the others that fit the bill, has been allowed to articulate the cause of regionalism and the redress of marginality nationwide. Yet it is neither the most remote, neglected, or oppressed, nor the least. Moreover, the intellectuals who have articulated this regionalism—perhaps without surprise—are hardly of the most oppressed or marginalized class. They are of the liberal professions—including journalism, psychiatry, and anthropology. More surprisingly, this regionalism asserts its host region's status as the main exemplar of a national essence.

With the exception of Machado de Assis's arguably regionalist portrayals of life in Rio de Janeiro, no other Brazilian regionalism has produced a single author rivaling the fame of Euclides da Cunha, Gilberto Freyre, or Jorge Amado, all of whom penned vivid tableaus of the Northeast. From the First Republic (1889–1930) to the Vargas era (1935–45; 1950–54), Northeastern literary Regionalism was central to bourgeois Brazilians' effort to rethink the role of race in the high culture, politics, and economic development of the Brazilian nation.

REGIONALISM AND TRANSNATIONALISM

The irony of Northeastern Regionalism is that it was actually a nationalist movement that denounced one kind of transnationalism while drawing energy from other kinds. It arose against the backdrop of the economic and political dominance of the Center-South elite, which was perceived as slavishly Europhile in its values and had, indeed, financed the costly transnational immigration of Italian workers in order to whiten the nation and circumvent the demands of native-born black and brown labor (e.g., Andrews 1991). Northeastern Regionalism also made capital out of the cosmopolitan sophistication of the transnational Jejes and Nagôs, which conferred luster upon the whole of the dark Northeast.

Since the 19th century, the central worry of many Brazilian intellectuals was that Brazil's racially mixed heritage would hinder its development. The pseudoscientific racism flowing (transnationally) out of Europe's printing presses had convinced much of Latin America that its black and Indian ancestors and citizens were its greatest encumbrance to modernization (e.g., Stepan 1991). The Brazilian bourgeoisie has long sensed that its country was closely watched by and judged in comparison to the apparently successful and "white" nations of the North—England, the United States, and France. Thus, in Brazilian Portuguese, the cosmetic appearances mounted to impress outsiders are described proverbially as *para inglês ver*—"for the English to see."

Nationalist romanticism enlisting the Indian as the emblem of Brazilian identity gave a dark symbolic face to the struggle for Brazilian independence from Portugal, achieved in 1822, and to subsequent imaginings about the character of the new nation (Skidmore 1974:6–7; Wafer and Santana 1990; Brookshaw 1986). But following the influence of pseudoscientific racism, the overthrow of the monarchy in 1889, and the rise of the oligarchical First Republic, the emblematic Indian gave way to lamentation about the poverty and atavistic ignorance of backland *mestiços*, as represented, for example, in Euclides da Cunha's famous *Rebellion in the Backlands* (*Os Sertões*, 1944[1902]).

One might date the beginning of Regionalism (with a capital R) from the First Brazilian Congress of Regionalism, held in Recife in 1926 (see Mendonça Teles 1986:343–44). And, though these Regionalists were typically *Northeastern* writers about the Northeast, military engineer and journalist Euclides da Cunha had come from Rio de Janeiro. He had offered his account of the army's assault on a mixed-race northeastern peasant community as a wake-up call to the Europhile Brazilian bourgeoisie. His *Rebellion in the Backlands* graphically reminded his fellow metropolitans that they were part of a *mestiço*, or mixed-race, nation—oppressed, hungry, and diseased but noble in their suffering. In *Rebellion*, the messianic peasant rebels are described as backward, unenlightened, and isolated, but the military assault on them becomes a metaphor for the Brazilian bourgeoisie's collusion in the sweeping away of all that is distinctive and native to Brazil. Cunha regrets these *mestiços'* isolation and unenlightenment, but these appear to be precisely what makes them truly Brazilian, rather than poor facsimiles of the European. Such is the redemptive bent of much of the Northeastern literary and artistic Regionalism that followed.

Resistance to the Europhile oligarchy of the First Republic would further enlist the imagery of blacks and *mestiços* during the Brazilian Modernist movement, conventionally dated from the Modern Art Week of 1922 in São Paulo. The 1920s also saw the beginnings of cooperation between the samba schools of Rio, on the one hand, and, on the other, composers and thrill seekers of the white middle class (Raphael 1981). In the late 1920s, when elite, white Carnaval clubs became critical of the First Republic government, they were shut down, ceding the most central public spaces to black Carnaval groups. These developments established the momentum for even weightier and more official changes in the 1930s.

In 1933, *The Masters and the Slaves* (*Casa-Grande e Senzala*), by native Northeasterner Gilberto Freyre, kicked off a series of magisterial and introspective monographs in which Brazilian intellectuals tried to pinpoint what distinguished their nation from a remarkably international set of comparison cases. Indeed, Freyre's Regionalism was oriented toward a pan-American and circum-Atlantic conversation about race, region, and nationality in the 1930s and 1940s.[3] Among the explicit aims of these writers was to valorize Brazil's difference from its neighbors and from the whiter "peoples of the North."[4] Chiefly through Northeastern Regionalist fiction and belletristic essays, the Brazilian bourgeoisie was coming to terms with its not-so-white Portuguese, Moorish, Indian, African, and, in a word, *mestiço* roots. Thus, Freyre and the other Regionalists of the 1930s recapitulated the earlier images of Indians, *mestiços*, and *caboclos* but increasingly placed the *negro* and the *mulata* at the literary center of the nation (e.g., Bastide 1978[1960]:22).[5] These dark figures would eventually become the chief exemplars of the *povo*, or the "folk."

However racially and culturally hybrid the nation might have been for the Regionalists, Freyre's *Masters and the Slaves* established the momentum behind the now-popular conviction that Brazil's distinctive cultural and political look is but a transnational fulfillment of the Portuguese national character and cultural essence—itself *mestiço*, expansive, culturally incorporative, and in love with dark women. Freyre would eventually describe the cultural and political formation that united Brazil with the rest of Portugal's tropical colonies and former colonies as a "transnational community," which he called "Luso-Tropicalism" (Freyre 1959[1945]:154).

Hence, despite the parochial sound of the term "Regionalism," its leading spokesperson was intensely interested in the "transnational" sociocultural formations of which the Northeastern region was the exemplary outpost. And far from being antinational, this Regionalism not only extended the efforts of the metropolitan bourgeoisie to clarify the Brazilian *national* character as a whole but also identified a *transnational* formation as the nation-state's roots and its enduring substance. The paradigmatic status of Northeastern society in the Brazilian nation-state lay in the age and solidity of its connection to a Luso-transnationalism. Freyre's praise of the overseas racial and cultural *mestiçagem* that resulted from the Portuguese national character was as revolutionary as his use of the term "transnational" in 1945 was prescient. Both concepts established Brazil's centrality to something great and brown, in contrast to the earlier bourgeois view of the nation as something darkly marginal to the great, white North.

On the other hand, while the heavily black population of the Northeast now appeared at the center of the Regionalist canvas, the *negros* and *mulatas* appeared largely as *patiens* to the Portuguese or Euro-Brazilian male *agens*, and as props for the real protagonist of Brazilian national history. Such is the characteristic role of dark people in the nationalisms and regionalisms of white or light-colored American creoles. For example, writes Freyre:

> In our affections, our excessive mimicry, our Catholicism, which so delights the senses, our music, our gait, our speech, our cradle songs—in everything that is a sincere expression of our lives, we almost all of us bear the mark of that [Negro] influence. Of the female slave or "mammy" who rocked us to sleep. Who suckled us. Who fed us, mashing our food with her own hands. The influence of the old woman who told us our first tales of ghost and *bicho* [beast]. Of the mulatto girl who relieved us of our first *bicho de pé* [a type of burrowing flea], of the pruriency that was so enjoyable. Who initiated us into physical love and, to the creaking of a canvas cot, gave us our first complete sensation of being a man. Of the Negro lad who was our first playmate. (Freyre 1986[1933]:278–79).[6]

Amid the intoxicating bouquet of *catinga*[7] with a dash of psychoanalysis, this passage sweeps us starry-eyed past the disturbing realities of race relations in Freyre's day. Freyre's magnum opus also sought to demonstrate that the slave plantations of his native Northeast were sites of such sensual interaction between white males, on the one hand, and their black nursemaids, their black pals, and their black and mulatto mistresses, on the other, that Brazil evaded the kind of racial purism and segregation that created the world-famous horrors of Jim Crow and an emergent Nazism. Thus, Freyre and his followers argued that Brazil, far from being a racially inferior nation, ought to be seen as a paragon of racial and cultural hybridity, or *mestiçagem*, and of "racial democracy"— that is, interracial harmony and conviviality.

Skeptical readers of Freyre's narrative will find copious evidence of racial inequality and racial discord. However, the author argues that the "consciousness of race" in this civilization is "practically non-existent" (Freyre 1986[1933]:3, 83), giving rise to a fundamentally mixed-race and harmonious society. Thus, the slave plantation society at its biological and spiritual foundation is surrounded with a halo of nostalgic innocence. Freyre creates an air of delight around what, to dark women, is the threat of rape, involuntary prostitution, and, at the very least, extremely unequal terms of negotiation under the slave regime that Freyre considers the roots of Brazilian society. Clearly, Freyre's history is not a nostalgia made for black people's appreciation. In Brazil, slavery ended later than in any other American country. At abolition in 1888, in the 1930s (when Freyre wrote), and today, Afro-Brazilians have experienced systematic and blatant job discrimination, remained disproportionately poor and sick, and suffered extreme underrepresentation in the making of state policy. Despite (or partly because of) the endurance of these realities, the ideology of "racial democracy" has also been extremely durable.

The justifiable suspicion of bourgeois romanticism amid the very real oppression the dark and of the poor motivated a major trend in the scholarship about Afro-Brazilian religions in the 1980s and 1990s. Motta (1994), Dantas (1982, 1988), Henfrey (1981), Wafer and Santana (1990), and others argued that the Euro-Brazilian dominant classes have "manipulated" the ideologies of Northeastern religions, imposing, particularly since the 1930s, certain fictitiously "African" standards in the name of "purity" in exchange for their protection and sponsorship of select temples.

Both Modernism and Regionalism rested on images of dark people's passivity and of the *povo's* inertia and isolation. So there is a further irony. Just as some regions are more regional than others, some of the *povo* are more *povo* than others. It was not the most inert, isolated, poor, or passive dark people who came to represent and speak for the dark *povo* of the Northeast. The best protected, the most lucratively sponsored, and the most nationally emblematic temples in the Northeast have not been the ones most easily manipulated by the Euro-Brazilian dominant classes. In fact, they themselves constitute a dominant class with transnational cultural references and sources of power that inspired *deference* from the Euro-Brazilian Regionalist scholars, journalists and artists alike.

We know of this influential family of temples because Northeastern literary Regionalism came about alongside another major genre of intellectual production in that region. The journalists and university intellectuals in the Northeast threw themselves wholeheartedly into the *ethnological and ethnographic documentation* of the real-live blacks, or "Africans," in their midst. Though committed to scholarly methods of little concern to the literary Regionalists—including the Columbia-trained anthropologist Freyre—these documentary observers penned a genre that one journalist and Candomblé initiate called "the black cycle" (*o ciclo do negro*).[8] Raimundo Nina Rodrigues (at the turn of the century), Manoel Querino (in the second decade of the century), and, in the 1930s and 1940s, Édison Carneiro, Arthur Ramos, Ruth Landes, and Donald Pierson documented volumes of what they saw and heard in systematic investigations of Bahian social life.

Just as Brazilian intellectuals feared that their country was racially and culturally inferior to the white "North," Northeastern intellectuals suffered under their region's reputation for racial and economic inferiority to the Center-South, which includes Rio de Janeiro and São Paulo. São Paulo had industrialized itself with the profits from its initially slave-based coffee plantations and had, over the course of the 19th century, surpassed the formerly prosperous and politically central Northeast in wealth, whiteness, and political power. However, Nina Rodrigues, Édison Carneiro, Arthur Ramos, and Gilberto Freyre publicized the idea that the black population of the Northeast was racially and culturally superior to the black populations of both the United States and the Brazilian Center-South, making the Northeast not so inferior after all. These writers knew that the Africans held captive in the Center-South had come mainly from West-Central Africa—the putative homeland of Candomblé's Congo and Angola nations. Thus the Sudanese, or West African, majority in the Northeast and the West African–inspired Jeje-Nagô Candomblé became a source of distinction and pride for the Regionalists. These light-skinned and largely Northeastern intellectuals personally knew the Jeje and Yorùbá/Nagô priests, merchants, and intellectuals of the Northeast and recognized their black genius as evidence of Northeastern dignity.

Freyre's *Masters and the Slaves* popularized these opinions on the national scene (Freyre 1986[1933]:299–315; also 281). Freyre cites a number of sources, but especially Raimundo Nina Rodrigues, as proof of the "superiority of the Negro colonizers of Brazil over those of the United States" (311; also 299) and the superiority of Northeastern blacks to those in the Center-South. He writes, "Pernambuco and Bahia received a higher grade of Africans than did Rio de Janeiro" (308). Freyre again cites Nina Rodrigues in identifying the African ethnic source of Bahian and Brazilian superiority: "The Sudanese [prominently including the "Gêges" and the "Nagôs"] ... were the predominating element in the formation of Bahian society" (Freyre 1986[1933]:300, 302, 304; also Dantas 1988:151–57).[9] Like Rodrigues, Freyre appears to have acquired his only direct knowledge of "Sudanese" culture and its "superiority" from a trans-Atlantic black traveler. Freyre's main informant was a literate priest of the Pernambucan Xangô religion known as "Pai Adão." In a letter to the translator of *The Masters and the Slaves*, Freyre describes this priest as "a personal friend of mine until he died in 1936 ... , who, though born in Brazil, had studied religion in Africa" (Freyre 1986[1933]:474n203; also 318).

The authors of Northeastern Regionalism repeatedly reveal their debt to such transnational black informants, as well as their excitement about these travelers' books, polyglot literacy, commercial acumen, sophistication, and chauvinism. The public prominence of these travelers in 1930s Bahia gave Regionalism and the allied ethnographic "black cycle" much of their credibility and some of their most memorable characters. In turn, these literary and scholarly movements paved the way for a new *Black* transnationalism, by the 1960s, to enter the pantheon of Brazilian national culture.

INDIGENISMOS AND NEGRIGENISMOS

New World-born whites have a long history of employing the images of the indigenous people they have oppressed or exterminated as emblems of their own political or artistic authenticity. The bourgeois elites who run the nation-state thus appropriate the cultural capital of the subaltern to legitimize their own sovereignty, as they did in Mexican *indigenismo* and the myriad U.S. uses of the Indian as an emblem of "American" identity. Alberto N. Pamies (1973:ix, xiii, xvii) coined the term "negrigenista" to describe pale Cuban and other Latin American nationalists who appropriated the image of their black subjects for similar purposes. As a *negrigenismo*, Fernando Ortiz's *Afro-cubanismo* finds counterparts in Brazilian Modernism, the "black cycle," and Euro-Brazilians' practice of the Umbanda religion. Westernized black elites employed a similar strategy in the Haitian Bureau of Ethnology. Jean Price-Mars affirmed the dignity of Haitian popular religion in the face of the U.S. occupation and its racist propaganda. In the 1960s, "President-for-Life" François Duvalier used priests and Vodou symbolism as means of political control. European nationalist "folklore" similarly represents the nation in images inspired by the doings of citizens from marginal classes, regions, or races—such as Gypsy flamenco in Spain and Breton crepes in France (see also Dietler 1994 on the role of the marginalized Celts in western European nationalisms).

Amid such nation-states' efforts to centralize authority and to establish the supremacy of a bourgeois bureaucratic culture, the elevation of the culturally marginal is the first irony, albeit a widespread one, in the emergence of the territorial nation. This irony became the norm in Latin America from the 1930s onward. Yet there is a contrast between the two behemoths of Latin America, in which it is difficult to say which case is the more predictable and which the more ironic. Whereas the capital of Mexican indigenism, Mexico City, was also the capital and eponym of the nation-state itself, the capital of Brazilian folkloric nationalism was in several ways *opposite* the political and economic capitals of the nation-state. The post-1930s capital of Brazilian culture is one of Brazil's poorer regions, it is the blackest, and it was long ago abandoned as the political capital of the nation-state. Brazil's romantic celebration of the politically and economically marginal or superseded region as a metonym of the nation is comparable to the elevation of the South in U.S. nationalism—above all in D. W. Griffith's film *Birth of a Nation* (1945) and perhaps in *Gone with the Wind* (1939) and Walt Disney's *Song of the South* (1946) as well. Consider also the depiction of the "Wild West" in the enormous genre of "cowboy" movies.

Mexico and Brazil contrast in a further way, which illustrates a widespread contrast *within* Latin American nations as well. Where the Indian prevails as a national symbol, the African usually does not. Indeed, like 19th-century Euro-Brazilian Indianism and Trujillo's 20th-century cultural politics in the Dominican Republic, some American indigenisms seem intent on denying the black presence and excluding black participation in the state. Where they coexisted, as in 1930s Bahia, the advocates of Afrocentrism were often the critics of Indianism. In the mid-1930s, an African-Bahian complained to Donald Pierson:

> These people think the *Indian* is worth so much! Brazil has been discovered for more than four hundred years, and there isn't a single book in Tupi like this [showing the Anglican prayer book printed in Yorùbá], or a single magazine like this [pointing to a periodical published in English in Lagos and containing several articles by Africans]. These people here call us fetish worshipers and say, "Aw, that's African nonsense!" It just goes to show that they don't know anything about what we have in Africa (Pierson 1942:272).

A similar defensiveness served Northeastern Regionalism in Freyre's text:

> Nothing is more absurd than to deny to the Sudanese Negro, for example, who was brought to Brazil in considerable numbers, a culture superior to that of the most advanced native. . . . the Negro slaves show that they came from more advanced stocks, and that they were in a better position than the Indians to contribute to the economic and social formation of Brazil. At times, in a better position than the Portuguese themselves (1986[1933]:281).

Which brings us to the third irony: though the regional and racially marginal institutions that have been elevated as emblems of the nation are often represented as the flowering of an organic, illiterate, and historically static lower-class culture, its indigenous spokespersons in the 1930s were in fact often highly literate, cosmopolitan, and innovative. The role of transnationalism in the making of Brazilian Regionalism and nationalism thus makes Brazil more immediately comparable to Cuba and Haiti than to Mexico, but a more comprehensive study of the near-simultaneous *indigenismos* and *negrigenismos* that swept Latin America in the 1930s might reveal a transnational impulse behind Mexican *indigenismo* as well. That these movements arose in resistance to transnational forms of Europhilia and eugenics, and did so within a few years of each other all across Latin America, suggests a measure of mutual inspiration or regional contagion. In more recent decades, the popularity of Mexican indigenism (in the form of U.S. Chicanos' adoration of Aztlán) and the *boricua* nationalism of Puerto Ricans in the United States seem to be overwhelmingly transnational and beyond the territories of the nations to which they bear allegiance.

Candomblé and the Brazilian Nation-State: The Regionalist Revolution

With the possible exception of Carnaval, Candomblé has now attracted more attention from anthropologists, sociologists, and journalists than has any other Afro-Brazilian institution. Alongside Carnaval and the black bean–based national dish called *feijoada*, it has, over the past half century, achieved the status of a national symbol in Brazil—in no small way because of Freyrean Regionalism and the allied "black cycle." Yet since the late 19th century, when Brazilian psychiatrist Raimundo Nina Rodrigues's studies of this religion inaugurated Afro-Brazilian studies as we know it today, Candomblé has been celebrated—with some exaggeration—as a uniquely pure manifestation of African culture in the Americas.

Since the colonial period, Candomblé and similar religions have variously been persecuted, tolerated, and encouraged—often all at the same time. As when it missionized other countries, the Roman Catholic Church in Brazil often encouraged the integration of pre-Christian iconography, celebratory practices, and ethnic identities into the local practice of Christianity (see Moss and Cappannari 1982:64; Bastide 1978[1960]:53; Russell-Wood 1974:579; Verger 1987[1968]:525; Mattoso 1986[1979]:129; Albérico Paiva Ferreira, personal communication, 29 December 1995 and 3 January 1996).

Not all members of the 19th-century Bahian planter class *favored* tolerance for African-inspired forms of celebration and worship. At times, however, the Brazilian state had its own motive for tolerance. Though the Conde dos Arcos, who served as governor of Bahia from 1810 to 1818, was too committed a Christian to endorse the practice of African religion in his province, he envisioned a divide-and-conquer strategy:

> Seen through the government's eyes, the [Afro-Brazilian dances known as *batuques* are] one thing; seen through the eyes of [slave-owning] individuals,

[they are] something quite different. The latter believe that the *batuques* infringe [on] their dominical rights, either because they want to employ their slaves in useful work even on Sundays, or because they want to station them before their doors on these days of rest as a way of showing off their wealth. The government, however, sees the institution of the *batuques* as something that *obliges* the blacks unconsciously and automatically to revive every week the feelings of mutual aversion instinctive to them since birth, which are nevertheless gradually extinguished in their common suffering. . . . If the different "nations" of Africa were totally to forget the furious resentment which by nature has divided them, . . . a tremendous and inescapable danger would descend upon and destroy Brazil. And who can doubt that suffering has the power to create fraternization among its victims? (Bastide 1978[1960]:55; see also Carneiro 1986[1948]:16–17).

Indeed, many conspiracies to mass flight and rebellion were hatched among members of one single "nation" or another during the early 19th century (Reis 1986:116; 1987:123; Verger 1981[1850]:227–28). And although many conspiracies were multiethnic, rival "nations" often betrayed each other's rebellions (Bastide 1978[1960]:56; Reis 1987). Whatever its degree of effectiveness, the Conde dos Arcos's policy of strategic tolerance lapsed under the subsequent provincial governor, and not a moment too soon for most of the province's planters.

In fact, the diverse officials of the Brazilian state were seldom coordinated in their policies toward African religious institutions. For example, in 1829, a justice of the peace ordered police to invade a Candomblé of the Jeje nation in the neighborhood of Salvador under his jurisdiction, despite the freedom of religion guaranteed in the constitution of 1824. There, the police not only stole a hat and various sacred objects but also extorted money from the worshipers. However, the priest, a freed African, sought the support of his *patron*, the Viscount of Pirajá, whose overseer paid the justice a visit in order to seek a remedy. When the efforts of the overseer failed, the justice received a query, much to his surprise, from the president of Bahia himself—the Viscount of Camamú (Reis 1986). As far back as the 1820s, documents reveal the active participation of whites in the activities of certain Candomblé temples (Reis 1986:118–19).[10]

Until fairly recently, most of the Northeast's temples of Afro-Brazilian religion operated under a cloud of insecurity because of intermittent extralegal and sometimes politically motivated persecution by the police, who interrupted ceremonies, stole valuables, extorted bribes, humiliated priests, and desecrated objects or impounded them for display in museums of criminology and pathology.[11] However, in 1937, a journalist and major expositor of the "black cycle," Édison Carneiro, helped organize the Union of the Afro-Brazilian Sects of Bahia, which was intended to guarantee the fidelity of Candomblé priests to African-centered traditions and intended to coordinate strategies for protecting them from arbitrary police intervention (Oliveira 1987:15). Also prominent in the early activities of the union was Àlvaro MacDowell de Oliveira, an Afro-Brazilian lawyer and the father of the children of Mãe Menininha, chief priestess of the Gantois temple (Pierson 1942[1939]:278, 317; Dantas 1988:190, 192ff.; Landes 1947:24). The union and its successors have always experienced the state's contradictory interventions, suffering, over time, just a bit more persecution than regulation and patronage.[12]

On the other hand, certain temples—particularly the "great" female-headed temples of the Nagô nation—have long enjoyed protection by the powerful and exemption from police persecution. Mãe Menininha of the Gantois temple recalled the persecution of many temples during the 1930s and 1940s by police commander Dr. Pedro de Azevedo Gordilho, but she also recalled the connections by which her own temple remained safe.[13] But the protected Candomblé houses were not simply the receiving partners in these relationships.[14] The current head priestess of the Opô Afonjá temple, Mãe Stella, remembers what sources also reported in the 1930s and 1940s: Opô Afonjá protected "black cycle" journalist Édison Carneiro, who was also a communist fugitive, from arrest by Vargas's police. He is said to have hidden in the shrine of Aninha's goddess Oxum.[15] Moreover, Mãe Aninha, the founder of Opô Afonjá, is said to have intervened with President Vargas to protect Candomblé temples generally.[16] (In a divergence of opinion that I will read more closely later, some Candomblé members alternatively credit the male priest Joãozinho da Goméia with this intervention.) If the events occurred even vaguely as they are described, these risky and bold political interventions by an allegedly "protected" house undermine the credibility of the view that Candomblé in the 1930s bought immediate safety at the cost of changes in their ritual protocols dictated by "the dominant class."

In hindsight, the uneven treatment of Afro-Brazilian temples in the 1930s was the sign of a transition under way, the rise of a new hegemony in the Brazilian nation. Indeed, these years occasioned a major transition in the ideology and shape of the Brazilian nation-state. This period followed the collapse of the Europhile First Republic and the end of costly efforts by the government of São Paulo State to whiten the population and diminish Afro-Brazilian bargaining power through the transnational importation of European labor. Though officially democratic, the First Republic had done little to represent the poor and colored majority of the country, or to enhance the quality of their lives.

In 1930, a group of junior army officers called the *tenentes*, or "lieutenants," overthrew the First Republic and replaced it with the dictatorship of Getúlio Vargas, whose rule over the subsequent fifteen years centralized government and incorporated numerous trade unions, commercial blocs, and social organizations as consultative bodies in the government. Vargas also drew popular support by establishing workers' rights to limited working hours, a minimum wage, social security benefits, and collective bargaining. Vargas also legislated that two-thirds of the employees of any enterprise had to be native-born. Moreover, the national constitution of 1933–34 restricted immigration (a move that surely would have benefited native-born labor generally), but the framers of that constitution had some expressly racist intents: they were particularly anxious to exclude nonwhite immigrants (Mitchell 1983), including the Africans whose movement and commerce had profoundly shaped the nations of Candomblé. Yet, judging by the communications recorded by Ruth Landes (1947) in the 1930s, the Africans of Bahia continued to "imagine" their livelihoods, religiosity, and community in transnational terms.

Vargas was the first head of state to abandon the view that state politics was an elite monopoly. He and his appointees carried out a parallel cultural policy. In the 1930s, the Vargas-appointed mayor of Rio began to subsidize the popular samba schools and deny funding to the white *grandes sociedades*, or exclusive Carnaval clubs. The samba schools became a well-regulated link between the state and the shantytown dwellers, or *favelados*.[17] The Vargas administration also invested in propaganda to improve the public image of the disproportionately dark *favelados*. Official magazines showed that the *favelado* was a worker like any other, not the *malandro*, or "con man," of the upper- and middle-class imagination. These same magazines propagated the image of Brazil as a racial democracy, where miscegenation was encouraged. Breaking with the policies whereby the previous government projected a white image of the country through a segregated navy and diplomatic corps, Vargas appointed a mulatto ambassador to France (Raphael 1981; Skidmore 1974:48, 125–36 [esp. 133–34]; Dzidzienyo 1985:136; Degler 1971:139, 148).

However, the legacy that represented blacks and mixed-race people as the deplorable antitype of the modern citizen did not automatically die. During the 1930s, local police persecution of Candomblé continued to be inspired by the aim to extirpate alleged African "barbarism" and "atavism" (see also Ortiz 1973[1906] on Cuba). Yet President Vargas himself is remembered as having intervened legally to protect the practitioners of Candomblé and Umbanda.[18] In all transitions there are apparent contradictions such as these. What is surprising is that both Vargas, at the helm of the state, and his Regionalist opposition—including both Gilberto Freyre and Édison Carneiro, both of whom he jailed—cleaved to a new image of the Afro-Brazilian and her relationship to the nation-state.[19]

Though Candomblé temples were treated unevenly by state officials, support by the Euro-Brazilian intelligentsia of the Northeast was more concerted. Three scholarly conferences called the "Afro-Brazilian Congresses" polished Candomblé's image as a national "folk" institution, In 1934, Freyre organized the first congress in Recife, where he brought priests and professors together to valorize Afro-Brazilian culture and end the persecution of Afro-Brazilian religions. Carneiro and Aydano de Couto Ferraz organized the next congress in Salvador in 1937. Ayres de Mata Machado and Joã Dornas Jr. organized a third one in Belo Horizonte just before World War II (see Gonsalves de Mello 1988[1935] and Freyre 1988[1937]; Bastide 1978[1960]:22). In 1935, Jorge Amado published his famous novel *Jubiabá*, about a wily Afro-Brazilian priest. According to Ruth Landes, virtually all the university professors in Bahia in those days studied and were personally connected to Candomblé (Landes 1947:7, 74; Dorival Caymmi, quoted in Barboza and Alencar 1985:194). This was the period when the black replaced the Indian as the indigenist trope par excellence in Brazil (Bastide 1978[1960]:22). And this transition happened at the hands of the Northeastern Regionalists. At the same time, the "black cycle" fueled a growing symbolic role for Candomblé in the national imaginary, as well as the exponential demographic growth of Candomblé and Umbanda, over the following decades.

But what made Candomblé particularly ripe for the symbolic role accorded it by the Regionalists? Judging by the ways they are personally quoted and consulted in the works of Raimundo Nina Rodrigues, Édison Carneiro, Ruth Landes, Arthur Ramos, and Donald Pierson, it was certainly not the priests' inertia, passivity, isolation, poverty, or folksiness that suited them to this role. They do not fit the stock characterizations of the brown and black roles in the nation popularized by Euclides da Cunha's *Rebellion in the Backlands* (1944[1902]) and Freyre's *Masters and the Slaves* (1986[1933]). They were neither the poor but noble *mestiços* nor "the Negro slave in the sexual and family life" of Freyre's Brazilian national protagonist (1986[1933]:278, 404).

Nor, obviously, were they suitable for such a symbolic role in Regionalism because of any form of biological superiority to other blacks. The Regionalists' strangely racial argument against racism and in favor of Northeastern regional dignity brackets the actual history of the Northeast's ongoing, transnational links with a cosmopolitan West Africa. However, it was those links that gave authority and credibility to the Northeastern Regionalists' struggle against that region's and the nation's suspicions of Northeastern racial inferiority. The first-class transnationalism of the Jejes and Nagôs even yielded a personal dividend for the light-skinned Regionalists of the Northeast, many of whom undoubtedly had Sudanese ancestors of their own.

Thus, the ethnographic "black cycle" reveals what Freyre takes for granted in his nationalist narrative. The "Africans" of Bahia shaped Bahian society with their own distinctly transnational set of priorities. The people who formulated the discourse of "purity" and so impressed Nina Rodrigues and the Regionalist intellectuals with it were hardly the dark "folk" of the purely literary Regionalists' imagination. They were literate international merchants and interlocutors in a circum-Atlantic, Anglophone-centered Black nationalism.

Freyre was fully aware of the ongoing transnational movement of African-Brazilian merchants and pilgrims. Yet they appear only as footnotes in two chapters called "The Negro Slave in the Sexual and Family Life of the Brazilian." Freyre's image of dark Northeasterners as a nameless "folk" shaped entirely by Portuguese men's promiscuous use of passive Indians, "negro slaves," and their cultures simply could not accommodate the biographies and words of these cosmopolitan Black travelers.

The impressive accomplishments of the African merchants and priests of that period lent credibility to Regionalist efforts to show that Northeastern Brazil might, on account of its Sudanese roots, be the racial equal or superior of Rio, São Paulo, and the United States. Thus, an alliance of mutual interest—if not always honest mutual comprehension—between nationalists and transnationalists established Northeastern culture as *the* classical Brazilian culture, preparing the ground for Candomblé itself to become a classic of Brazilian national civilization.

THE NATIONALIZATION OF CANDOMBLÉ?
ON THE POST-REGIONALIST EXPANSION

At the height of the Northeastern Regionalist project, the 1930s, urbanites in the Center-South created a watered-down version of Candomblé called "Umbanda." By the 1950s, two decades after Freyre's landmark publication, both of these religions had begun a vast demographic expansion and entered the canon of Brazilian national culture. Throughout the 20th century and the beginning of the 21st, the migration of Northeastern workers and priests into the more industrialized Center-South has given Candomblé a notable presence in Rio and São Paulo. Moreover, the construction of a superhighway linking Salvador to Rio brought prosperous southern workers and tourists into Candomblé's native territory.[20] In the 1960s, the growth of the petroleum industry in

Bahia brought to Salvador numerous migrants from the interior of Bahia and other Northeastern states, swelling the local Candomblé temples with new supplicants in search of health and community.[21] Thus, Afro-Brazilian religion became not only a homogenizing influence and idiom of community among dislocated migrants but also a major tourist attraction, whose economic potential few state governments were unwilling to exploit.[22] Not just lively entertainment, it also offered the potential for social communion across classes, a repeated theme in the talk of priests and journalists.[23] Moreover, Candomblé was, to borrow a phrase from Claude Lévi-Strauss, "good to think with." That is, as a putatively autochthonous expression of the national "folk," it became a Rorschach, useful in eliciting and authenticating a contradictory array of editorial pronouncements about the ideal character of the nation. Whereas under the First Republic, Candomblé had been branded the racialized antitype of the nation's civilized future, it increasingly became, after the 1930s, the prototype of what the nation should ideally become.

In the minds of many of their Euro-Brazilian advocates (not altogether unlike their foes), Afro-Brazilians walk on their hands. Candomblé and other Afro-Brazilian institutions are seen in a Turnerian sense as "anti-structural" (V. Turner 1969). Carnaval, for example, is often made to typify Afro-Brazilian culture and is reduced to a carnivalesque inversion of "daily life," notwithstanding the level of year-round financial coordination, social hierarchy, and, particularly in Bahia, transnational politics of Black identity that shape the samba schools, *afoxés*, and *blocos Afro* (pace, e.g., DaMatta 1991[1979]; V. Turner 1983:103–24). Similarly, many bourgeois Brazilians see what they want to see in Candomblé.

Thus, Candomblé and its priestesses have been convenient journalistic canvasses for escapist white fantasies and political allegories about *communitas* (V. Turner 1969) and democracy during times of dictatorship; about Dionysian release as opposed to the "Cartesian rationalism" of white society; about love and "real" spirituality, in contrast to the leftist politics alleged to prevail in certain sectors of the Brazilian Catholic Church; about socialism amid rabid Brazilian capitalism; about matriarchy in a patriarchal society; and even about the religion as primitive therapy for the ills of modern life.[24] With the best of intentions, Candomblé's intellectual and journalistic admirers, like those of the Carnaval, regularly misread Candomblé as the model of every cultural or political reform they desire, as the inversion and antitype of every sort of social oppression and psychological distress suffered by Euro-Brazilians. These opportunistic Euro-Brazilian readings of the Candomblé reproduce the spirit and geographical focus of Freyre's argument that Northeastern plantation life demonstrates the "racially democratic" nature of Brazil and, as the reader will see, Landes's argument that the Northeastern Candomblé demonstrates the gender egalitarianism of Brazilian society (see chapter 5 of this volume).

It is perhaps the *highly multiple and conflicting self-representations and the imposed expectations* shaping national life at economic peripheries, such as Brazil, that give psychiatrists such an important role in articulating "national culture" in the third world.[25] Politically, economically, and culturally, Brazil bridges the European and the non-European, the colonizer and the colonized, the hierarchical and the egalitarian, the sadistic and the loving, the Francophone and the Anglophone, the Portuguese and the Coastal—the images of which are enshrined in rival canonical imaginations of Brazilian nationality. Psychoanalysts and activists Albert Memmi and Frantz Fanon are famous for inspiring emergent nationalist movements in North Africa, but they postdated major precedents in Haiti and Brazil, where the image of the possessed person presents a uniquely powerful image of the multiple personality of the nation itself.

Haitian psychiatrists and other physicians with psychiatric training played a central role in the canonization of "Voodoo" as a Haitian national tradition in the early and mid-20th century. François Duvalier, Louis Mars, J. C. Dorsainvil, and Jean Price-Mars described Vodou as an authentic and essential characteristic of the Haitian folk, such that, even if spirit possession was in some sense pathological, it became, especially in the foundational work of Price-Mars, grist in the struggle against U.S. racism and colonial domination. Vodou proved that Haiti had a distinctive culture and historical legacy of its own, worthy of appreciation in its own terms (see, e.g., Price-Mars 1983[1928]; Dorsainvil 1931).

Like the Brazilian psychiatrists, these Haitians offered a nativist reaction to a transnational racism. In Brazil, however, race and racism became an idiom for the inequality among regions of Brazil as well. In Pernambuco, psychiatrist Ulysses Pernambucano de Melo (Dantas 1988:174–78) and, in Bahia, psychiatrist Raimundo Nina Rodrigues established fundamental new principles for the interpretation and regulation of social life in the Northeast. To this day, psychiatrists are nearly as common around the Candomblé as are anthropologists, another professional group intimately involved in making Candomblé a Brazilian national symbol.

In the 1950s and 1960s, a long history of rumors that powerful politicians and generals frequented Afro-Brazilian religious houses accelerated into credible journalistic reports.[26] In the 1970s, most Brazilian newspapers and national news magazines tardily turned from simply reporting the alleged ritual crimes of priests to chasing them for quotes and explanations, not only about Candomblé gods and rituals but about the doings of important priests and African visitors, divinatory predictions of Brazil's future, and the religious conduct of white Brazilian politicians. There were also the multiplying temple federations, which used the newspapers to dictate new policies and standards of ritual practice. Candomblé thus gained some recognition as an agent in, rather than the antipodes or the folkloric past of, the nation-state.

By the 1980s, more than a few Brazilians of all colors would avow that Candomblé's enormously popular offshoot, Umbanda, is the only "genuiunely Brazilian religion" (D. D. Brown and Bick 1987:85). Nonetheless, as a movement to "re-Africanize" has swept through Umbanda since the 1980s, numerous Umbanda temples have converted to the ritually more formal and demanding Candomblé, and Candomblé has become, all the more clearly, the holy of holies in this new temple of Brazilian national identity.[27] True to Freyre's Regionalist impulse, this central religious institution in Northeastern society has become an emblem not of the region's withdrawal from the national society but of the region's status as an exemplary bastion against alienation and marginalization for people of numerous regions, races, classes, and genders in Brazil.

Since the 1930s, the conditions of trans-Atlantic commerce and social functions of "African" ethnic identity in Bahia have also changed. From the late 19th century to the 1930s, the "Africans" of Bahia were in many ways a classic trade diaspora, whose members are united by shared religious and commercial interests in links to their West African coreligionists and trade partners. The interruption of the West African trade during World War II appears to have had lasting effects. Trans-Atlantic merchants and pilgrims no longer visibly dominated the Candomblé elite between the late 1940s and the early 1980s. In this generation, prosperous priests of the Candomblé still retail Candomblé products, but mainly those imported by others or cultivated locally.[28] Some priests were restaurateurs or state functionaries. Mãe Stella of Opô Afonjá was a university-trained nurse (Omari 1984:104n7), while the chief of the Gantois temple, Mãe Cleusa, was an obstetrician.[29] These are the heirs to the literate, polyglot, and wealthy "English professors" and transnational merchants who lifted up the Nagô and Jeje nations in the early 20th century.

However, a higher proportion of the financially comfortable Candomblé affiliates now than in the 1930s seems to consist of the Euro-Brazilian professors, psychiatrists, lawyers, and artists themselves. Their point of entry would have been the post-Freyrean nationalist discourses still evident in much bourgeois journalism and fiction—such as the antipodal logic of Candomblé as primitive folk therapy for the ills of modern civilization and of Candomblé as a religion of nature and environmentalism. These readings of Candomblé usually go uncontested.

Indeed, few priests will question them because psychiatrists and the administrators of the nation-state regard them as good reasons to fund Candomblé. However, such sponsorship, far from altering Candomblé standards, has enabled many temples to restore their direct, transnational connections to the Coast.

THE TRANSNATIONAL NETWORK REBUILT AT THE NATION-STATE'S EXPENSE

From the 1960s onward, military men and civilians, democrats and dictators have agreed in their public deference toward the Candomblé. From the early 1960s onward, politicians and generals sought the blessings of Mãe Menininha, chief of the Gantois temple.[30] In the Center-South, some military officers participated in Umbanda and secured protection for it from the time of that religion's foundation in the 1930s until the 1960s.[31] These supporters of Umbanda clearly avowed their *belief* in that religion (D. D. Brown and Bick 1987:87), but it is tempting to regard ruling-class support for the Candomblé as patronizing, as the pursuit of a photo op with the *povo*, or "folk." On the other hand, some of those photo ops show no small deference to Candomblé ritual protocol and trust in the priestesses' goodwill. One photo from the 1980s shows a priestess manipulating the head of a kneeling Antônio Carlos Magalhães, "the uncontested political chief" of Bahia and, many say, of all Brazil.[32] By the 1980s, cabinet ministers, mayors, and governors constantly visited Candomblé temples, jockeying to display their largesse toward the "great" houses.[33]

This largesse ultimately subsidized transnational communities. During the late 1940s and 1950s, the Nagô priests maintained contact with their African compatriots largely through the voyages of itinerant whites, such as historian and photographer Pierre Verger and diplomat Antônio Olinto, and through the occasional African sojourns of Afro-Brazilian priests—namely, Balbino of Ilê Axé Opô Aganjú in Bahia and Pai Adão of Recife (verger 1995:16; Dantas 1988:181n; Bastide 1978[1960]:77; Freyre 1963:418). However, since 1963, the Brazilian federal government and the municipal government of Salvador have financed or facilitated numerous overseas voyages by Candomblé priests and by the descendants of the trans-Atlantic travelers of an earlier age. Among these government-sponsored travelers were Romana da Conceição, Pai Deoscóredes (Didi) dos Santos, Mãe Olga de Alaketu, Dr. George Alakija, Beatriz and José Elles da Rocha, and Mãe Stella.

The governments of Brazil, Nigeria, and the People's Republic of Bénin have used Jeje and Yoruba genealogical ties to fortify the moral, political, and commercial links among their nation-states. Indeed, the first ambassador that Nigeria sent to Brazil was a da Rocha, the descendant of a proverbially wealthy family of 19th-century returnees to Lagos and cousin to a prolific Brazilian sublineage that includes my friend, Candomblé initiate Beatriz da Rocha. During the 1970s and 1980s, the Nigerian government sent numerous students and professors to Brazil, some of whom taught Yorùbá and conducted scholarly research on Bahian Candomblé.

But the Nigerian federal government is not the only Nigerian actor to build on these trans-Atlantic connections. Apparently as a result of the communications conveyed between Brazil and West Africa by white travelers, the king of Ọ̀yọ́ conferred upon Mãe Senhora of Opô Afonjá the palace chieftaincy title "Ìyá Nàsó," the same title borne by the founder of the Casa Branca temple a century earlier.[34] In 1968, the king of Ọ̀yọ́ named Pai Didi dos Santos the "Bàálẹ̀ Ṣàngó" ("chief of Ṣàngó's community"—in Brazil, Didi is not only the most important priest of the *egum* masquerades in Brazil but also the son of Mãe Senhora, and an important traveler, researcher, and artist in his own right). The king of Kétu gave him a further title—*Bàbá Mọ́gbà Ọ̀gá Oníṣàngó*, "Most Respected NonPossession Priest of Ṣàngó" (Santos 1988[1962]:6). Similarly, in the early 1980s, the king of Ifẹ̀ made a Black North American priest, Ọfuntọla Oseijeman Adelabu Adefunmi I, his vassal-king in the United States (Adefunmi 1982:vi).

Many recent trans-Atlantic journeys have been financed and coordinated by UNESCO, the Ford Foundation, the University of Ibadan and Ọbafẹmi Awolọwọ University in Nigeria, the Brazilian Ministry of Culture, and the Federal University of Bahia.[35] Some of this travel has been intended to strengthen Brazil's specific links with Nigeria and the Republic of Bénin. These efforts complement Brazil's ongoing pursuit of leadership among "nonaligned" nations. This African-Brazilian "traveling culture" (Clifford 1997) also serves Nigeria's aspirations to leadership in the transnational Black world. The second World Black and African Festival of Arts and Culture (FESTAC), held in 1977 in Lagos, hosted a sizable Afro-Brazilian delegation.

The idiom of transnational Black unity in the Anglophone world was itself undergoing changes due to new population flows after World War II. The U.S. Black Power movement of the 1960s resurrected and popularized Africa-centered idioms of Black political unity, which received a further boost from the Hart-Cellar Immigration Reform Act of 1965. That act radically increased the number of Anglophone Caribbean immigrants admitted to the United States and therefore encouraged the pursuit of a least common cultural denominator between them and native-born blacks (see Kasinitz 1992:26–27, 255n2). That least common denominator has been the sartorial, aesthetic, and religious imagery of a shared African legacy. The immigration of Cubans into the United States, particularly after the 1959 revolution on the island, made the Yorùbá-inspired Cuban gods, or *orichas*, available as spiritual guides and further emblems of pan-African unity. And Black North Americans spearheaded a movement to expunge the Catholic saints from their Cuban borrowings. What makes this traveling culture Black with a capital B is the premise that its practitioners are united by a political and cultural identity, an imagined community, rather merely an exogenously perceived phenotypical likeness. Blackness itself is a self-conscious transnation.

The Anglophone Black cultural transnationalism of the 1970s penetrated Brazil massively in the form of North American "soul" music and Jamaican reggae, amplified by enthusiasm over the new independence of Africa's Portuguese colonies. In the 1980s, Nigeria launched a series of World Conferences of Orisha Tradition and Culture, which articulated Ilé-Ifè, New York, Bahia, Havana, Port of Spain, and São Paulo as the multiple capitals of a single transnational religion. At each conference, Black cultural nationalists and *òrìṣà* worshipers from three continents once again met face-to-face, just as they had in Lagos at the turn of the last century.

It was soon thereafter that Mãe Stella announced her view that Candomblé is a self-sufficient religion that ought to be freed of its Roman Catholic saints and its links to the Catholic Church. Her campaign in the 1980s was undertaken with major transnational inspiration and support. At the point when she began her campaign against syncretism, she had, with the financial support of the Nigerian and Brazilian nation-states, entered into frequent contact with the priests and professors of the Yorùbá-majority University of Ifè, in Nigeria. Her central role in this communication with Africa enhanced her prestige in Brazil and built on the fifty-year-old precedent by which Mãe Aninha had done the same under the tutelage of Martiniano do Bonfim.[36] After numerous international planning sessions and two international plenary summits of *òrìṣà* worshipers and scholars, Mãe Stella publicly announced her position at the Third World Conference of Orisha Tradition and Culture, in New York City. Her message was clearly intended to reform not only Afro-Brazilian Candomblé but all the *òrìṣà* traditions in the Americas. Nor were foreign governments and the circum-Atlantic *òrìṣà*-worshiping community her only transnational audience and sources of support. The program met with the vocal support of Roman Catholic Church officials as well, including the Bahian archbishop of the time, Cardinal Lucas Moreira Neves.[37]

Thus, the Brazilian state has facilitated Candomblé priests' participation in black Atlantic organizations and alliances centered on Black Anglophone nationalism and has helped to enhance among these organizations the emblematic status of the *òrìṣà*. *Vodun* worshipers, *nkisi* worshipers, black Caribs from Central America, practitioners of Afro-Jamaican religions, Kardecist *espiritistas* from the Caribbean, and a general public of uninitiated Latinos and Black North Americans were found in numbers at the International Congresses of Orisha Tradition and Culture. But the institutional solidity of Candomblé and the Quêto/Nagô nation's relationship with the Brazilian state have made Brazil an anchor, an exemplar, and a growing exporter of materials and ideas to Black cultural nationalists worldwide. Bahia has become a major destination for Black North American and Afrophile *nuyorican* pilgrims and heritage tourists.

The sacred transnationalism of the Black Atlantic triumvirate has inspired some rethinking of national identity in the People's Republic of Bénin as well. Not only was a House of Brazil built in the Beninese city of Ouidah to parallel the House of Bénin in Salvador, but the Beninese government has also now established an official annual *vodun* holiday, on January 10.[38] And in 2003, the Nigerian head of state appointed Wande Abimbọla to a ministerial post, as special adviser on traditional affairs.

Though subsidized by nation-states, the restoration of black Atlantic travel networks after the 1960s has also restored the old Lagosian-style link between *òrìṣà* worship and Black nationalism, cutting across a nation that Freyre "imagined" constitutionally devoid of a "consciousness of race" (1986[1933]:3, 83). The current head priestess of Bahia's Opô Afonjá temple—the most purist and Africa-centered of Bahia's "great houses"—is heeded and her person lionized not only within the Candomblé but also among the Black nationalists of the Unified Black Movement in Bahia, Rio, São Paulo and elsewhere who regard Candomblé as the most dignified common cultural denominator of Afro-Brazilians. Journalists often ratify the Black nationalists' conflation of Candomblé with the entire Black race, calling Mãe Stella, for example, the "priestess of a race."[39] Stella's declaration of a self-determined African religion autonomous from Roman Catholicism is celebrated as the model for a new Afro-Brazilian political activism.[40] Because of her, many more hundreds of middle-class Afro-Brazilians without prior exposure to the Candomblé have "rediscovered their roots." Along with music, the *orixás* have become a primary symbol of Afro-Brazilians' connection to an international community of the African diaspora.[41]

Neither Freyre nor the framers of the 1933–34 constitution found Afro-Brazilians' identification with an international African diaspora congenial. One framer regarded foreign blacks generally as malcontents who, if allowed to immigrate into the country, would inspire Afro-Brazilians to revolt (Mitchell 1983:18–19). And Freyre repeatedly spoke out against Black North American influence on Brazilian music and politics. Quite contrary to these Brazilian nationalist interests, Mãe Stella and the Afro-Brazilian activists who admire her do find identification with non-Brazilian Blacks congenial, persuasive, and largely consistent with their religious priorities.[42]

Though these self-consciously and transnationally Black leaders pursued ends well beyond the control or the interests of Brazil's ruling classes, they were convenient ambassadors in Brazil's pursuit of leadership in the nonaligned world and in pursuit of markets for Brazilian industrial manufactures. They also served as fictitious emblems of the dictators' and politicians' respect for the "folk" (Dzidzienyo 1983). The symbols of pan-African cultural and political unity never stopped being available as symbols of Brazilian national identity. Hence, the priests and scions of black Atlantic travelers have attracted some of the highest honors of the Brazilian state. In 1977, President Geisel and Generals Hugo de Abreu and Golbery Couto e Silva conferred upon priestess Olga de Alaketu the title of "knight" (cavaleiro) of the Order of Rio Branco.[43] One article reports that she was appointed to the Order of the Cruzeiro do Sul as well.[44] At the invitation of Antônio Carlos Magalhães, she dined with the likes of President João Figueiredo and the Catholic primate of Brazil, Avelar Brandão Vilela, at the presidential palace in Brasília.[45] In 1989, a descendant of Brazilian returnees to Lagos, Angélica da Rocha, received from President Sarney the medal of the Order of Rio Branco, the highest honor that a non-Brazilian can receive from the government (Beatriz da Rocha, personal communication, 1 January 1996). And, on the occasion of her seventieth birthday, in 1995, Mãe Stella received the Maria Quitéria Medal from the city council of Salvador.[46] The citywide celebration included a series of lectures, concerts, exhibits, and public gatherings that demonstrated the near-unanimity with which Northeastern politicians, intellectuals, professionals, and artists endorse these towering symbols of Northeastern dominance in Brazilian national culture. And journalists recognize that to ignore them is to ignore what moves a major section of the Bahian and Brazilian population.

In the contemporary Candomblé, priests most often remember the days of persecution in terms of its most recent legal forms. From 1967 to 1976, Bahian priests had been obliged to register with the Bureau of Games and Customs (Delegacia de Jogos e Costumes) at the Secretariat of Public Security—in effect, the police (Barbosa 1984:70). However, with the support of the right-wing political party (ARENA) and in coordination with other state governors, Bahian governor Roberto Santos explicitly leglislated an end to the requirement of police registration in 1976, for which he earned the enthusiastic and enduring affection of the Candomblé priesthood.[47]

In fact, such restrictions had already been forbidden by successive federal constitutions, which guaranteed freedom of religious practice. However, since the mid-1970s, Bahia and most other jurisdictions of the Brazilian nation-state have been anxious not only to guarantee Afro-Brazilian religions' *equal* protection but to interpret the law and conflicts of interest among groups *in their favor*. In 1972, despite ridicule from the conservative newspaper *Estado de São Paulo*, the government guaranteed retirement benefits (from the Instituto Nacional de Previdência Social) for priests of the Afro-Brazilian religions.[48] Other measures were intended to produce more dramatic results on the visible landscape of Brazil.

In 1985, Mayor Mário Kerétsz, author Antonio Risério, and singer Gilberto Gil founded Salvador's Gregório de Mattos Foundation expressly to upset the Eurocentric priorities of the Brazilian "artistico-intellectual class" (Risério and Gil 1988:233). Thus began the greatest series of public and corporate financial investments in the Candomblé up to that date. What makes them truly significant is not that they represent vast sums of money but that, in a series of struggles among the rival factions of the governing class, the partisans of the Candomblé have *repeatedly* won and the dividends for the Candomblé—both symbolic and monetary—have continually *increased*. Recently, moreover, there has been a *vast* increase.

The foundation funded the renovation of the paradigmatically Black Pelourinho neighborhood, the purchase of *trios elétricos* (trucks bearing mobile stages and sound systems) for use in popular musical and theatrical performances, and other projects. Of greatest interest to Candomblé was Project Bahia/Bénin, which financed the travel of Candomblé priests to the People's Republic of Bénin and arranged for the city to donate a centrally located building to house the House of Bénin, a museum of Candomblé and Beninese culture. It opened in January 1987. Further moves were made to clean up and protect Saint Bartholomew Park, which Candomblé members use to perform ablutions and to deposit sacrifices. Most significantly, the foundation's Project Terreiro funded extensive renovations of the Bogum temple, and with the city's encouragement, the petrochemical company Copene financed the fabulous renovation of the Gantois temple of the late Mãe Menininha. Though Copene disbursed the funds, Mãe Menininha's son managed the renovation project. Casa Branca, Ilê Axé Maroketu, Ilê Axipá, Opô Afonjá, and Luiz da Muriçoca's Ilê Axé Ibá Ogum temple realized smaller but nonetheless significant benefits from Project Terreiro.[49] Note that all the temples benefiting are of the Nagô nation, except for one Jeje house, and that twice as many female-headed temples as male-headed ones benefited. The selection was made to achieve "the right kind of publicity," building on the publicity surrounding Mãe Menininha's death, and thus to further the integration of "*negromestiça* culture" into the national canon (Antonio Risério, interview, 19 December 1995). However, the next mayor, Fernando José, abandoned the Gregório de Mattos Foundation and its projects altogether.[50]

CANDOMBLÉ'S SACRED ENVIRONMENTALISM

Environmentalism is only the latest transnational movement with which Candomblé has entered into a dialogue, and with some degree of fitting opportunism by both Candomblé priests and Regionalist intellectuals. There have long been various references in the Candomblé to the veneration of "nature" (*a natureza*), both as a term for one's personal "nature" and as a construction of nonhuman flora, fauna, and minerals (e.g., Carneiro 1991:174–79; Mãe Aninha, quoted in Pierson 1942[1939]:294). These days, Mãe Aninha, the founder of the African purist Opô Afonjá temple, is regularly quoted as saying something like, "Don't even pluck a single leaf or snap the smallest branch from a tree, for it is like breaking an arm. A tree is like a person, and must be respected as such" (see also Pierson 1942[1939]:293).[51]

Yet in the discourse that I have heard in and around the Candomblé since I began my involvement in 1987, public descriptions of Candomblé as the worship of nature have appeared far out of proportion with such references in the earlier literature. For example, in the run-up to Eco '92 (the international environmentalist conference in Rio), São Paulo's Mãe Sílvia convened the conference "Ecology, Citizenship and African Religions and Cultures." To whit, on 25 April 1992, about eighty Afro-Brazilian priests from São Paulo, Minas Gerais, and Rio Grande do Sul gathered at the Archive of Afro-Brazilian Memory and Life (Acervo da Memória e do Viver Afro-Brasileiro) in São Paulo to make sure that their enemies would not represent the Afro-Brazilian religions as antiecological, in the sense that practitioners might be accused of littering the environment with ritual detritus and carelessly razing the forests with their votive candles.

This episode suggests, first, that, at the geographical peripheries of Afro-Brazilian religion, such as São Paulo, not all priests are confident of their claims on the ecological discourse and, second, that, under the circumstances, such priests know how to organize strategically to assert their rights. Much to their surprise, they found fertile ground at the Rio conference to assimilate their religions to a range of new and old spiritualities sharing and emphasizing their ecological sympathies. The power of what Brazilian journalists call the international "ecological wave"[52] makes such talk into great apologetics—not apologetics *alone* but persuasive and legitimizing talk, nonetheless, about a sometimes-stigmatized religion.

Yet there might be reasons to balk at this discourse. First, it flirts at times with the European modernist and racist notion that blacks are somehow closer to "nature" and are, for better or worse, wilder, more sexual, less refined, and less self-controlled. Indeed, one hears such stereotypes repeated by the same people who endorse the Candomblé. More important, any unprejudiced look at Candomblé reveals it as the worship of fairly *un*natural kings, queens, and knights. Much of this religion's extraordinary beauty lies in the highly *un*natural human craftsmanship in clothing, jewelry, ceramics, dance, and drumming. So, why the selective attention in labeling? First, calling Candomblé a religion of "nature" is more consistent with European and New World stereotypes about black people; second, it has helped to legitimize the Candomblé in materially profitable ways.

Gregório de Mattos Foundation cofounder Antônio Risério lately describes himself as a "Green," and there were hints of an appeal to environmentalism in Risério and Gil's 1988 explanation of their project. However, in the wake of Eco '92, an appeal to environmentalist discourse enabled the mayoral administration of Lídice da Mata to invest still greater financial resources in the Candomblé. The city called its new program Project Gardens of the Sacred Leaves, inspired by the innumerable religious and medicinal purposes to which Candomblé practitioners put plants. Between 1993 and the beginning of 1996, the municipal Secretariat of the Environment had either invested or earmarked more than $80,000 in the renovation of temples, erosion prevention around temples, temple decorations during the festival season, the planting and protection of sacred trees, filmmaking, and the construction of plazas commemorating deceased Candomblé priests. Over $130,000 more had already been allocated, and a total expense of more than $500,000 was anticipated.[53] The rationale for this commitment of public funds is that Candomblé is a religion of "nature worship" and is therefore a locus of indigenous environmentalism, which, in the opinion of the organizers, should be encouraged and used as a conduit for further environmental education.[54]

Though the diversity of Candomblé's sacred nations is marginally better represented among these beneficiaries than among those of Project Terreiro, male chief priests benefited little more than in the past. Only three out of the eleven houses benefiting from the current project have been male-headed. The social and ideological priorities of the Northeastern intellectual class in the 1990s are remarkably consistent with those of the Regionalists and the ethnographers of the "black cycle" in the 1930s, as well as those of a transnational feminism that had deeply influenced them in the 1930s. This last transnational wave is the subject of the next chapter.

If one embraces the reasoning of Anderson and Appadurai, Candomblé's transnational orientation seems to contrast with the territorial logic of the nation-state. Yet, in striking parallel to the geographical metonym of the national imagined community, the priority of the priestesses has long been on the control of territory.

THE LAND OF TRANSNATIONS AND THE NATION-STATE'S LARGESSE

Though they see themselves as transnational, diasporic communities, the Jeje and Nagô nations are powerfully attached to land and to places, such that we should be cautious about describing diasporas as "deterritorialized" (e.g., Appadurai 1996), a term that suggests the diaspora's unconcern about collective and individual landed property in both the homeland and the diaspora. On the contrary, many diasporas passionately seek property ownership in the host country, build houses in the homeland, and value the sense of an ethno-national territory that makes return to the homeland possible in case of strife in the diaspora. To my mind, diaspora is therefore better described as a *pluralization* of territorial identities than as an abandonment of them.

Indeed, land remained a central icon of the dispersed population's belonging in both the homeland and the diaspora. The subdenominations of, for example, the Jeje nation are called its "lands" (*terras*), and every temple is described metonymically as a *terreiro* (yard) or a *roça* (free farm). This last term was once used to describe the land where slaves cultivated food for their own personal use, profit, or manumission (Barickman 1994). And the Candomblé temple that most identifies itself as a pure extension of its West African religious homeland—Ilê Axé Opô Afonjá—has also been the most adroit at acquiring and defending its landed property in Brazil. Hence, the numerous local conflicts over land further illustrate the irony that the most transnational phases of Afro-Bahian culture have also been the greatest beneficiaries of the nation-state's favoritism.

The "great" Nagô and Jeje houses make their concerns well known to the press. One concern dominated all the others from the late 1970s onward—that is, the integrity of their sometimes-sizable urban landholdings. The practice of Candomblé requires privacy and access to plenty of undeveloped land and water sources; thus, most of the oldest houses initially established themselves in areas distant from the city. Sometimes, as in the case of Opô Afonjá, they bought land. In other cases, such as that of Casa Branca, the priests simply squatted on initially outlying land that had appeared unused. They claimed large areas, on which they collected sacred herbs, deposited offerings, and disposed of ritual detritus.

Though critical to the practice of Candomblé, such land claims were often insecure. For example, as the city's population swelled, outsiders laid claim to the undeveloped temple lands[55] or sought to reactivate earlier claims. Even the long-term squatters' rights of some important houses, like Casa Branca, were legally vulnerable to the rival claims of the titleholders.[56] In another set of cases, temple real estate was, again perhaps legally, claimed by rival groups of familial heirs to the founder[57] or by internal rivals to the head priestess.[58] Still other temples were threatened by official plans to use Candomblé lands for the benefit of the general public, such as a footbridge that would carry pedestrians safely over a major highway. The proximity of that bridge to one temple—the Casa de Oxumarê—would have compromised the secrecy of the temple's rites.[59]

However, in each of these cases, between 1977 and 1992, the Bahian state stood up for Candomblé practitioners against other social classes and interests—including a gasoline franchise, a wealthy absentee landowner, unorganized and landless immigrants to the city, and rival legatees who wished to dismember temple properties. As far as I can tell, in none of these cases was either claimant manifestly wrong in her or his claim. These were conflicts in which the state might have been expected to mediate evenhandedly or, if not in favor of the right, then in favor of the national collective or of the strong.[60] We will now look at another case, where it was clearly the strong who won out.

What sympathizer with "the folk" could flatly deny the virtue of giving untitled and undeveloped bush to homeless people? In Bahia, Saint Bartholomew Park is a beautiful and unoccupied stretch of hills, waterfalls, woods, and streams. For generations, it had been used primarily for Candomblé rituals, such as spiritual "cleansings" and the discarding of ritual detritus, such as food offerings contaminated with the malign influences drawn out of supplicants' bodies and the shaved hair of initiands, which is normally deposited in streams. When, in 1982, Mayor Renan Baleeiro proposed that the park be parceled out to eight hundred needy families, a team of intellectuals and artists led by novelist Jorge Amado—the leading Regionalist author of the late 20th century—successfully campaigned to derail the plan. Thirteen months later, it was still unclear where else the needy families might live.[61]

Jorge Amado was not manifestly wrong. We need not lament for eight hundred needy families who do not yet even appear to have been selected. No poor people had been led on and then disappointed. One might ask why, if not to spite members of the Candomblé, a different site had not been chosen in the first place. I do not know, but most stretches of habitable land that large are much farther away from the city, exacting insufferable transportation costs from poor people who are likely to have to work in town. Furthermore, Salvador does need parkland, even if it is attractive for use only by Candomblé practitioners. If nothing else, Saint Bartholemew Park generates oxygen for the surrounding non-*candomblecista* population, which was precisely the argument of the Greens in the municipal government who followed through on Amado's campaign in the late 1980s and 1990s. They agreed that an overcrowded city needs foliage and open air. Moreover, if Amado is correct, this site has been used for these religious purposes for more than three hundred years, and in a religion regularly described as the "worship of nature," does the absence of human architecture make the park any less worthy of protection than, say, a church is in a Christian community?

These cases do not demonstrate the wrongness of any of the rival claimants, or the evilness of the Candomblé. They simply show that the members of the Candomblé are in no way reducible to an indistinct "folk," with interests equivalent to those of all the other simple, common people. Nor are they vulnerable, powerless victims. In fact, labeling this entire range of threats to their interests "land speculation" (*especulação imobiliária*), the Candomblé priests gained public sympathy, monopolized journalistic and official sympathy, and won in virtually every case.[62] However, representing themselves as "the folk," the people of the Candomblé appeared as a collective Brazilian victim of the capitalist onslaught. Partly by persuading journalists to publish exaggerated reports of their temples' age, priests succeeded at proving to the government and the public the truth of their dignity and foundational character in Bahian life.[63] As a result, the state and national governments repeatedly protected Candomblé real estate through an official process of consecration and preservation called *tombamento*.[64] The temples thereby became what we would call "historical landmarks," immune to private alteration or dismemberment.[65]

Though the priestesses' motives may seem pedestrian in their materiality—protecting real estate from other groups with legally and morally sustainable claims—the success of their claims might have meant life or death for some sacred communities and their trans-Atlantic nations. Fortunately for them, it is not for that reason alone that Northeastern artists, intellectuals, and politicians have taken up the priestesses' cause. Overshooting the priestesses' own mark, Candomblé's Euro-Brazilian advocates have now redefined Brazilian national identity itself. They have compelled the central government to make revolutionary changes in the official definition of the national legacy—the Patrimônio Nacional.

Most students of the Candomblé recognize the Casa Branca do Engenho Velho, or Ilê Axé Iyá Nassô, as the oldest Candomblé temple in Brazil.

Its initiates, in turn, founded two of the other oldest—Gantois and Opô Afonjá. Though they operate independently, whatever and whoever enhances the fame of one amplifies the fame of the others. Thus, the renown that Casa Branca brought from the 19th century and from the attentions of early ethnographers has ballooned further with the fortunes of, in turn, Opô Afonjá's Mãe Aninha, Gantois's Mãe Menininha, and Opô Afonjá's Mãe Stella. Casa Branca is the mother of great mothers and, as journalists put it, the "womb [or origin] of the Nagô Nation" (*matriz da Nação Nagô*).[66]

When the current temple building was erected in 1850, its occupants claimed a large area of land for their use. However, their newer neighbors have now occupied all but 6,804 remaining square meters. The roots of a potential disaster took shape in 1925, when a powerful family claiming ownership of the land began charging rent. Though the Bahian newspaper *A Tarde* reports that the family's title to the land was undocumented,[67] the history of rental payments proved the right of the heir and current titular owner, then living in Rio, to demand the evacuation of the land. Given the rapid development of the surrounding area, he knew that the land had much more profitable uses. The threat of eviction prompted Afro-Brazilian and/or Northeastern anthropologists, a Catholic monk, famous artists, politicians, Bahian Carnaval clubs, and Mãe Menininha herself to demand the official expropriation of the land from the absentee owner, as well as the removal of the temple's least scenic new neighbor, a gasoline station that allegedly posed a threat to nearby sacred trees.[68]

In order to save the "womb of the Nagô Nation" from demolition, two successive mayors separately declared it a municipal historical landmark. It was then declared a public utility and the land expropriated from the absentee owner. Most important, in 1984, the Secretariat of the National Historical and Artistic Patrimony conceded to the demands of Afro-Brazilian and Northeastern celebrities to declare it a part of the National Patrimony and immediately allocate tens of thousands of dollars for initial renovations. This status as a national historical landmark had previously been granted only to "churches, palaces and manors." Hence, the national secretary of culture and presiding officer of the consultative council, Marcos Vinicius Villaça, admitted that deliberations on this case had been "polemical": only three of six council members voted in favor of the innovation, which journalists described as "historic" and "a revolutionary decision."[69] Just as important, articles in numerous newspapers made this decision a highly public event through this medium par excellence of the Brazilian "imagined community"—the newspapers.

In the Northeast, this honor was reserved for a female-headed temple. In the Center-South, a temple founded by a male priest passed muster, even though, by that time, his niece had inherited the throne. But again, the relatively pedestrian needs of the priesthood inspired a revolutionary change in the historical charter of the Brazilian nation-state. In 1992, the state of São Paulo would enshrine the much newer Axé Obá Ilê temple as a state historical landmark. Founded in 1974 as an Umbanda temple, Axé Obá Ilê displays in its architecture and altars the impressive personal wealth of its founder, a dark mulatto who made his fortune in the meat trade. When he died intestate and without children in 1984, several of his nieces and nephews made claims on the estate. To prevent the dismantling of the temple, moves were made in 1988 to enshrine it as a historical landmark.

According to the president of the Council for the Defense of the Historical, Archaeological, Artistic and Touristic Patrimony of São Paulo (CONDEPHAAT)—Edgard de Assis Carvalho—the grounds for the temple's eligibility in the state constitution were not clear, and it took two years to collect the six expert opinions needed to approve the measure. Fortunately for this religious institution, the claims of the rival heirs were stayed and the temple continued to function pending its actual *tombamento* (enshrinement) in 1992.[70] Again, the Candomblé and its advocates had the clout in the 1980s and 1990s to fend off major rival interests and threats to its existence. In the process, they modified well-established standards of the Brazilian nation's historical self-definition and compelled the reinterpretation of the constitutional laws of the Brazilian state.

On the one hand, the nationally enshrined Candomblé temples appear to have won the victory of acceptance within a nation-state that had once persecuted them. On the other hand, some Candomblé priorities have been undermined. The "great houses" rest atop long histories of efforts to keep themselves *autonomous* from the Brazilian nation-state. Incorporation into the National Patrimony entails not only honor and material advantage but also the renewed risk of state regulation and a public visibility that is anathema to this secretive initiatic tradition. Few priests, however, regret being represented as exemplars of the nation-state's ideals. The resulting prestige is effective in other imagined communities as well.

WHO'S IN CHARGE HERE?

To challenge the premise that Euro-Brazilian elites have "manipulated" Candomblé is not to deny that this religion has been affected by its dialogue with bourgeois nationalists. In interviews with various newspapers, some priests have enunciated priorities suspiciously consistent with

Regionalist expectations about authentic and decently folkish religion. They have publicly disavowed commercialism, harmful magic, and "immediatism," or the pursuit of immediate material benefits instead of the virtue of long-term piety in itself.[71]

Despite these disavowals, articles in Center-South newspapers document, with little criticism, the commercial dimension of Afro-Brazilian religions.[72] After all, the provision and practice of these religions is the means by which many Brazilians make a living. Moreover, to justify freedom of religious choice, scholars and journalists in the 1980s regularly invoked the terms of the emergent international hegemony of neoliberal economics. In dozens of publications, the term "the religious market" (o mercado religioso) was used to describe the ease with which Brazilians moved from one confession to another as it suited their needs.[73] Candomblé is often more consistent with the neoliberal capitalism of the Center-South than with the pastoral nationalism of the Northeastern Regionalists.

Indeed, some priestly enunciations have militated against Regionalist values. Since the 1970s, priests interviewed in Brazilian newspapers almost uniformly condemn "folklorization" (folclorização)—that is, the display of Candomblé ritual objects, music, and dances in Carnaval parades and in nightclub shows, all of which fascinate tourists and generate revenue for the Northeastern state governments.[74] Mãe Stella's well-publicized efforts to eliminate "syncretism" (sincretismo—i.e., the integration of Roman Catholic saints and church visits into Candomblé practice) oppose the cultural and racial mestiçagem, or hybridity, endorsed in Freyrean nationalism. And, in what is proverbially the "largest Catholic country in the world," the growing denial of Catholic Church authority over Candomblé is at best disorienting to Northeastern bourgeois convictions.[75] Contrary to the representations of the recent critical ethnography, the best-documented and seemingly best-"protected" Candomblé houses have sidestepped the most explicit interests of their alleged "protectors."

The priorities of powerful factions in the Brazilian bourgeoisie at times harmonize with those of the Candomblé priesthood, but it is not always clear who is "manipulating" whom and who is garnering the benefits. State officers might imagine themselves engineering a new public conceptualization of the national heritage or engineering a more environmentalist nation, but the houses that these officials are most anxious to draft into such projects can afford to be choosy about their involvements. The accumulated cultural, political, and financial power of the "great houses" of the Nagô nation has often enabled them to go their own way, in patent disregard of the incentives provided by the Regionalist bourgeoisie and the nation-state. For example, when Opô Afonjá's

Mãe Stella was offered the opportunity to have her temple made into a Brazilian historical monument, she refused, unwilling to allow the state any regulatory authority over whatever changes, repairs, or renovations she or her successors decided in time to undertake.[76]

Though Project Terreiro director Juca Ferreira had hoped to use state largesse as an occasion to educate the priestesses about ecology, one priestess called the project "Juca's *cachaça*," or Brazilian rum—dismissing it as a product of the potential benefactor's typically Brazilian enjoyment or addiction rather than the need of the Candomblé community. She surmised that, at some level, the donor was in pursuit of his own irrational and indisciplined emotional pleasure. Whereas Motta, Dantas, Brown and Bick, Fry, Henfrey, and others see Candomblé temples as dependents of the "dominant class," this priestess seems to see the state bureaucrat as the dependent party.

Such a priestess had the clout not only to reject the state's offerings but also to make demands of the state according to her own priorities and value orientation. For example, she secured support in the early 1980s for a Candomblé-sensitive public school on the grounds of her temple (which has been built) and for a commercial bakery (which had yet to be built as of 1995). The priorities of powerful factions in the Brazilian bourgeoisie and state at times harmonize with those of the Candomblé priesthood, but it is not always clear who is "manipulating" whom.

DIVIDE AND CONQUER: THE CANDOMBLÉ'S STRATEGY OF RULE

The efforts of the "dominant class" to suppress Candomblé or incorporate it into the nationalist project have been too uncoordinated to prevent the priests from sustaining alternative, crosscutting communities of their own imagining. For example, politicians and patrons who visit, protect, or fund Candomblé temples in exchange for votes and other forms of support are in competition with politicians and patrons who neglect to do so. Police persecution of Candomblé continued well into the 1970s, but documented cases of white elite or state cooperation with Candomblé have abounded since the 19th century (e.g., Reis 1986; Rio 1976[1906]; Bastide 1978[1960]:55; D. D. Brown and Bick 1987:83). Such cooperation sometimes met with subtle or violent dissent by other powerful Euro-Brazilian factions.[77]

Brazilian law, even as seen through the nation's press, does little to convince the reader that all the territorial nation's citizens experience the same form of homogeneous sovereignty or egalitarian community (pace Anderson (1991[1983]). In 1982, Mayor Renan Baleeiro of Salvador signed a decree again affirming the applicability of the federal constitution to Candomblé temples, this time to establish their exemption from state taxes.[78] In 1985, it was still news to Mãe Nicinha that her temple was tax-exempt, as she hoped that tax exemption would follow from municipal legislation just being proposed by the mayor that year.[79] By 1989, in the face of chronic nonenforcement of existing legislation, the state constitution of Bahia would promise extraordinary measures to guarantee respect for and preservation of Afro-Brazilian religion. Article 275, in chapter XV, reads:

> It is the obligation of the State [of Bahia] to preserve and guarantee the integrity, respectability and permanency of the values of Afro-Brazilian religion especially:
>
> I—to inventory, restore and protect the documents, works, and other goods of artistic and cultural value, the monuments, fountains, flora and archaeological sites linked to Afro-Brazilian religion, the identification of which will be left to the temples and the Federation of the Afro-Brazilian Cult;
>
> II—to prohibit the state agencies responsible for promoting tourism from exhibiting, commercially exploiting, broadcasting [veiculação], captioning [titulação] or prejudicially handling the symbols, expressions, music, dances, instruments, paraphernalia, clothing and cuisine, strictly linked to Afro-Brazilian religion;
>
> III—to assure the proportional participation of representatives of the Afro-Brazilian religion, alongside the representatives of other religions, in whatever commissions, councils and agencies are created, as well as in events and promotional activities [promoções] of a religious character;
>
> IV—to promote[,] in the teaching of geography, history, communication and expression, social studies and artistic education[,] an adequate acknowledgment of the Afro-Brazilian historical reality in state-run institutions at the primary, secondary, and university levels.[80]

Like newspapers, law is a bourgeois mode of imagining the national community. It, too, displays the inconsistencies, and indeed the heteroglossia (Bakhtin 1981), that priests have often been able to exploit. Over the course of the 20th century, Candomblé received such momentous but inconsistently enforced concessions from elite white Brazil precisely because the image of Candomblé accommodated diverse and sometimes opposite programs by the Euro-Brazilian elite. Both military dictators and democratic politicians have publicly knelt and kissed the hands of the "great" black priestesses.

"White" Brazilians are not always possessed of the superior symbolic capital in this negotiation. For example, we have seen that the Bahian

Nagôs' British passports and command of English were major capital in 19th-century and early 20th-century Bahia, and that Djedji identity in the colony of Dahomey was heavily subsidized by French territorial competition with the British. In an age of increasing Anglophone U.S. domination, it is significant that, after World War II, French scholars and a French-trained Argentine scholar—Bastide, Verger, and Elbein dos Santos—became the most persuasive literary spokespersons for Candomblé's distinctness, integrity, and virtue in Brazil. The two-century-old contest between the French and the Anglo-Saxons for global hegemony has been highly productive in the nationalist imaginings of the black Atlantic world. In Bahian religion and scholarship, as in the U.S. academy, French intellectual products provide powerful symbolic ballast against Anglo-Saxon dominance in the Americas (see also Landes 1947:35).

CONCLUSION

The state official most often remembered for his persecution of the Candomblé in the 1930s is "Pedrito"—police commander Dr. Pedro de Azevedo Gordilho.[81] He was persistent in his predatory "hunt" (caçada) for Candomblé practitioners, invading and desecrating numerous temples in Salvador.[82] For that reason, but not that reason alone, his name is found in the archives of the Tribuna and Tarde in Salvador, of the Estado in São Paulo, and of Bahiatursa. It comes up in the speeches, interviews, and writings of Candomblé officials, which are quoted repeatedly by contemporary journalists. While Candomblé members remember Pedrito's malice and his horrific deeds, they also recall him in playful song:

> Sexta e sábado
> Domingo é meu,
> Cadê Pedrito?
> O gato comeu.

> Friday and Saturday
> Sunday is mine,
> Where is Pedrito?
> The cat ate him.[83]

This predator who was himself eaten embodies both the shared convictions and the controversies that Candomblé and its advocates have now made central to the debate over the content in Brazilian "national culture."

Pedrito personifies the nation-state's deepest antagonisms toward the Afro-Atlantic nations on its territory. And yet the Candomblé "ate" him.

The ditty suggests that Candomblé brought Pedrito to heel. The culture-specific nature of this conquest is conveyed in a double entendre. *Gato* is slang for "guy," and *comeu* means "ate" or "penetrated sexually," suggesting that someone penetrated Pedrito. Working-class Bahians tend to regard anal penetration as a form of humiliation and domination, to which spirit possession itself, as we shall see in the next chapter, is dangerously similar. Male priestly authority in Brazil has been variously described and condemned in these very terms by elites who, beginning in the 1930s, have questioned the "African purity" of male priestly leadership.

This question of male leadership appears to be at stake in newspaper and oral accounts that seem unable to settle the tandem questions of who conquered Pedrito and whether it was Mãe Aninha, Pai Joãozinho da Goméia, or Pai Procópio who secured Vargas's commitment to end police persecution of Candomblé. The non-Candomblecista journalists report as certainly that it was a *woman* as leading male Candomblé officials report that it was a *man*.

One journalist's story goes as follows: "The police-soldier Pedrito, commanding a detachment, invaded the temple [of Aninha] during a festival in homage to Ogum [god of war and iron]. . . . the *filho-de-santo* [Aninha's initiate], following an order given by Mãe Aninha with her eyes, incanted some words in *nagô*. Ogum possessed the policeman and [Pedrito], possessed, turned against his colleagues, setting them in flight from the temple.[84] In the eating of Pedrito and the statist forces he represents, Candomblé has played the role of the eater or penetrator. Another version of these events is less detailed but, like a campaign, appears in more numerous sources. It is promulgated in multiple forums by Jeová de Carvalho—a poet and male official of the Jeje Bogum temple—in his soi-disant "Reinvention of the Kingdom of the Voduns," but it is also seconded by male scholar and Opô Afonjá official Vivaldo da Costa Lima. This story reports that the infamous "hunter" of priests ultimately became an *ogã*, or male advocate and honorary official, of Pai Procópio—a male priest of Jeje ancestry.[85]

By both reports, the state, in the person of its most radical official enemy of Candomblé, has been converted by ritual means into a servant of the gods. Yet the stories contrast in their identification of the responsible agent. Was it a male priest linked to the respected but moribund Jeje nation? Or was it a priestess—the great ancestress of Nagô sacred imperialism? As a mythic charter, the tale of Aninha's coup seems more consistent with today's reality that the Candomblé priestess far outranks the Candomblé priest in Brazilian national culture. However, the press has been and still is a powerful instrument in the making of that reality. In the 1980s, Jeová and Vivaldo implicitly lent their efforts to resurrecting the legitimacy of male leadership in the African purist Candomblé.

But their efforts fly in the face of what has, over the past half century, become a local form of common sense to the contrary.

The logic of "African purity" and Yoruba superiority may not have originated from the Euro-Brazilian ruling class, but the efforts of the Euro-Brazilian bourgeoisie to secure a position of esteem in the transnational culture of nationalism did dictate at least one powerful transformation in the Candomblé priesthood after the 1930s. Chapter 5 documents the transnational efforts and nationalist motives by which the Candomblé priesthood was declared a female monopoly, and centuries of male priestly leadership in *òrìṣà* and *vodun* worship were reclassified as a pathology. In this way, the ethnographic "black cycle" of the 1930s left a powerful mark on the leadership of Candomblé.

Para Inglês Ver

SEX, SECRECY, AND SCHOLARSHIP

IN THE YORÙBÁ-ATLANTIC WORLD

> "I know by now that women are [in Bahia] the chosen sex,"
> I said to Édison. "I take it for granted just as I know in our
> world that men are the chosen sex."
> —RUTH LANDES, ethnographer

> Yorùbá society did not make gender distinctions and instead
> made age distinctions.
> —OYERONKẸ OYEWUMI,
> Yorùbá-American sociologist (1997:157)

NOT ONLY DITTIES, like Pedrito's, but also clichés reveal social history. The Brazilian expression *para inglês ver* (for the English to see) describes acts of subterfuge and self-camouflage—presenting a facade to outsiders and dominant parties who might respond with contempt or punishment if they knew the truth. One story reports that the expression originated during the 19th century, after the British outlawed the maritime slave trade to Brazil in 1830 and slave traders developed means of camouflaging the slave ships to avoid capture by the British navy. The false appearance of an innocent maritime commerce was "for the English to see" (see esp. Fry 1982:17). That the English are chosen as the paradigm case of the critical outsider evokes both the specific power of the Anglo-Saxon gaze in the political history of Brazil and the general fact that national identities have always taken shape in a transnational context.

CANDOMBLÉ AND THE AFRO-ATLANTIC DIALOGUE

Certain massive 20th-century changes in the gendered leadership of Candomblé would be difficult to explain if we failed to examine a series of transnational dialogues—involving Afro-Brazilian priests alongside state officials and an international community of scholars. I will argue that the Candomblé religion owes not only much of its international fame but also the gendered transformation of its internal leadership to Ruth Landes's *City of Women* (1947), in which the author offers Candomblé as a living and time-honored example of matriarchy, available to inspire the opponents of sexism in her own native society, the United States. Landes's scholarly gaze and intervention, however, were not the first. In the 1930s, Brazilians inaugurated a series of scholarly conferences uniting scholars and priests, making it increasingly important for us to acknowledge the role of scholars' "official" pronouncements and cover-ups in the ongoing transformation of the Yorùbá-Atlantic religions. Yet Landes's international exposé became influential precisely because Candomblé has been a convenient template for other scholarly and political agendas, including Gilberto Freyre's Regionalist nationalism, which we have already seen in action, Melville J. Herskovits's effort to redeem African Americans from what he calls the "myth" that we lack a cultural past and are therefore inferior to whites, and Oyeronkẹ Oyewumi's diasporic Yorùbá nationalism. Such agendas have shifted not only the reputation but also the *practice* of òrìṣà worship in the diaspora, and this transformative dialogue continues in the 21st century.

Not only scholars and literati but also political leaders of local or international standing have presented Candomblé as a metonym of their imagined communities and struggled to redefine it in the service of their interests. Yet the popular realities of Candomblé and the heterogeneous interests of its devotees always resist such agendas. For example, the hottest touchstone in the transnational debates over the meaning of Candomblé and the communities it authenticates is a cultural persona that is as normal to Candomblé devotees as he is anathema to the normative vision of the nation-state—that is, the "passive homosexual," or, in Candomblé parlance, the *adé*.[1]

The *adé* priest, like his counterparts in the African-inspired religions of Pernambuco, Rio, Cuba, and Haiti, is regarded as normal and eminently respectable by most devotees but has, since the 1930s, been summarily dismissed by nationalist and transnational feminist scholars as either "untraditional" or nonexistent. This chapter is intended not simply to correct an empirical error but also to demonstrate the role of national and transnational scholars in creating what is now widely regarded as the primordially African tradition of "cult matriarchy" in Candomblé.

The main historical riddle addressed in the present chapter is this: How did the Jeje-Nagô Candomblé come to be widely regarded as traditionally "matriarchal," while male priests play a numerically equal and often politically dominant role in Candomblé's counterparts and antecedents elsewhere in Brazil and around the Atlantic perimeter? The number and authority of male priests equal or exceed those of priestesses in Ọ̀yọ́ religion, Pernambucan Xangô, Cuban Ocha, and Haitian

Vodou. How did the "cult matriarchate" (Landes 1947) become the standard of legitimate practice among scholarly observers and government sponsors in Brazil?

I will argue that the dialogue between priests and non-Candomblé elites did bring about this one radical change in Candomblé practice, one apparently unintended by the priests identified as its exemplars. However, Euro-Brazilians were not the power behind this foremost case of elite manipulation. It was, I will argue, a class of transnational scholars, of which Margaret Mead and Ruth Landes have become the most famous exemplars—social scientists who identify in distant societies ideal models for the destruction of Western sexism. In short, Ruth Landes helped to make real in Brazil her own subaltern North American vision of a primitive matriarchy. Since the 1990s, sociologist Oyeronke Oyewumi has reinterpreted Yorùbá-Atlantic religion in a similar program to challenge the sexism of the West and puncture the myth of Western racial superiority.

The "City of Women"

Ruth Landes became a great foremother of feminist anthropology by declaring a unique status for women in Candomblé. Her 1947 title, *The City of Women*, advanced the opinion that, by tradition, women were uniquely suited to serve the African gods. Landes added her voice to an ongoing tradition of privileging the Jeje and Quêto/Nagô nations over the Angola nation, the Congo nation, and the *caboclo* worshipers, on the grounds of the Quêto/Nagô nation's alleged African purity and, therefore, its unique authenticity. For Landes and her many North American admirers, the Quêto/Nagô Candomblé, in particular, has inspired great hope as a shining example of "matriarchy." Thus Landes named Candomblé and its Bahian host city the "City of Women."

On the other hand, during the 1930s and 1940s, Landes's research embarrassed some Euro-Brazilian nationalists for two reasons. First, in a country ambivalent about its demographic and cultural blackness, it was embarrassing that she studied Candomblé at all. However, her agenda paralleled, in some ways, the new antiracism of Gilberto Freyre and Melville J. Herskovits, fellow students of anthropologist and cultural relativist Franz Boas. Landes studied Afro-Brazilian religion not as a racial flaw to be hidden but as proof of the richness of a transnational African legacy. More important, she added the antisexist proposition that Candomblé demonstrated the potential for women's equality elsewhere in the world. Her argument, however, required her to broach the second embarrassing topic—Brazilian sexuality.

The final two paragraphs of the book summarize how Landes saw her own relationship to the guardians of Brazil's international reputation:

> When I left Rio for the United States, Brazilian friends escorted me to the boat, and one of them said, half teasing but with a certain defiant patriotism, "Now you can tell them that no tigers walk in our streets."
>
> I nodded, and added: "I'll tell them also about the women.... Will Americans believe that there is a country where women like men, feel secure and at ease with them, and do not fear them?" (Landes 1947:248)

In her study of Candomblé in Brazil, a country that she knew to be highly sexist, she felt that she had found evidence of what she called a "cult matriarchate," in which women ruled in the religious affairs, and therefore the most important affairs, of Bahian blacks (Landes 1940:386–97). Actually, argues Healey, she had constructed a primitivist cliché (Healey 2000[1998]:87–116). In her search for the antipodes, for a primordial alternative to the lamented condition of her home audience, Landes had created an otherworldly Bahia, of which she declared, "I know by now that women are [in Bahia] the chosen sex.... I take it for granted just as I know in our world that men are the chosen sex" (Landes 1947:202). Like Margaret Mead in Samoa, Landes had silenced or distorted a great deal of the evidence at her disposal (Freeman 1983; Mead 1928). For example, to account for the significant number of men leading Candomblé temples at the time of her visit, she claimed that, no matter how numerous they might be, they did not count, because they were violating "African tradition" due to their own personal psychological problems and due to the alleged ritual laxity of the women who had initiated them. Indeed, Landes effectively founded the Brazilian tradition by which a temple's commitment to excluding men from the possession priesthood became a significant measure of its "African purity."

Landes ignored the significant number of leading male priests in the Nagô nation and instead identified the male priesthood chiefly with a variant of the Nagô religion in which Indian spirits, or *caboclos*, were also worshiped prominently. But given that a priestess of the Nagô nation is credited with founding the *caboclo* cult (Landes 1940:386–97, esp.391), that the Nagô *orixás* remained preeminent even in the *caboclo* worshiping houses (Landes 1940:391–92), and that virtually every Nagô temple also worshiped the *caboclos* (Carneiro 1986[1948]:54), the categorical separation that Landes drew between the female-dominated Nagô temples and the male-dominated *caboclo* temples seems suspiciously a priori and inconsistent with the self-declarations of the temple leaders themselves (cf.Landes 1940:393 with Carneiro 1986[1948]:52).

In fact, across all nations, including the supposedly traditionalist Nagô and Jeje nations, male leadership in the Candomblé had been an old phenomenon (V. Lima 1977 [1971]:106n19; Motta 1976). Throughout the 19th century, men significantly outnumbered women in the Bahian Candomblé priesthood generally, and men were common in the priesthood of the supposedly all-female priesthood of the Nagô and Jeje nations as well. Indeed, the increase in *female* leadership was the more recent phenomenon (Harding 2000:71–74, 77, 103; Carneiro 1986[1948]:57, 104–9; Butler 1998:193, 195; Wimberly 1998:82–85). The *degree* to which male leaders outnumbered female leaders appears to have declined somewhat (from 88 percent to 58 percent) between the first and second halves of the 19th century (Harding 2000:72), but so did the proportion of men in the general Afro-Bahian population following the mid-century cessation of the slave trade. Equally important was the increasing willingness of Brazilian-born captives (as opposed to the African-born) to participate in these religious practices. Despite such shifts, and contrary to the consensus among scholars since the 1930s, it is extremely unlikely that men were ever excluded in fact or in principle form Candomblé's 19th-century leadership.

Butler believes that a tradition of exclusively female temple leadership began in the Casa Branca, or the Ilê Iyá Nassô temple, in the mid–19th century (which, one might add, is relatively late in the documented history of Jeje and Nagô religion in Brazil) and spread due to the fast-growing prestige of that temple among scholars and bourgeois elite sponsors (Butler 1998:193–209; Butler, personal communication, 3 December 2002). Yet, as we shall see, the evidence of an exclusively female leadership—statistically or in principle—is ambiguous even in Casa Branca and its scions before the 1930s.

In the 1930s, male chief priests (*pais-de-santo*) still equaled or outnumbered female chief priests (*mães-de-santo*) (Corrêa 2000:245; Carneiro 1986[1948]:104; 1940:272; Herskovits 1966[1955]:230).[2] Nonetheless, since the publication of Landes's work (1940:386–97; 1947), the scholarly advocates of Jeje and Nagô superiority have come to speak with one voice on the matter: in the Candomblé priesthood generally, "women are the chosen sex." However, Landes and her companion Édison Carneiro recorded, albeit dismissively, copious evidence against their own interpretive models. Yet Landes's tendentious interpretation of this evidence clearly changed the minds and conduct of Candomblé's leading bourgeois allies and, consequently, the conditions of that religion's reproduction in Brazilian society.

GENDER AND THE DIALECTICS OF "PURITY"

V. Lima (1977[1971]:106n19) believes that Landes's revision of Candomblé tradition arose directly from the prejudices of Martiniano do Bonfim, but Martiniano's words are ambiguous. Indeed, Martiniano is the first local authority quoted in Landes's *City of Women* (1947) on the matter of gender roles and sexual morality in Candomblé. At the end of a nostalgic diatribe against the inability of young or middle-aged priestesses to keep their minds off of money and sex, even in the most esteemed Nagô temples, Martiniano notes that the less esteemed *caboclo*-worshiping temples even failed to keep "men" from dancing in the midst of the priestesses during their sacred performances. Martiniano complained:

> Anyhow, I don't believe much in these young women who run the terreiros [temples] nowadays. They want to make money, and they want to get men. Most of them are too young to be dedicated to the gods. After all, Menininha [chief priestess of the prestigious Gantois temple of the Nagô nation] is only forty-two or -three, and the blood is hot in her! . . . Nowadays in the middle of their services they are thinking of other things: the man they saw last night or the one they will see next night, or the men who are watching them dance. That is hardly right when they are supposed to be praising the gods! And these new temples of the caboclo "nation"—my God, they are crowding everything else out, they are throwing our traditions away! And they allow *men* to dance for the gods! (Martiniano do Bonfim, quoted in Landes 1947:32; italics in original)

For Landes, this last phrase of Martiniano's becomes the opening volley in the argument that, by tradition, men simply do not belong in the Candomblé priesthood. However, Landes's worries do not appear to be Martiniano's. In the context of his worry about inappropriate thoughts of sex and money during sacred ceremonies, Martiniano seems to suspect that men on the sacred dance floor would inspire even *more* thoughts of sex and money in the priestesses, that the sacred dance circle had become like a dance hall, tainted with financially interested sexuality (cf. Landes 1947:196–98, 200). Later, I will discuss the further ambiguity in the term "men": not all males are "men" (*homens*) of the sort that would bring such a temptation to the priestesses.

However, in favor of Landes's interpretation, the sentence "And they allow *men* to dance for the gods" does literally concern the legitimacy of *their* dancing, not the threat they pose to the *priestesses'* concentration. Yet a careful reader would have several reasons to doubt that Martiniano is being quoted faithfully. First, Martiniano's lengthy prior and subsequent words are unrelated to such a worry. Second, a person of even Martiniano's apparently limited exposure to the religions of the Lagosian hinterland in the late 19th century would have known that many West African òrìṣà possession priests, who necessarily danced for the gods, were cross-dressing men. Indeed, they were the most publicly visible priests.[3] One cannot but suspect that Landes's reportage of unrecorded speech, inspired by a strongly partisan agenda, resulted in some inaccurate or *biased* quotation. Martiniano's apparent endorsement of spatial segregation and role differentiation between men and women almost certainly obeyed motives other than the affirmation of "cult matriarchy." After all, he was a male priest, and one whose influence, furthermore, exceeded that of any priestess in his time.

As we have seen, the forms of "purity" that Martiniano endeavored to "restore" to Bahian Candomblé often followed priorities foreign to the Ọ̀yọ́ traditions implicit in the "great" temples' avowed origins. Instead, he often advocated the priorities of a racial nationalism that emerged in reaction to British colonial racism in Lagos and in dialogue with Black responses to racism in the United States. The Lagosian reaction was imbued not only with the values of the African religions that became emblems of Yorùbá dignity but also with the British nationalist and Protestant idioms of England's dialogues with its colonies (see Comaroff and Comaroff 1997; 1991 on the South African case).

During the last quarter of the 19th century, Martiniano had been educated by Anglophone Africans in the Presbyterian Faji School in Lagos, where he certainly would have been exposed to Victorian gender norms and to Afro-Atlantic reactions against European stereotypes that blacks and other southern peoples are oversexed, or excessively "hot." Such reactions, including Martiniano's, are often even more puritanical than the values and behaviors of Victorian whites, at least partly because some of the "men" at Candomblé services were prosperous Euro-Brazilians who had come in pursuit of potentially exploitative sexual dalliances with dark women (e.g., Landes 1947:145).[4] Hence, Martiniano's apparent sexual puritanism and, if he is being quoted accurately, his dichotomous thinking about priestly gender roles seem far more indebted to Victorianism and Anglo-Atlantic nationalism (including its racial variants) than to the Ọ̀yọ́ religions of spirit possession.

Landes's main guide through the world of Candomblé, however, was Northeastern lawyer and journalist Édison Carneiro. However, his interpretation remains at odds with Landes's, in that the dancing men are not singled out for his opprobrium; even more than Martiniano, he criticized male and female priests on the same grounds.

As quoted by Landes, Carneiro and his Yorùbáphile friends Mãe Menininha, Eulalia, and Manoel identify three main phenomena that distinguish proper priests from improper ones: (1) the use of hot combs, which desecrate the part of the possession priest's body that hosts the possessing god (Landes 1947:38, 157–59, 196, 238); (2) modern greed, trade, and profit orientation (157–59, 192, 193); (3) the lack of initiation (157–59); and (4) the practice of malevolent sorcery (182–83, 192). Most of these phenomena converge around the negative symbolism of "heat."[5]

What is most surprising, in the light of Landes's argument that the priesthood is ideally a female monopoly, is that Carneiro and his friends implicate priestesses and priests equally in these alleged failings. The reader witnesses as much testimony attributing a decline in the quality of the priesthood to the admission of inappropriately young, moneygrubbing, and concupiscent women as to the admission of "men," whose inappropriateness as priests is similarly blamed on their concupiscence and pursuit of money (e.g., Landes 1947:38, 202).

If, from the time of Martiniano's intervention, some emergent discourses made the marginalization of male possession priests conceivable, that marginalization was demographically unverifiable and hardly commonsensical in the 1930s. It was, however, highly evident by the 1990s. The cause of this transformation, I submit, is not to be found in Ọ̀yọ́ tradition.

Rather, at least three rival imaginations of community with firm roots in a circum-Atlantic dialogue engaged in a "struggle for the possession of the sign." In Landes's text, three interpretations of the Candomblé dance space, or *barracão*, become emblematic of three rival and overlapping constructions of the collective self. Through the drama of the *barracão*, Martiniano imagined a transnational Yorùbá community uncontaminated by Brazilian national money and miscegenation. Carneiro imagined a Northeastern regional community anchored by an unchanging, premodern innocence. What we see most clearly in the text, however, is Landes's own imagination of a transnational community of women who are sufficiently alike in substance that the "matriarchy" or gender equality they experience in one place is equally possible in other places.

GROUNDS FOR DISMISSAL: THE NATION-STATE AGAINST THE *ADÉ*

The credibility of Landes's novel claim that Candomblé is a matriarchy, despite demographic and historical facts to the contrary, relied on a deus ex machina, embodied in a specific antihero that neither Martiniano do Bonfim nor Édison Carneiro had previously singled out for disapproval. Thus, the greatest casualty in this struggle for the possession of the sign was the *adé*—antihero of the territorial nation and of Landes's "cult matriarchy" alike.

Landes's argument of images harmonized with the transnationally exported logics of North American and European nationalisms. As George Mosse (1985) shows, homophobia is a common adjunct of nationalism. In Landes's appeal to a counterfactual nostalgia equally typical of nationalisms, she dismissed the male presence as the result of a recent corruption. As if to confirm that the male presence was recent and nonnormative,

Landes reported the widespread view that all male possession priests were *adés*, or, in Landes's psychiatric parlance, "passive homosexuals"—that is, men who are penetrated during sexual intercourse. Thus, without reference to any indigenous discourse, Landes inaugurated the scholarly tradition of diagnosing male Candomblé priests as diseased and therefore alien to any legitimate cultural tradition (see also Bastide 1961:309; Ribeiro 1969:122; Corrêa 2000:246–48, esp. 246nn24, 25; Healey 2000[1998]:88). However, as Landes herself reported, the *adés'* sexual identity seems not to have troubled the other priests or adherents of Candomblé. As priests, these men were, observed Landes, "supported and even adored by those normal men of whom they were before the butt and object of derision" (Landes 1947:37; 1940:393).

Landes's revelation of these sexual matters particularly discomfited her Brazilian scholarly colleagues, who clearly were more attuned to transnational standards of national respectability and more concerned to guard the open secrets of the Brazilian nation than were the priests and subjects of Candomblé's sacred nations. Sometime Brazilian state functionary and culture broker extraordinaire Arthur Ramos flatly denied Landes's claims about a "cult matriarchate" and about a significant homosexual presence, and, in retaliation for Landes's divulgence, he cooperated with Melville J. Herskovits in virtually shutting down her academic career (Corrêa 2000:246–48, esp. 246n; 2002:14; Fry 2002: 25–26).[6]

Though Carneiro had started out as Landes's guide, his attitudes, or at least his public discourse, about the male Candomblé possession priest changed remarkably over the course of his dialogue with Landes. In 1936, before Landes's sojourn in Bahia, Carneiro wrote of the possession priests: "In Bahia, those priestesses are called daughters-in-saint. In the olden days (and even still today), the men could be sons-in-saint too. It is noteworthy that they had to dance, during the grand festivals, wearing women's clothing" (1991[1936]:56; also 58, 60, 91). Throughout his 1936 publication, he casually describes male and female possession priests engaged in the same ritual duties, doing so with equal legitimacy and with equal deference from the public.

In the midst of his professional and personal relationship with Landes, Carneiro is quoted expressing a subtly different range of ideas and feelings, which appear to entitle "abnormal" men, as we shall see later, to a respected and beautiful role in the possession priesthood. Rather than condemning Bahia's cross-dressing male priests, he opines, "Sometimes they are much better-looking than the women" (Carneiro in Landes 1947:37). Carneiro and the young Mãe Menininha actually praise the "solid reputation" of two priestesses whom Landes lambasted for the alleged aberration of initiating men (Landes 1947:158, 192). And when Carneiro says that "men are not entitled to [be possessed] unless they are abnormal," he implies that "abnormal" men *are* somehow entitled to be possessed (Carneiro quoted in Landes 1947:201–2). In another conversation, Landes asks Carneiro which god ruled the head of a highly esteemed and allegedly homosexual male priest named Bernardino: "'Just what you might expect,' Édison answered with a certain satisfaction, as though the rules of the cult were working out properly, 'it is Iansã, the bisexual goddess. Bernardino is always consistent.' He smiled" (Landes 205; also 203).

The "rules" appear to have been working out perfectly, but Landes simply could not, or would not, conceive of those "rules" in their own terms. Apparently embarrassed by the pathologizing gaze of this powerful transnational visitor, Carneiro would, within a few years correct his earlier position of acceptance and admiration for the *adé* priests.

We cannot say with certainty whether his was a true change of heart or a dissimulation "for the English to see." However, he presents his first full about-face in the Anglo-American *Journal of American Folk-Lore*, in an article paired with one by Ruth Landes. The venue of this publication leaves little doubt about the motive and the origin of this change. Over the course of their acquaintance, Landes seems to have changed Carneiro's mind, or at least made him answer to a transnational culture of national respectability. Having declared in 1936 to a Brazilian reading public that the cross-dressing male possession priesthood had originated "in the olden days" and even continued today, Carneiro chose later to tell the U.S. reading public,

> It seems that formerly the *candomblé* was a woman's business.... The ascendancy of women dates from the introduction of the *candomblés* in Bahia, with the establishment of the Nagô house of [Casa Branca do] Engenho Velho about 1830.... As against so many "mothers," we know of the existence of only a few "fathers," like Bambuxê and "Uncle" Joaquim.... Despite the superior importance of women in *candomblé*, today the number of "fathers" and "mothers" is equal. (Carneiro 1940:272)

Thus Carneiro had literally reversed the course of history. This English-language publication is the first in which Carneiro systematically lambastes the male possession priests, condemning them for "giving themselves up to homosexuality, where they take the passive role, dropping into the small gossip typical of lower-class women." Of Bahian public opinion toward the "homosexual" priests, Carneiro tells the United States, "Criticism is always more venomous about 'fathers' than about 'mothers,' labeling them insincere, dishonest, and evil" (Carneiro 1940:273). In the defense of his country's international reputation, his own venom even exceeds that of Landes (cf. Landes :37; 1940:393).

By 1948, in a publication for the Museum of the State of Bahia, Carneiro was prepared to denounce the male priesthood before the Brazilian national public. He did so in *Candomblés da Bahia* (1948), a decade after his collaboration with Landes had begun and after Landes's two major publications on the subject, in 1940 and 1947. Following Landes, he established the canonical view in Brazilian and Brazilianist folklore studies, and even in the historical consciousness of many priests today, that male priests are uniquely disreputable, that their numerical predominance is recent, and, therefore, that they are unrepresentative of Northeastern and Brazilian society (Carneiro 1986[1948]:103–9).

A Northeastern mulatto himself, Carneiro had a personal and political stake in dignifying the Nagô and Jeje nations and making theirs the standard by which the typical folk religion of his region was judged. Thus, nostalgia for the Jeje-Nagô-centered, putatively matriarchal and innocent prehistory of Candomblé came to unite the spokespersons of two imagined communities—Northeastern Regionalist Édison Carneiro and transnational feminist Ruth Landes.

Yet the copious detail in Carneiro's 1948 volume frequently undermines his own argument that the Nagô nation is exclusively or in principle matriarchal. For example, Carneiro credits the 19th-century African-born priest Bambuxê (Báṁgbóṣé), a man, with initiating Aninha, the future chief priestess of the prestigious Opô Afonjá temple. Aninha's disappointment that another man, Joaquim Vieira, did not succeed the recently deceased chief priestess of Casa Branca, the mother church of Bahia's preeminent family of Quêto/Nagô temples, is given as the reason for Aninha's secession from Casa Branca and for her role in the founding of Opô Afonjá. Finally, even though the histories recounted nowadays at Opô Afonjá seem to leave no doubt that Aninha founded that temple and was its first chief priest, Carneiro reports that Joaquim himself had been that temple's first "chief" (1986[1948]:57; also Butler 1998:195).

Carneiro also mentions in passing numerous eminent male possession priests of the Jeje and Nagô nations who were alive during his time, such as Eduardo Mangabeira, Procópio, Manuel Falefá, Manuel Menêz, Cosme, Antônio Bonfim, and Otacílio (1986[1948]:106, 119–23).[7] Despite the contrary evidence that he himself recorded, all of Carneiro's synoptic statements about the "tradition" seem aimed to satisfy the same partisan notions of gender respectability that Landes had invoked. Not even the most gymnastic and speculative argument was barred in the effort to dismiss the male priests and thereby guarantee the international respectability of an "authentic" Northeastern tradition left by their absence.

Landes's transnational feminism and Carneiro's efforts in the Regionalist project thus conspired to keep a secret. Yet Carneiro had not thought of Candomblé's "passive homosexuals" as much of a secret before 1938. The *adés* first became a secret amid the conflict between North American Ruth Landes and Brazilian academic gatekeeper Arthur Ramos. A close observer of Bahia and a close friend of Landes's, Carneiro could not deny the newly embarrassing reality, as Ramos had done, but he was in a position to marginalize that reality authoritatively. Deeply informed by Landes's interpretive biases, Carneiro's *Candomblés da Bahia* has been reissued at least six times since 1948. There could hardly be a richer testament to the longevity of Landes's profound influence on the national reading public and imagined community of Brazil.

On the Throne of Two Realms: the Ascending Queens of Nation and Transnation

From the 1930s onward, the priestess became an object of public talk to the same degree that her *adé* antitype became an object of silencing. While the scholarly intervention of Landes made the Candomblé priestess into the most widely cited and credible reference in the cross-cultural pursuit of "matriarchy," the dialogue among Freyre, Carneiro, and Landes inspired an elaborate popular discourse about the Black Priestess at the heart of the Brazilian nation-state. Since the time of Landes's intervention, the priestesses of Candomblé have remained a touchstone not only of transnational feminism but of Brazilian national identity as well.

Perhaps the most internationally famous transnational expositor of Candomblé was the existentialist philosopher, sociologist, traveler, and feminist foremother Simone de Beauvoir. Her autobiographical *Force of Circumstance (La Force des choses* 1968[1963]) devotes fourteen pages to the description of her 1960 experience in Northeastern Brazil, focusing on her encounter with Regionalists Gilberto Freyre and Jorge Amado, with the anonymous priestesses of Candomblé, and with Pierre Verger and Vivaldo da Costa Lima, intellectuals who made a profession of studying Candomblé. Beauvoir even documents Verger's role in bringing religious items from Africa to the Brazilian Nagô priestesses.

Beauvoir represents Candomblé less as a feminist idyll than as a noble form of cultural resistance, the next-best spiritual alternative to Cuban-style revolution for a people racially and economically oppressed (Beauvoir 1968[1963]:523–37). Yet Beauvoir brings an implicitly feminist reading to Candomblé. She had first encountered the religion through Amado's novel about a male Candomblé priest, *Jubiabá* (1935), and her subsequent understandings were mediated by Amado himself and Roger Bastide's *Les Religions des Afro-Brésiliennes* (1960, published in English as *The African Religions of Brazil* [1978]), an impressive sociology of religion, one of whose main flaws is its failure to analyze gender. By contrast, Beauvoir's own ethnographic portrait lingers on the female membership of the religion, and especially its female leadership.

Sally Cole's 1994 reedition of Landes's *City of Women* crowned the book a classic of feminist ethnography and declared it a precursor, in its candid revelation of the politics of representation, to the reflexive ethnographies of the 1980s. Kim D. Butler's *Freedoms Given, Freedoms Won* (1998) and Rachel Harding's *Refuge in Thunder* (2000) underlined Candomblé's power as an exemplar of black women's power in particular. Even as Candomblé's reputation crossed more and more borders with the growth of the transnational feminist movement, its role in the gendered self-image of the Brazilian territorial nation continued to expand as well.

In the national public sphere, the priestesses of Candomblé benefited enormously from their superficial likeness to a major character in the Freyrean mythology of the Brazilian nation—the beloved, Mammy-like nursemaid called the "Black Mother" (*Mãe Preta*) (Silverstein 1979). Once a scourge of the eugenicist nationalism of the First Republic, she became perhaps the ranking heroine and an object of nostalgic adulation in Freyre's *Masters and the Slaves* (1933) and eventually the object of a national cult. She receives star billing in Freyre's two chapters titled "The Negro Slave in the Sexual and Family Life of the Brazilian," where she is credited, psychoanalytically, with the foremost influence on the collective psychology (and especially on the libidinous desires) of Euro-Brazilian men and, therefore, of the Brazilian nation. I cited Freyre's words more succinctly in the last chapter, in a passage where his further words leave no room for ambiguity:

> Every Brazilian, even the light-skinned fair-haired one, carries about with him on his soul, when not on soul and body alike.... [t]he influence of ... the female slave or "mammy" who rocked us to sleep. Who suckled us. Who fed us, mashing our food with her own hands. The influence of the old woman who told us our first tales.... The psychic importance of the act of suckling and its effects upon the child is viewed as enormous by modern psychologists, and it may be that Calhoun is right in supposing that these effects are of great significance in the case of white children brought up by Negro women (Freyre 1986[1933]:278–79).

The Black Mother became a powerful icon of Regionalist thought, which, much like southern regionalism and nationalism in the United States, included the sentimentalization of plantation slavery. For the white bourgeoisie of the mid–20th century, the Candomblé priestess, in her usual 19th-century *baiana* skirt, surely tapped into the modernist nostalgia for the premodern—the longing, in the midst of an urbanizing and industrializing society, for a bucolic past world of personalistic and emotional attachments, a world where everyone knew his or her place, an infantile world where a warm and caring servant-mother provided all of one's needs on demand. In a mid-20th-century world of labor unrest, the communist threat, and lumpen crime, such an image acquired a special appeal.

The dignity of the Black Mother was guaranteed by her personal attachment to the lord of the manor, and the poverty and exclusion of the black masses were sublimated in her image (e.g., Freyre 1986[1933]:278; Giacomini 1988:63; Andrews 1991:215–17, 231, 302, 330–31).

Lopes, Sigueira and Nascimento (1987:128) describe the image of the Mãe Preta as "masochistic," but it has also been a source of pride for Afro-Brazilians and of empowerment for the Candomblé. Writes George Reid Andrews,

> Since early in the [20th] century, São Paulo's black organizations had venerated several mythic figures symbolizing black contributions to Brazilian history and the forging of the Brazilian nation. The most prominent of these figures was the Mãe Preta, the Black Mother, who . . . was often cited in the black newspapers as a symbol, not just of the sacrifices that black people had made for Brazil, but also of the powerful ways in which Euro- and Afro-Brazilians were linked in a common destiny. (Andrews 1991:215)

These organizations established and annually celebrated the Day of the Black Mother on 28 September—the anniversary of the passage of the quasi-abolitionist Law of the Free Womb (1871)[8]—and later on 13 May, the anniversary of the complete abolition of slavery in 1888. The official importance of the Black Mother for the state was recognized in 1955, when the city of São Paulo erected a statue of her in front of the city's oldest historically black church. The city's Candomblé priests took a leading role in this celebration during the 1960s. By 1970, the mayor and archbishop of São Paulo were attending regularly, and, in 1972, President Médici himself attended along with "every major politician in Greater São Paulo" (Andrews 1991:215–16).

Though President Geisel refused to attend, on the grounds that the celebration represented "reverse racism," and the North American–inspired Black militancy of the 1970s rejected the Black Mother as a symbol of submission (Andrews 1992:216–17), the statue is now the object of much ritual devotion. Supplicants light candles around it, and Candomblé priestesses dance before it. The area around the church and the statue remains a site of frequent antiracist political protest, often coordinated by members of the Unified Black Movement (Movimento Negro Unificado). During these protests, Candomblé priestesses have been known to sprinkle the Black Mother statue with perfume and to distribute popcorn, a means of spiritual cleansing closely associated with Omolú, the Candomblé god of pestilence and healing (see Andrews 1991:215–17, 231, 302, 330–31).

The dignification of the Black Mother and her association with the Candomblé are not, however, restricted to São Paulo. The Black Mother erected in Rio in 1969 was modeled on Mãe Senhora—the then-recently deceased chief priestess of Bahia's Opô Afonjá temple. One journalist recorded the following scene at a huge public gathering of Umbanda practitioners near the statue, held in Rio under the patronage of Riotur (the state tourist agency) and several politicians: "The participants in the parade directed themselves to the statue of the black mother of Brazil, . . . modeled on the image of . . . Mãe Senhora, descendant of African kings. The black mother holds on her lap a little black boy. Beside the statue, taking care of it and welcoming those coming to pay homage, is an impressive woman dressed in a more African fashion." This attendant introduced herself as Ajubanã, or "assistant to the chief priestess" (Cacciatore 1977:154). Having vowed to "dress up and guard the black mother" for seven years, she was then in the third year of her vigil.[9]

Mãe Senhora's successor as chief of Opô Afonjá, Mãe Ondina, was officially awarded the title "Black Mother of Brazil." However, the late Mãe Menininha has been the greatest beneficiary of this Black Mother imagery. In the words of journalists, novelists, songwriters, and state tourist agents, she appeared to personify the nostalgic, infantile, and narcissistic dreams of the privileged class in a postslavery society. She was self-sacrificial, self-effacing, long-suffering, generous, constantly available, free of malice, plump, "simple" (simples), possessed of "wisdom" (sabedoría), and always "sweet" (doce).[10] She was equally maternal to blacks and whites.

Mãe Menininha first became the object of attention from populist Bahian politicians and tourists from Rio in the mid-1960s. However, in 1971, Dorival Caymmi composed his famous song "Prayer of Mãe Menininha" ("Oração de Mãe Menininha"), which magnified her already great popularity many times over. As of 1985, it had been recorded by Caymmi himself in 1972, 1984, and 1985; Gal Costa and Maria Bethânia in 1972; Clementina de Jesus in 1973; Paul Mauriat and his orchestra in 1974; and Osmar Milito in 1980.[11] These numerous recordings established the emotional tenor of subsequent journalistic coverage of Menininha and of some highly dissimilar priestesses who came in her wake (see Barboza and Alencar 1985:24, 26, 148–89, 213). Musicians, Carnaval performance groups, and journalists lavished over two decades of national attention on Mãe Menininha, representing her as the sweet, maternal nursemaid writ large, who awaited the visits of her high-class children in an enormous house, spotlessly clean, with a permanent yard, a large kitchen, lots of food and entertainment, and always a kind word. Politicians were said to *pedir colo dela*—"to ask to be held on her lap."[12]

I must emphasize that I have said nothing about Mãe Menininha's actual character or behavior. I have never met her and have no reason to doubt her kindness. I am observing only the nearly predictable terms in which her image was appropriated in the bourgeois nationalist cult of the Black Mother during the 1970s and 1980s. So great were her fame and charisma that some called her a "living *orixá*." Thousands of politicians, intellectuals, artists, and workers honored her at her funeral in 1986.[13]

Mãe Menininha's successor as the most talked-about and respected Candomblé priestess in the country is Mãe Stella, and she has inherited this vocabulary of reference. Some journalists automatically describe her "sweetness" (*doçura*). Now, as if by reflex, Brazilian journalists describe even the toughest Candomblé priestesses in the clichés characteristic of the Black Mother image—self-sacrificial, self-effacing, long-suffering, generous, constantly available, and so forth. The son of Dorival Caymmi, Danilo, proclaimed on the occasion of Mãe Stella's seventieth birthday, "She is our mother. She is the umbilical cord that unites us."[14]

Today's journalistic descriptions of Candomblé's Black Priestess are similarly tinged with nostalgia for an innocent past, in which the Black Mother is equally maternal to blacks and whites. Just as important, she is always described nostalgically as "one of the last" of a dying breed.[15] On the occasion of Mãe Menininha's funeral in 1986, even the *New York Times* reproduced this discourse carefully and respectfully, alongside the observation that the deceased priestess was "perhaps the best-known woman in all of Brazil." Thus, in Brazilian nationalist nostalgia, the Black Mother of the Candomblé is warm and caring, but journalists offer the further assurance that she is also "firm" (*firme*) with her masses of black children and has always "commanded [them] with an iron hand" (*comandava com mão de ferro*).[16] This dimension of her "matriarchal regime"[17] seems a godsend to the Brazilian privileged classes.

Equally associated with the white bourgeois wish list and with the Afro-Brazilian imagery of the *orixás'* power is the repeated imagery of the priestesses as "queens" (*rainhas*).[18] No one has attracted this description more than Mãe Stella—"Queen of the Yorùbá Nation," "Queen of Axé Opô Afonjá," and "Lady Queen and Mother of Axé Opô Afonjá," who "reigns absolutely" over her community.[19] Such imagery is particularly striking after more than a century of republicanism in Brazil and nearly two decades of hard-won and stable electoral democracy. This royalist vocabulary suggests some bourgeois Brazilian comfort with the idea that the country's "lower classes" ought to be governed by some system other than the presidential democracy that gives the bourgeoisie so much freedom and potentially so much veto power over state political programs. Such are the reasons that national imaginations of community never entirely supersede imperial ones.

These are the terms of the Candomblé priestess's ongoing incorporation into the narrative of *mestiço* nationalism, and of the Candomblé priest's relegation to the closet. These terms appear to be a significant reason why, since the 1960s, the city government of Salvador, the Bahian state, the Brazilian federal state, businesses, and the national media outlets have lent vastly more moral support and funding to the female-headed temples than to male-headed ones, and why such support and funding are subject to such fanfare in the Bahian press (see chapter 4).[20] The Brazilian bourgeois sentimentalization of the Black Mother has indeed opened up a symbolic space for black female assertion in Candomblé, but, in general, it represents a space significantly more selective than the already small space allowed to black men, and those black women who have successfully exploited it did so only by way of their own genius at pushing its boundaries.

Despite empirical evidence of their "double marginalization" and of their avowed grievances toward whites, black female Candomblé priestesses have been far more successful than their black male counterparts at making alliances with light-skinned, prosperous, and Western-educated men, many of whom classify themselves as white. These women have somehow managed to attract not only journalistic and ethnographic attention but also material resources to their temple communities, even in a society as thoroughly patriarchal and racist as Brazil. For journalists who know nothing about the ritual protocols that insiders regard as one key to success, the other insider's criterion of success has become the focus of journalistic commentary—the talent of the priestesses at cultivating "friends on all social levels" (*amigos em todas as camadas sociais*). Indeed, successful priestesses publicly broadcast their interest in doing so.[21]

Thus, for Afro-Brazilians, the power of these queens and mothers is more attractive than coercive. Like male priests, they provide healing, shelter, and entertainment. To a far greater extent than most male priests, these well-connected priestesses can also arrange employment and protection for their followers. In death they draw even larger crowds than in life. Though Candomblé initiates tend to avoid funerals, fearing that the hand of the deceased will drag them under as well, the funerals of leading Quêto priestesses are legendary for their size (see, e.g., Landes 1947:36; Pierson 1942[1939]:316–1721).[22]

Whatever their personalities, ritual competencies, or grievances, dark-black and middle-aged to elderly women in Brazilian society can wittingly or unwittingly take advantage of a well-established cultural image in Brazil, one that is especially useful in attracting those seeking the literal or

metaphorical "snuggly embrace" (*aconchego*) of the black "matriarch."
In the 1990s, this phenomenon found its latest exemplar in Mãe Stella,
who, like a series of Quêto women since Mãe Aninha—particularly Mãe
Senhora, Mãe Olga de Alaketu, and Mãe Menininha—has ridden the momentum of her illustrious predecessors and, by the force of her own character, has augmented it.

To summarize, both Landes and Freyre have profoundly reshaped the
allegories by which the Brazilian bourgeoisie represents itself to itself
and to others on the transnational field of national respectability. And
Landes has placed the Candomblé priesthood within the force field of
Western feminism, powerfully affecting the options and life chances of
Afro-Brazilian men and women.

Mãe Stella is the first priestess offered the opportunity in the national
press to explain women's preeminence in her religion. Mãe Stella characterized the Candomblé in the precise words of Landes, as a "matriarchate." She then offered three historical and psychological reasons. First,
the Candomblé was initially "brought [to Brazil] by three ladies, Iyás
Deta, Kala and Nassô, three people from the kingdom of Xangô, who
had the courage, even with all the repression, to do their Candomblé."
Second, since female domestic slaves had free time and both the competency and the option to cook, they were also uniquely able "to continue
practicing their original religion." Finally, she detailed a unique qualifica-
tion in the very nature of women: "I think that a woman always has a
special little maternal way of taking care of things. Men too take care,
but it is not the same thing." Thus, she reflected on what draws a following to any given priest: "That doesn't mean that the man lacks the capacity to be a priest [*babalorixá*]; it's that the woman is the mother *figure*,
and when people come into a Candomblé temple, they are looking for
more of that snuggly embrace [*aconchego*]. Women have the capacity to
offer more tenderness. It's just that" (emphasis mine).[22]

Mãe Stella appropriately identified this "figure" as an imago rather
than as a real person, for the characteristics highlighted by journalists and
other bourgeois noninitiates are often at odds with the personalities and
ritual roles of those observed. For example, many of the chief priestesses
of the Candomblé are not snuggly or affectionate at all, and more than a
few are childless by choice. For someone not in search of a conventional
"mother" figure, the description of their manner that might first come to
mind is "executive." Moreover, while plenty of male and female initiates
can cook very well, Candomblé-related "domestic services," as Carneiro
put it, are generally preferentially assigned not to just any woman but to
women consecrated to female gods (Leão Teixeira 1987:44–45; Azevedo
Santos 1993:52–54). In other words, many women are disqualified in
principle, and many men are qualified in practice. Herskovits's view that

women predominate because their time and earnings are more dispensable to the consanguineal family than are men's is surely more a midcentury white and bourgeois projection than a reality of most Afro-Brazilian women's lives.

Such allegorical images of black and Indian women in journalism, folklore, and anthropology have been a central element in New World creoles' imagination of their national communities. And such images—ranging from la Malinche in Mexico, Pocahontas and Sacagawea in the United States, and the lascivious Mulata in Cuba and Brazil to the "Mammy" and Black Matriarch in Brazil and the United States—can become the prisons or the tools of the real people so represented. Since the 1930s, a few priestesses have acquired a mighty set of allies in the overlapping imagined communities of the Brazilian nation and transnational feminism. And, to the same degree, all the male priests of the Candomblé have acquired a powerful set of enemies.

However, this argument must not be mistaken for the plainly false claim that all transnational feminists or all Brazilian nationalists are homophobic. Such a misreading would misconstrue the nature of imagined communities and their leadership. Demagogues are the most extreme example of leaders who, by making bold, vivid, and distinctive pronouncements, capture the imagination of large populations and thereby rearrange their sense of where the boundaries of the community lie, how its internal hierarchies are arrayed, and what the shared purpose of the community is. However, not even the most capable demagogues win the unanimous agreement or consent of their target populations. Rather, they become centers of gravity in target populations that often have multiple centers of gravity. Gilberto Freyre, Arthur Ramos, Édison Carneiro, and Ruth Landes are hardly demagogues, but Freyre, Ramos, and Carneiro are important centers of gravity in the Regionalist imagination of the Brazilian national community, just as Landes is an important center of gravity in the transnational feminist imagination of community. While her homophobia is accidental to the entirely rightful feminist aspiration to gender equality (and, once detected, anathema to the principles of most feminists I know), her particular discursive strategy and the Regionalists' fear of the transnational gaze have deeply compromised the status of the *adé* priest in Northeastern Brazil.

Landes's work has now outlasted Mead's as a measure of potential success in the transnational movement to combat sexism. However, Landes paid the price of a nearly jobless career for mobilizing the *adé* as the antitype of the Black Priestess.[24] Nonetheless, once built on this foundation, the City of Women proved durable and harmonious with the enormously influential Regionalist project of Freyre and Carneiro. Carneiro was the first to translate Landes's vision to the Brazilian public, and he did so as an agent of the state's image-making apparatus—the Museum of the State of Bahia.[25] His *Candomblés da Bahia* (1948) endeavored to rescue the transnational reputation of Bahia and Brazil by excoriating the male priesthood as a class.

The *adé* remained under a pall of silence until an accelerating sequence of publications between 1961 and the 1980s. However, few of these works have been published in languages other than Portuguese, rendering them inaccessible to Anglophone scholars who might otherwise read Landes's work more critically. Yet until the mid-1980s, even Portuguese readers would still have the impression that the presence of *adés* in the Candomblé priesthood resulted from either their psychological abnormality or their symbolic liminality within Brazilian national society. Their presence or their ideal absence was still "imagined" within the framework of transnational feminism or Brazilian nationalism. Let us now consider the *adé*'s historical and symbolic connection to another transnational community—the African diaspora.

MOUNTED MEN: NIGERIAN MALE E̩LÉGÙN AND NEW WORLD PASSIVES

There are no reliable statistics on how many Candomblé priests are what Landes called "passive homosexuals," or what Candomblé adherents call *adés*.[26] Nor does the following explanation concern their actual numbers. Rather, I seek to explain why so many members and cognoscenti of Candomblé assume—with or without statistical accuracy—that male initiates of the possession priesthood are normally *adés*, why many Afro-Brazilian men who love men feel at home in Candomblé, and why Candomblé-inspired terminology dominates the argot of gays all over urban Brazil.

Today there are numerous explanations of *adés*' alleged prominence among Candomblé possession priests. But before we can understand them, we must identify the set of semantic contrasts of which the "passive homosexual" is a part in Brazil. On the one hand, English-speaking North Americans tend to distinguish sharply between those men who engage in sex with other men ("homosexuals") and those who do not ("heterosexuals"). On the other hand, Brazilians are far more likely to distinguish men who penetrate others during sexual intercourse (*homens*, or "[real] men") from those who are penetrated (*bichas, viados*, or, in Candomblé language, *adés*; Fry 1986:137–53).[27] Brazilians share the basics of this pattern of classification with many peoples around the Mediterranean, as well as much of pre-19th-century Europe, Native America, and most of the rest of the world (Trexler 1995). Contemporary European and Anglo-American prison populations and sailors seem no exception. Even when the Bahians I know use the term *homos-sexual*, most are referring only to the party in sexual intercourse who is assumed to be habitually penetrated, or "passive." Of course, the real be-havior of both *homens* and *bichas*, or *adés*, is regularly more varied than what is stereotypically attributed to them, and the normative assumption that the "active" party is dominant both in the sexual act and in the non-sexual dimensions of the social relationship is often more fantasy than ma-terial reality (e.g., Kulick 1998). However, local *ideological* assumptions and expectations tend to link habitual male "passivity" with transvestism, feminine gestures, feminine occupations, and the social subordination of the penetrated party.[28]

So why do many Brazilians think there is a connection between the possession priesthood and so-called sexual passives?[29] Following Victor Turner and Mary Douglas, Peter Fry argues that the homosexuals' *limi-nal* status in Brazilian national society suits them symbolically, in the Brazilian popular imagination, to professions dealing with "magical power."[30] Lima moves in the direction of acknowledging what is *normal* about homosexuality in Candomblé ideology: both Afro-Brazilian religions and Brazilian Kardecist Spiritism, he argues, have shown themselves more generally tolerant than the Roman Catholic Church (D. Lima 1983:167ff.). This is the explanation most widely heard among today's Candomblé practitioners. The next most frequent indige-nous explanation is recorded by Birman (1985). She reports that men whose heads are governed by female divinities and sex-changing gods—like Iançã, Oxum, Oxumaré, and Logunedé—are expected to share in the female dispositions and desires of those gods (Birman 1985:2–21; also Leão Teixeira 1987:48; Landes 1940:395).

Fry also notes the advantageous flexibility enjoyed by *bichas*, or "pas-sives," in the performance of social roles normally reserved, in the wider society, primarily for one sex or the other. That is, they can acceptably do the cooking and embroidering necessary for the temple and yet, in a similar religion in Belém do Pará (where Fry conducted his research), re-tain the social advantages of men in transactions with the "world of men"—of police, judges, doctors, lawyers, and politicians, "whose ser-vices they themselves may use or broker to clients for their own advan-tage" (Fry 1986:147–49). In the Bahian case, men's advantages over the "great" Nagô mothers in this regard are not so evident. What is more evident, as Leão Teixeira observes, is that homosexual men bring to the Candomblé three other advantages that they share with all men: (1) higher average earnings; (2) the ritual license to perform all the ritual duties normally restricted to men, such as the sacrifice of four-legged an-imals, the care of the all-important Exú (god of sex, mischief, and com-munication), Ossaim (god of herbal medicine), and the Eguns (ancestral spirits); and (3) immunity to the restrictions placed upon menstruating women, such as exclusion from the shrine rooms.[31] A woman consecrated to a male god is eligible to receive a further initiation (mão de faca) that entitles her to sacrifice birds, but, while menstruating, she cannot even do that. For Leão Teixeira, these facts suggest that priestly competency itself is coded male.[32] Priestesses, too, seem to question the "matriarchate." For example, Mãe Stella declares the servile status of women consecrated to female orixás and emphasizes that in the "African tradition" of her temple, the god Xangô has always been the real "boss" (chefe). [33]

These facts lead me to two main points. First, what Landes called Candomblé's "cult matriarchate" is not a fact given simply by "tradition"; nor is it simply a lie. It is a plausible but interested and contested construction of "tradition." It is also the product of various imagined communities' fear of outside judgment or pursuit of an ideal model of community for themselves. And despite the pronounced homophobia of many contemporary third world bourgeois nationalists (including a number of prominent Anglophone African elites [e.g., Sweet 1996:84–202]), one would be hard-pressed to locate the precolonial West African precedents for the homophobia that Landes, Ribeiro, and Bastide have presented as psychoanalytic proof of male priests' inferiority. The homophobia that denormalizes adés, or "passives," in the Candomblé priesthood has its origins not in an aboriginal Africa but in emergent nationalisms and a particular brand of transnational feminism of the mid–20th century.

The second main point is that the proliferation of latter-day explanations of the prominence of adés in the Brazilian Candomblé, Cuban Ocha, and Haitian Vodou priesthoods and the well-documented history of Candomblé adherents' comfort with adés in this role appear to share a common historical root. That root is evident between the lines of Landes's informants' testimony in the 1930s and most clearly implied by my own comparative field research between Brazilian Candomblé and its West African "homeland." A West African logic of "mounting" and its attendant transvestism converged in Brazil with a Brazilian logic of sexuality and social hierarchy, thereby helping us to understand why, in Brazil, male Candomblé possession priests are widely believed to be bichas, adés, or "passive homosexuals" (Matory 1988:215–31; 1991; 1994b). Thus, mine is an argument about the local "reinterpretation," to borrow Herskovits's term, of cultural forms that appear in diverse but historically connected places.[34]

Ọ̀yọ́-Yorùbá people formed not only a plurality of the African captives taken to Bahia in the 19th century but also the founding priests and priestesses of Bahia's most influential temples—including Casa

Branca. No African ethnic group has influenced Candomblé more than the Ọ̀yọ́ people. In West Africa, Ọ̀yọ́-Yorùbá worshipers employ multiple metaphors to evoke the nature of people's relationships to the gods. Like Brazilian *candomblecistas*, they might call any devotee of an *òrìṣà* the "child" (*ọmọ*[Yorùbá]; *filho* [Portuguese]) of that god. In both traditions, motherhood and fatherhood are used as metaphors of *leadership* in the worship and activation of the gods. For example, a senior male West African Yorùbá priest of, say, Ṣàngó might be addressed as Bàbá Olórìṣà (Father [or Senior Male] Owner of the Òrìṣà); a senior priestess would be addressed as Ìyá Olórìṣà (Mother [or Senior Female] Owner of the Òrìṣà").[35] In Brazil, the male head of a Candomblé temple is called a *pai-de-santo* (father-in-saint), while a chief priestess is called a *mãe-de-santo* (mother-in-saint).

Yet the Yorùbá terms that mark out the priest's competency to embody the god through possession-trance and to act as his or her worldly delegate rely, above all, upon allied metaphors of marriage and sexuality. According to Édison Carneiro, as we shall see, these metaphors were very much alive in the Brazilian Candomblé of the 1930s, and they were present in local understandings of both male and female participation in the priesthood.

Most Ọ̀yọ́-Yorùbá possession priests in West Africa are women.[36] The numerous male possession priests, on the other hand, cross-dress. But their cross-dressing requires a culture-specific reading. They dress not as "women" but as "wives" or "brides" (*ìyàwó*)—a term that otherwise refers only to the women married to worldly men. Novices to the priesthood—whether male or female—are designated metaphorically as *ìyàwó*, meaning "brides" or "wives" of the gods (Matory 1991, 1994b). The term implies the productive subordination of the "wife" to the real and metaphoric "husband" (*ọkọ*).[37]

Bahians call novices to the priesthood by a cognate of the Ọ̀yọ́ term— *iaô*. However, the unanimity with which Bahians understand the word *iaô* to mean "wife" or "bride" has declined since the 1930s. In the 1980s and 1990s, I normally heard it glossed as "wife," often as "son" or "daughter," and occasionally as "slave" of the god.[38] I have also regularly heard the marriage metaphor invoked to explain other ritual conventions. For example, Quêto/Nagô and Jeje priests are ideally possessed by only one god; on other occasions, by contrast, an Angola priest might be possessed by several. One Quêto/Nagô purist priest I know therefore derided priests who receive more than one god by calling them "promiscuous" (*promiscuos*). Thus, at the beginning of the third millennium, the Yorùbá meaning of the term *iaô/ìyàwó* continues to shape the logic of Candomblé speech, recruitment, and practice.

For months after the initiation, male and female novices among the West African Ọ̀yọ́-Yorùbá wear women's clothes: *ìró* (wrap skirts), *bùbá* (blouses), and *ọ̀já* (baby-carrying slings); on ceremonial occasions, they also wear *tìróò* (antimony eyeliner), *làálì* (henna for the hands and feet), delicate bracelets, earrings, and so forth. As mature priests, or *ẹlẹ́gùn*, women and men braid their hair and follow the latest styles in women's coiffures. But on ceremonial occasions, they also continue to don *tìróò* eyeliner, henna, and delicate jewelry. Many uninitiated Yorùbá *women* do these things, but male possession priests are virtually the only men who do so. In Ìgbòho, the Ọ̀yọ́-Yorùbá town where I conducted my principal West African field research, both the strip-weaving of cloth and barkeeping are considered female professions.[39] So, almost predictably, the only male strip-weaver and the only male barkeeper in the town were Ṣàngó possession priests.[40] Yet, West African Ṣàngó priests present themselves ritually, sartorially, and verbally not as women but as wives of the gods, *by analogy to the female wives of earthly husbands.* This extended metaphor includes a further important term.

Indeed, the most pervasive and dramatic gendered symbol in the possession priests' metaphoric representation of their relationship to the gods—from the initiation onward—is the web of metaphors implicit in the Yorùbá verb *gùn*, meaning "to mount." The most common term for "possession priest" (*ẹlẹ́gùn*) contains this root and means "the mounted one." *Gùn* refers to what a *rider does to a horse* (hence, possession priests are also called "horses of the gods" [*ẹṣin òrìṣà*]). The term *gùn* also refers to what an animal or a brutal man does sexually to his female partner (and possession by Ṣàngó is often spoken of as a brutal act).[41] The term *gùn* also refers to what a god—especially Ṣàngó—does to his possession priests. And Ṣàngó's is the most influential possession priesthood not only on the Bight of Benin but, to an even greater extent, among the *òrìṣà* worshipers of Brazil, Cuba, Trinidad, and the United States.[42] However we translate the verb *gùn* into English, the term *montar* in Caribbean Spanish and Brazilian Portuguese and the Haitian Kweyòl term *monte* (all cognates of the English verb "to mount") encode the same three referents—sex, horsemanship, and spirit possession—and have a long history of usage by worshipers in Cuba, Brazil, and Haiti.

Let me illustrate how Afro-Latin Americans—such as the priests and cognoscenti of the Bahian Candomblé—consciously construed these West African metaphors in the 1930s, at the time of Landes's research in Bahia. These are the words of Regionalist ethnographer and long-term Candomblé affiliate Édison Carneiro early in his acquaintance with Ruth Landes:

> Sometimes they call a priestess the *wife* of a god, and sometimes she is his *horse*. The god gives advice and places demands, but often he just *mounts* and plays.

So you can see why the priestesses develop great influence among the people. They are the pathway to the gods. But no *upright* man will allow himself to be *ridden* by a god, unless he does not care about *losing his manhood*. . . .

Now here's the loophole. Some men do *let themselves be ridden*, and they become priests with the women; but they are known to be *homosexuals*. In the temple they *put on skirts and mannerisms of the women*. . . . Sometimes *they are much better-looking than the women*. (Landes 1947:37; emphasis mine)[43]

This parlance is highly consistent with the West African, Ọ̀yọ́-Yorùbá symbolism of spirit possession I observed among Nigerian Ṣàngó priests of both sexes in the 1980s, save one important detail: the reluctance of "real men" to be possessed in the Brazilian Candomblé.

Sex was not an infrequent topic of conversation among male friends of my age-group in Ìgbòho, and, no matter what their age or sex, the Ṣàngó priests in the town were vocal and ribald in their humor about the matter. Yet I never became aware of any commonly used vocabulary in Ọ̀yọ́-Yorùbá language to distinguish "upright men" from a category of men who are "homosexual" or somehow like women. I have never heard any West African òrìṣà priest speak of himself or his fellow priests as anything like a "homosexual" or as engaging in same-sex intercourse. I argue simply that the Afro-Brazilians have *reinterpreted* West African metaphors of spirit possession in the light of Brazilian gender categories. For many Brazilians in the 1930s and now, submission to a god's agency has seemed analogous to sexual "passivity," or the experience of being penetrated during intercourse. In other words, to Brazilians, a physically mountable man seems highly qualified, in a symbolic sense, to be mounted spiritually. The metaphor-ridden "loophole" by which Édison Carneiro told Landes that men entered the Candomblé possession priesthood in the 1930s was virtually identical—in both its terms and its emphases—to the *dominant* logic of the Ọ̀yọ́-Yorùbá Ṣàngó priesthood that I observed in the 1980s and others had observed since that West African priesthood was first written about (see Matory 1994b:171). Carneiro's own publication (1991[1936]) also suggests that neither this "loophole" nor the male presence in the priesthood was truly new in the 1930s.

PENETRATION AND THE CHANGING NEGOTIATION
OF "RESPECT" IN THE CANDOMBLÉ

Spirit possession and vocabulary are potent symbols not only of priestly relations with the gods but also of priestly social status in the world. Thus, in pursuit of collective respectability, the Quêto/Nagô nation has, over time gradually euphemized its African-inspired vocabulary of spirit possession, expurgated its ritual protocols, and increasingly declared itself a female monopoly. Similarly, in the pursuit of greater individual authority and respect, individual priests endeavor to minimize the frequency of their possession performances, and the numerous male chief priests often volunteer the information that they are sexually impenetrable.

Thus, given the relative ideological commitment to "African purity" in powerful sectors of the Bahian Candomblé, transformations since the 1930s would have been difficult to predict. The first unpredictable transformation was an ironic instance of Quêto/Nagô influence on other nations, whereby Angola and *caboclo* priests are nowadays even more likely than Quêto/Nagô priests to call themselves "horses" (*cavalos*) and to say that they are "mounted" (*montados*) by the gods. Such vocabulary no longer conveys the dignity that the "great" Quêto/Nagô houses seek (see also Wafer 1991:102). Preferentially, the god is said, for example, to "manifest" (*manifestar*) or "incorporate" (*incorporar*) in the priest; or the priest is said to "turn into the saint" (*virar no santo*), "fall into the saint" (*cair no santo*), or "turn around with [*rodar com*] the god or saint." Similarly, spirit possession by the important but morally lowly god Exú now occurs exclusively in Angola temples, despite that god's origins among the ancestors of the Yorùbá and the Quêto/Nagô (Omari 1984:16).[44]

It is difficult to interpret the further major change—the evident success of lesbians in today's elite Quêto Candomblé. In the 1930s, Carneiro reported that the other priestesses of the prestigious Casa Branca temple had halted one lesbian's succession as the chief priestess of that temple. However, it is not perfectly clear whether she was disqualified on the grounds of her being a lesbian (the explanation Landes favors), of her having incestuously chosen a priestess of the same temple as her partner, or of her being the child of two "respectable mulattoes" who therefore obstructed her full training in Candomblé (Carneiro quoted in Landes 1947:46–47). Since I entered the world of Candomblé, I have never known lesbianism to be as common a subject of insider conversation as is male homosexuality, perhaps precisely because Landes and Carneiro did not make it an analytic issue big enough to require popular or nationalist reaction in Brazil.

In recent decades, reputedly lesbian priestesses have been appointed without controversy to the chief office of some of the African purist houses. Of course, the Candomblé is more tolerant of variety in sexual preference than is, say, the official Catholic Church. Also, the lack of a husband and of children might allow a lesbian more time than a heterosexually attached woman to obey the frequent *resguardos*, or periods of abstinence, that are necessary for consistent participation and apprenticeship. Indeed, Candomblé members have long expressed the view that the priesthood is practically and symbolically inconsistent with marriage to a man. One is tempted to seek the history of this opinion in the same semiotic conditions that structured local understandings of male recruitment to the 1930s: since lesbians (*monas de aló*, in Candomblé parlance) are not habitually penetrated by mortal men, they are exclusively reserved—and thus ideal—for penetration by the gods. Conversely, a woman who does not wish to or is not able to marry finds in Candomblé a community and a form of respectability that does not require her to marry.

Clearer still is the simple fact that Candomblé members regard women in general and "passive" men as the *normal* possession priests. What these two social categories share is the symbolically loaded fact of their penetrability. Conversely, both submission to sexual penetration and submission to spirit possession are considered behaviors inconsistent with being a *macho* or "man" (*homem*), and when a *macho* shows the slightest signs of getting possessed, he is automatically suspected of being an *adé*. At the first sign of an oncoming possession (*barravento*), males who wish to preserve their reputation as "men" will quickly flee the *barracão*, or sacred dance space, and, no matter how often they experience this and other forms of calling, they go to great lengths to avoid initiation as possession priests. Thus, at the conjunction of European and African discourses about sex and power, the *barravento* implies these men's involuntary subjection to the penetration and will of an exogenous Other, which is both a precondition to eligibility for the possession priesthood and a great threat to those who regard masculinity as their only ticket to some semblance of authority in daily social interactions.

Thus, success as a Candomblé leader involves cultivating certain forms of self-discipline that, in this same idiom of penetration, demonstrate his or her ability to dominate *people* and reverse the subordinating implications of the chief priest's penetrability. Therefore, priests dramatize their ascent in the temple hierarchy by diminishing the frequency of their possession episodes and denying their physical penetrability. The performance of the "passive" male priest reveals an ambiguity. If the metaphoric source of his spiritual penetrability (i.e., his alleged submission to physical penetration) implies his subordination to the *macho* in the real world, how does he keep control over the *machos* amid the daily workings of his temple? In daily social life, it is clear that "passive" males are not consistently submissive to their *macho* sexual partners. Indeed, their often quick-witted, fast-talking, determined, and decisive manipulation of their long-term *macho* lovers seems the more characteristic scenario. Hence, "domination" by the *macho* is the ideal image, something like an erotic fantasy, played out in the ritual service of the gods. Indeed, Martiniano describes Candomblé's preeminent god,

Xangô, as a "rapist" (Landes 1947:203). Such fantasies can be continually reinvested with an aura of reality at the height of public ceremonial, when the *macho* drummers and singers induce the chief priest to lose self-control by "calling the saint" into his head.

Conversely, many chief priests in the Northeast and the South feel that their respectability as the conscious leaders of their temple communities depends on establishing a certain distance from the "passive" role—by avoiding the term, by concealing their sexual activities from their congregations, by denying their role as "insertees" in sexual intercourse, and/or by avoiding penetration by any potential rival.

Many male chief priests volunteer the information that they have never been penetrated.[45] The information is probably often true. I have no reason to believe that all male Candomblé priests are in fact "passives"; several among my close friends definitely are not. Whether literally true or not, such denials highlight the priests' awareness that their alleged penetrability implies a level of real-world submissiveness that is inconsistent with their leadership of other human beings. Thus, while this reputation for sexual "passivity" is perfectly consistent with their privileged relationship to the gods, priests demonstrate their dominance over human communities by *denying* that they are ever sexually penetrated.

Ultimately, the ritual and social authority of a chief priest, like that of a chief priestess, depends on the parallel abilities to dominate others, to avoid being dominated by others, and to restrict even the occasions when the god will dominate him. Nothing conveys junior and subordinate status in the Candomblé better than the tendency to get possessed frequently and at inappropriate moments. Analogously, few things inspire more critical commentary than a profligate priest or priestess and one who allows others to use his or her sacred ceremonies as occasions for assignation. In common, these qualities indicate and encourage *a falta de respeito* (the lack of respect). And "respect" is the key to the proper, hierarchical functioning of the Candomblé. Thus, public questions about the importance of "passives" and about the likeness between sex and possession are sometimes met with protests such as "Candomblé has nothing to do with sex." Indeed, ritual activity tends to require sexual abstinence (*resguardo*), which proves not that Candomblé has nothing to do with sex but that priestly competency is defined in sexual imagery, and priestly success depends on disciplining its projection.

This study illuminates the welcome accorded to men who love men in Haitian Vodou and Cuban Ocha as well. In common, these religions embody a scientific morality, which observes and balances among the diverse powers, possibilities, and causal phenomena in social life, rather than judging, censuring, and prescribing normative limits (see also Kramer 1993[1987]; K. Brown 1987). Moreover, the cultural logic of spirit possession in these contexts undeniably widens the normally narrow space of agency for numerous categories of subaltern actor (Lewis 1989[1971]), but especially for categories whose subordination is, in the wider society, constructed dramatically in terms of their "passive" role in sexual intercourse. Such symbolic phenomena, however, are far from unique to the Yorùbá-Atlantic religions. Indeed, it would be difficult to find a religion anywhere in the world where images of marriage and intercourse, as well as restrictions on their practice and exposure, are not central metaphors of human-divine relations and equally central metonyms of community among worshipers.

A YorÙbÁ-Atlantic Controversy

I first proposed this explanation for the locally perceived normalcy of *adés* as possession priests and therefore as the heads of Candomblé temples, in 1988 (Matory 1988:215–31; 1999d). My published work has elicited the surprise of self-discovery among gay activists and scholars, as well as the deep offense of one widely read Yorùbá scholar. Our debate and the responses of both New World priestesses and the 20th-century Yorùbá diaspora in the United States illustrate the ongoing centrality of sex and secrecy in the making of overlapping imagined communities. This case demonstrates that Regionalist, nationalist, and international feminist communities are not the only ones that are continually transformed, in a cosmopolitan context, by the gaze of other imagined communities and by the nationalist silencing of certain homegrown realities (Herzfeld 1997). As communities, the African diaspora in general and the new microdiasporas within it are constituted by certain open secrets, and they can be reconstituted by reselections and rereadings of what secrets need to be defended.

My argument about the Yorùbá antecedents of the *adé* called forth multiple imagined communities and inspired public clarification of their boundaries. Were it not for the increasingly vocal homophobia of Protestant Anglophone African bourgeoisies and the hot-button nature of sex as an object of cultural intimacy among nationalists, my argument would be not only better substantiated but also little more controversial than Herskovits's view that "shouting" in Black North American churches is a "reinterpretation" of African spirit possession (Herskovits 1958[1941]:211–16). It would be little more controversial than explaining how porcelain soup tureens replaced calabashes and earthenware on Cuban and Brazilian orisha altars, or how the *idòbálẹ* and the *ìyíkàá* salutes in Cuba and Brazil reinterpret similar gestures in West Africa (see introduction and figure 8).[46]

I have never said or believed that the West African transvestite priests were or are in any sense homosexual (Matory 1994b:208; 1991:22, 520–21, 538; 1988; pace Oyewumi 1997:117).[47] While many have embraced my argument as logical and empirically sound, some others have found it easy to misinterpret, either as (1) proof that homosexuality is as widespread and natural in Africa as it is in the West (Stephen O. Murray 1998:100; personal communication, 1996), or as (2) a defamation of authentic, "traditional" Yorùbá culture (Oyewumi 1997:117). The first misinterpretation is beyond the scope of my argument and of the evidence that I present here.[48] The second is the subject of what follows—a case study in the making of an open secret.

One Yorùbá scholar in the United States, sociologist Oyeronke Oyewumi, read my argument and then, in print, accused me of describing the West African possession priests as "drag queens" and "actual if not symbolic homosexuals" (Oyewumi 1997:117). Oyewumi is clearly less interested in summarizing my argument than in expressing her deep offense and her own preference to classify "homosexuals" as anathema to this new diasporic nationalism.

Oyewumi's caricature of my argument was but one link in her thesis that there is no gender whatsoever in authentic Yorùbá culture (1997:31, 78, 157).[49] She argues, in sum, that colonization by the West is the origin of all the sexism and, indeed, of all the gender conceptions that exist in Yorùbáland today, and that, because English language continuously marks the gender of its human referents, conventional scholarly discourse in English consistently misrepresents the gender-free culture of the Yorùbá. Ungendered features of Yorùbá language are taken as proof that Yorùbá culture in general was once and still is, in its essence, both nonsexist and free of any gender differentiation. Therefore, Yorùbá and non-Yorùbá scholars who see gender in Yorùbá cultural history do so simply because they have falsely translated the gender-neutral terms of the Yorùbá language into the gender-specific terms of the English language. Such scholars are thus both victims and agents of Western imperialism (Oyewumi 1997:16).

In evidence, the author cites the extensive gender coding of pronouns, names, kinship terms, and occupational terms in English, as well as numerous Yorùbá pronouns, kinship terms, and occupational terms that, in her opinion, do not encode gender—such as òun (she or he), omo (child), ègbón (senior sibling or cousin), oba (monarch), Ìyá Olónje ("Food Vendor" [lit., "Senior Female Owner of Food"]), and Bàbá Aláso ("clothier" or "weaver" [lit., "Senior Male Owner of Cloth"]).

Oyewumi spends much of her argument explaining away or concealing the gender coding that actually does appear in these and other

Yorùbá terms and social practices. For example, when evidently old gender-marked aspects of Yorùbá language are addressed at all, they are excused by various arguments that would obviously be absurd if applied to languages and cultures more familiar to the reader. For example, *bàbá* ("father" or "senior man") and *ìyá* ("mother" or "senior woman") are said to indicate not only sex but *also* adulthood; therefore they are not gendered, argues Oyewumi. Does it follow, then, that the terms "father" and "mother" in English are not gendered?

Oyewumi argues that the term for "bride" or "wife" (*ìyàwó*) is ungendered because it refers to both the female brides of worldly husbands and possession priests regardless of sex. Does the fact that the church is called the "bride of Christ" in English then imply that the English term "bride" is also ungendered?

In English, as in Yorùbá, one could recite an endless list of gender-free references to people without ever proving that the language or the culture is gender-free. Could one reliably infer from the gender-neutral English terms "I," "you," "we," "they," "parent," "cousin," "sibling," "child," and "president" that Anglo-Saxon or Western language and culture are in their essence or once were free of gender and of gender hierarchy? I think not. But this is the logic of Oyewumi's linguistic argument that Yorùbá culture, in its deep past and in its present essence, is completely without gender.

There are clearly words in Yorùbá for "male" (*akọ*), "female" (*abo*), "man" (*ọkùnrin*), and "woman" (*obìnrin*). The terms of address and reference for parents, senior relatives, senior strangers, and people of almost every occupation indicate the referent's gender—as in *Bàbá Ayọ̀* (the teknonymic "Father of Ayọ̀"), *Bàbá Ẹléran* (butcher), and *ìyá mi* (Mommy). Most professions in Yorùbáland have and have long had vastly more of one sex than another practicing them, and virtually all social clubs (*ẹgbẹ́*) are segregated according to sex.

Certain Yorùbá religious and political titles are strongly gender-marked, despite their infrequent adoption by a person of the other sex, such as *babaláwo* (a type of divination priest [lit., "senior male who owns the mystery"]); *baálẹ̀* (nonroyal quarter or town chief [lit., "father of the land"]); and *baálé* (head of residential compound [lit., "father of the house"]). But as far as I know, a man can never be an *ìyálé* (eldest wife of the house [lit., "mother of the house"]). It should be noted that these last two terms—*baálé* and *ìyálé*—are *etymologically* distinguished from each other *only* by the gender of the referent. Yet in real social life the persons described as "fathers of the house" rank far higher in the house than do the people called "mothers of the house."[50]

Oyewumi's argument neatly parallels the claim that Brazil is a "racial democracy" (expressly opposite in character to the United States and the rest of the Euro-Atlantic world) and that foreigners' analyses of race and racism in Brazil result from the imposition of an imperialist North American logic (see Freyre 1986[1933]; Turner 1985:78–79; Skidmore 1985; Azevedo 1975). Of course, Oyewumi's argument is far more extreme. While denying the existence of racial discord in Brazil, Freyre concedes that social rank in the slave system was at least flexibly correlated with color (e.g., Freyre 1986[1933]:xiii); Oyewumi admits of no such subtlety. Both Oyewumi's argument and Freyre's dramatically remind us of the cross-cultural variation in the interpretation of human phenotypes (a point that may have been surprising to the Brazilian general public in the 1930s but is hardly news to the scholars who have studied gender since the 1970s).[51]

However, the work of Oyewumi and Freyre also alerts us to a genre of nationalistic allegory that is common in a transnational world, where scholars and other workers in the diaspora articulate some of the most emotionally powerful and politically persuasive images in the national imaginaries of the homeland. The Brazilian Freyre, too, formulated his influential sociomoral allegory during and following his sojourn in the United States. Both seek to affirm the sharpest conceivable contrast between their home and host societies. Both arguments rely on the construction of an idyllic past time beyond immediate scrutiny. They equally invoke a sense of national honor around the conspiracy to conceal contrary facts that every insider knows.

A (CULTURALLY) INTIMATE GATHERING OF PRIESTS AND SCHOLARS

Since Gilberto Freyre organized the First Afro-Brazilian Congress in 1934, dozens of such conferences have brought together priests and scholars intent on rethinking and reorganizing òrìṣà religion, and reflecting on its significance for the crisscrossing imagined communities of the Atlantic perimeter. Several such conferences have had momentous effects, largely because they have helped to establish which priests' practices are normal, which are best silenced, and who legitimately speaks for the group.[52] For example, the 1937 congress organized by Édison Carneiro in Bahia culminated in the organization of the Union of Afro-Brazilian Sects, the first organization to regulate priestly conduct and to unite the Bahian temples and their supporters against police repression. In 1981, Wande Abimbọla and Marta Moreno Vega organized at the University of Ifẹ̀, Nigeria, the first World Conference of Orisha Tradition and Culture. Thus, for the first time in history, a conference brought together scholars and priests of òrìṣà religion from Brazil, Cuba, Puerto Rico, Trinidad, the United States, and Nigeria. A dozen such conferences have followed, albeit under an increasingly factionalized leadership. As the leader of one series of conferences, Abimbọla is now regarded by some as the paramount leader of the global òrìṣà-worshiping community. Such an understanding of Abimbọla's role is clearly contested. However, until now, no one else has to my knowledge ever been credited with such authority. This is cultural history in the making.

It is against this backdrop that events at the 1999 conference at Florida International University acquire their historical significance. Titled "Òrìṣà Devotion as a World Religion: The Globalization of Yorùbá Religious Culture," this conference brought together dozens of U.S.-based Nigerian, Cuban, Puerto Rican, native U.S. American, and Brazilian scholars with priests from equally diverse geographical and national origins. On and off the dais, priests and scholars debated over whether whites and Westernized-looking Yorùbás could legitimately speak for Yorùbá tradition, whether Yorùbá was the only language in which Yorùbá religious concepts could be discussed, whether "each group [i.e., Cuban and Cuban-inspired *santeros*, Brazilian *candomblecistas*, Nigerian Yorùbá people, Trinidadian Shango practitioners, etc.] should speak for itself,"[53] and whether certain scholarly disagreements should be settled publicly or privately. University conferences are not simply forums where truth is worked out through debate; they are also stages where *social* priorities are debated and dramatized. Officially authorized speakers have diverse priorities, and so do audiences.

Oyewumi's extempore presentation at this conference urged caution in translating Yorùbá concepts into English terms. She gave examples of bad translations, such as glossing *ọba* as "king" (when, in truth, it means "monarch"), and insisted that the term *ìyàwó* (bride) was ungendered, since it describes not only married women but also junior òrìṣà possession priests of either sex. Based on this evidence, she reaffirmed her conclusion that Yorùbá language, and therefore Yorùbá culture, is devoid of gender.

The talk's logical and empirical inadequacies notwithstanding, two Trinidadian priestesses and an African-American priestess in attendance stood up to applaud it. Oyewumi's nostalgic reconstruction of an ideal Yorùbá past and essence held great appeal for New World priestesses who would resist the patent sexism of American societies (including the forms of gender inequality strongly evident in New World traditions of òrìṣà worship), for diasporic Yorùbá people anxious to subvert North Americans' tendency to regard Africa and its cultures as inferior, and particularly, according to Mọlara Ogundipẹ, for Yorùbá men happy to be exonerated of sexism.[54]

Dozens of West African Yorùbá scholars have written with sharpness and clarity about gender and gender relations in Yorùbá religion and culture generally (e.g., Abiọdun 1989; Awẹ 1977, Fadipẹ 1970[1939]; Okediji and Okediji 1966; Ogundipẹ-Leslie 1985; Olupọna 1997; Abimbọla 1997). Yet, in this public forum, the Yorùbá scholars present chose to circle the wagons. None publicly challenged Oyewumi's inferences, and a number of them offered examples apparently intended to support her case, such as the occasional affectionate use of the term *bàbá*, which normally identifies fathers and other men a generation or more older than the speaker, in addressing the speaker's patrilateral kin, and the similar use of the term *ìyá* for matrilateral kin. In fact, far from demonstrating the absence of gender for Yorùbá culture, such parlance reveals that Yorùbá people genderize whole categories of kin that few Americans consciously think of in terms of gender. Those Yorùbá scholars present who disagreed, restricted their self-expression to private conversations (for further details, see Matory 2003). Many of these scholars have already contributed significantly to the academic study of gender in Yorùbá culture and have chosen other venues to express their opinions. Hence, these observations are intended not to impugn the quality of these scholars' published work but to show that scholarly conferences have long been and still are important theaters for the imagination of community in the African diaspora.

Large North American professional organizations are also venues of such imagining. For example, without consulting any Africanists, much less Yorùbánists, the Sex and Gender Section of the American Sociological Association (ASA) awarded Oyewumi's *Invention of Women* its Distinguished Book Award in 1998 (anonymous selection committee member, personal communication, 19 November 2001). Thus, a new form of nostalgia and silencing has united the Sex and Gender Section of the ASA in common cause both with Yorùbá long-distance nationalism and with New World priestesses who are aware of the gender bias that has long been a part of their New World òrìṣà traditions and are ready to do something about it. This nostalgia is new in some details but is logically similar to the reasoning that united Brazilian Regionalism and nationalism with Landes's brand of transnational feminism. It is not clear how fast, how commonly, or how deeply this new alliance of ideological forces will affect the practice of òrìṣà worship, but every subsequent conference of scholars and priests is likely to add authority to these motivated representations of the shared past. These representations, in turn, acquire the credibility to structure new communities and hierarchies in the present, just as the historical revisions of Landes and Carneiro did in Brazil after the 1930s.

Life imitates scholarship, as in the case of many African Americans who, since the 1960s, have modeled the program of conscious cultural pan-Africanism known as "Afrocentrism" upon the African-centered

cultural history of Melville J. Herskovits. The logic of "survivals" articulated in his *Myth of the Negro Past* (1958[1941]) has become the rationale behind the adoption of forms of dress, cuisine, worship, marriage,
music, dance, and politics long ago abandoned or long avoided by Black
North Americans. It is now, similarly, news that the priests and followers of the Yorùbá-Atlantic traditions frequently own and read books
written by university-trained scholars about those traditions—such as
Juana Elbein dos Santos's *Os Nàgô e a Morte* (1976) and Pierre Verger's
Orixás (1981) in Brazil; Lydia Cabrera's *El Monte* (1954) among
Cuban-inspired adherents of Ocha; and Robert Farris Thompson's *Flash
of the Spirit* (1983) among U.S. orisha devotees. The degree to which
these books become catechisms or procedural guides is variable, but it is
clear that many priests use the information that scholars bring, particularly when those scholars possess a credible claim to information from
the African ancestral "homeland."[55] Thus, the opinions of even West
African and West Africanist scholars with no priestly credentials can be
enormously influential in the transformative projects of New World
priests and priestesses. Our scholarly analyses are often employed as
models of African "tradition," which can be used to include and elevate
particular segments of New World religious communities. Therefore,
our analyses can also be used to marginalize particular segments of
those communities and to delegitimize existing practices. Our influence
can be powerful, whether we are committed to studying our field sites
honestly, or we are committed to misrepresenting them as allegories of
some ideal that we are recommending to an audience unable to check
our facts. Thus, African, Africanist, and Afro–Latin Americanist scholars
can be influential not only in priestly projects but also in the political
projects of first world scholarly communities with little knowledge of
Africa and little intrinsic interest in Africa's complex truths.

A further form of transnational cooperation and concealment has taken
shape around Western feminists' good-willed but sometimes ill-informed
protests against "female genital mutilation" in Africa. Since the mid-1990s,
several female illegal immigrants from Africa, including at least one Yorùbá
woman, have won asylum in the United States based on the assertion that,
if repatriated to their homelands, they or their daughters would be subject
to forcible genital surgery. Despite the extreme doubtfulness of the two
southern Nigerian cases I have examined, both applicants won asylum. One
African woman invited to a televised roundtable about the Igbo applicant's
asylum case refused to participate, telling the convener that, although she
did not believe the asylum seeker's claim, she could never take sides against
her "African sister." Various imagined sisterhoods such as Euro-American-
led feminism are among the transnationally imagined communities that
crosscut the "brotherhood" of the American territorial nation, and all these

continue to struggle over the meaning of the gendered black body and which parts of it need to be kept secret.

CONCLUSION

In sum, neither the Candomblé "cult matriarchate" of Landes nor the primordially "genderless" Yorùbá society of Oyewumi is a neutral, or careful, report of cultural history. They are inventions in the service of overlapping imagined communities. Yet they also do a disservice both to Yorùbá cultural history and to the thousands of male priests who have built institutions, housed the poor, and healed the sick in the Brazilian Candomblé. Nonetheless, one cannot avoid the fact that these inventions move people and change history. The debates and transformations that I have detailed here reveal not only the pitfalls of tendentious scholarship but also the fact that long-distance, transnational dialogues have continually reshaped even "traditional" cultures and religions like Candomblé.

Moreover, transnational social movements and diasporas, like nation-states, propagate secrets and defend the intimate zones that are created around those secrets. The facts that the leaders of imagined communities choose to make secret can be surprising—such as the facts of gender and racial inequality in Brazil and gender inequality in Yorùbáland, not to mention the open secret of intergenerational class inequality in the United States; it is difficult to fathom how an outsider could *fail* to see them. However, any fact that a community can be persuaded to discuss privately and to silence in the company of outsiders can serve the same community-defining function—a function that Herzfeld (1997) calls "cultural intimacy." Indeed, an old imagined community can be reinforced or a new community imagined into being by the forceful assertion that some fact of its life needs to be hidden from a larger encompassing group or forbidden for outsiders to speak of. In these cases, populations that once felt like a part of other groups can suddenly circle the wagons around a newly contrived secret and thus separate themselves strategically from larger or crosscutting groups.

We have so far located Candomblé at the intersection of multiple transnational, national, and regional forces. Let us now consider how these processes have converged in the life of one man, Pai Francisco Agnaldo Santos, and one enduring friendship—ours. This compound story both recapitulates the themes of the preceding chapters—tradition, transnationalism, and matriarchy—and illustrates the challenges and passions through which they are embodied in the accidental complexities of real human lives.

CHAPTER SIX

Man in the "City of Women"

> If you just sing Jeje songs at your festival,
> you'll end up singing alone. . . .
> So, we have to sing Quêto songs too.
>
> —PAI FRANCISCO, 1992

LIKE ME, Pai Francisco lives in a place where multiple old and new transnational flows converge. Therefore, our friendship is new, the novel convergence of two sites of convergence—one African-Brazilian, the other African-U.S. On the other hand, our friendship recapitulates the old and truly circum-Atlantic circuits that long ago united Africa, the United States, England, France, Cuba, and Brazil—during the Lagosian Cultural Renaissance, the Négritude movement, and Afro-Cubanismo, as well as the circuits that have lately wrought the Yorùbá Cultural Renaissance now sweeping African-American and Latino immigrant communities in the United States (Matory forthcoming a). These transnational flows in no way demonstrate the lapse of inequality among the ethnically, racially, economically, and internationally diverse actors involved. Nor do they foretell the collapse of the nation-state. Rather, they demonstrate the forms of dialogue that powerfully unite the Afro-Atlantic world in an age of Anglo-American global dominance.

Next to hip-hop and Rastafarianism, Anglo-Yorùbá culture has become the most important reference point of Afro-Atlantic cultural unity generally.[1] In the 1970s and 1980s, the Nigerian oil boom, the Brazilian federal government's pursuit of leadership and commerce in the non-aligned movement, and the expansion of U.S. foundation and university support for research on black culture together reinvigorated a powerful circuit of exchange among Nigeria, Bahia, and the United States, which further undermined the rival leadership claims of the numerically preeminent Congo/Angola nation. Equally undermined were the purist Jeje-Arará nations around the Atlantic perimeter, reversing post-1930s trends that had created a greater reputation for the Rara tradition in Haiti and might have created an equal reputation for the Jeje nation in Bahia. At the same time, the feminization of Candomblé's leadership has proceeded rapidly. The male majority reported by Harding (2000) in the 19th century and by Carneiro (1986[1948]) in the mid–20th century has disappeared not only in the Quêto/Nagô nation but in every nation of Candomblé (Santos 1995).

This chapter illustrates the transnational forces that not only shaped Pai Francisco's life and his exceptionally successful struggle for a place of respect in the City of Women but also made our friendship possible. Here I tell the story of how old and new transnational flows have shaped both our lives and the perspective from which I have, throughout this book, retold the history of Candomblé.

It was August 1987 when I first met Pai Francisco Agnaldo Santos, who is not only my best friend in Brazil but also a *pai-de-santo*, or male chief priest, of Candomblé, a religion often regarded as the best proof that African culture has "survived" in the Americas.[2] Indeed, Pai Francisco knows an extraordinary array of songs in the Jeje and Nagô languages—songs that he and his priestly colleagues agree are purely African. Yet Pai Francisco's acclaim flows not only from his unparalleled command of the complex liturgy of his trans-Atlantic nation but also from his surefooted dance across the overlapping imagined communities that continually redefine its meaning and unsettle the grounds of priestly legitimacy.

"ANAGONIZATION"

Pai Francisco is usually known as "Francisco of Xangô," since his divine father is the god of fire and thunder. However, when it is important to affirm the dignity and autonomy of his nation, he reminds people that he worships not the *orixás* of the Quêto/Nagô nation but the Jeje *voduns*. Thus, properly speaking, he is a son of Sobô. Nonetheless, he usually calls his temple by a Yorùbá name. At the times that Pai Francisco is anxious to assert the fidelity of his temple to "purely" Jeje practice, he calls the temple by one of several Jeje names.

At the root of this shifting nomenclature is a deep fear. Pai Francisco fears that the substance of his power is draining away. We have now seen some of the various transnational flows that shaped the birth, death, and resurrection of this nation. Like the Quêto/Nagô nation, the Jeje nation has in some ways flourished as a result of its propagation of "African purity" and its access to the transnational means of demonstrating it. However, being outnumbered, outsponsored, and outpublicized by the Quêto/Nagô nation, the Jeje nation has found it difficult to resist forms of Quêto/Nagô influence that result in a loss of its own specifically Jeje "purity."

Bahians have a vivid name for the nemesis of Pai Francisco and many others' mighty struggle: "anagonization" (*anagonização*). I first heard the term spoken by a priestess of the Angola nation in conversation with Pai

Francisco. In their biographical narratives, numerous priests recount their conversion to the Quêto nation or express their frustration over having forgotten the distinctive words or songs of their own non-Quêto nations. In representing the juggernaut expansion of Quêto/Nagô religion, the term "anagonization" resurrects the complete Yorùbá word for the Dahomean borderland peoples—*ànàgó*—and, through the addition of a Portuguese suffix, makes the process sound like a form of "agony" (*agonia*).[3] That is precisely how the swallowing up of their nations feels to those who have risen to the top of priesthoods in non–Quêto/Nagô nations. The prestige of Quêto detracts from the value of their lifelong investment in other nations. They must both adopt Quêto vocabulary and liturgy and submit to the demeaning charge that their own nations are now "mixed" (*misturadas*). They thus concede defeat according to the most central tenet of spiritual health, sanctity, honor, and dignity in today's Candomblé—"purity" (*pureza*).

Despite this trend toward substantive homogenization among nations through the adoption of Quêto practices, the logic of "purity" inspires equal and opposite assertions of national distinction. The potential of anagonization to privilege the priestesses of the great Quêto houses as the heads of the entire Candomblé has also reinforced the incentive of other priests to avow and emphasize the distinctness among nations. Similarly, since the world conferences of *òrìṣà* worshipers began in the 1980s, the potential for Nigerian Ifá diviners to declare themselves sovereigns of the global *òrìṣà* community has led Bahian Candomblé, Cuban Ifá, Cuban-inspired Ocha, and Trinidadian Shango priests to reassert the autonomy of their traditions and the legitimacy of their divergences from West African traditions. By contrast, the white converts who have in recent decades made São Paulo into a wealthy capital of Candomblé have chosen to submit themselves to the Nigerian *babaláwo*s. Their strategic choice is reminiscent of Mãe Aninha's in the early decades of the 20th century. On the grounds of their race and their recent conversion from Umbanda, the white priests of São Paulo might have chosen to submit themselves to authorization by the high priests of Bahian Candomblé. They would thus always suffer by comparison to the model and seniority of Bahia. Instead, they have chosen to jump the queue, seeking their authorization and models directly from the African source.

Members of the African-Brazilian nations that are smaller or less prestigious than Quêto endeavor self-consciously to use the distinctive terminology of their own nations, even if communication with outsiders requires them then to translate into Quêto/Nagô language. The Jeje and Angola nations maintain their own names for various gods that are best known by their Quêto names: the Jeje Leba and the Angola Bobinjira, or Pombagira, correspond to the Quêto messenger god Exú; the Angola

Roximucumbi corresponds to the Quêto Ogum; the Jeje Agangatolú and the Angola Kibuco correspond to the Quêto hunter god Oxossi; the Jeje Caviano, Badê, or Sobô and the Angola Luango or Zazê correspond to the Quêto thunder god Xangô; and so forth.[4] The Jeje Dã or Besseim and the Angola Angorô correspond to the Quêto Oxumarê. Of course, each god is correlated with a variable range of Roman Catholic saints as well, but the Quêto pantheon is now the definitional standard, or, as Risério (1988) put it, the "metalanguage," of Candomblé. The òrìṣà occupied the same "metalinguistic" role in the documentation and canonization of West African "Djedji religion" by the Brazilian-sponsored fathers of the Society of African Missions in the mid–19th century. Though scholars recognize the Quêto nation's debt to Jeje vocabulary and ritual conventions (e.g., Verger 1970[1957]:21; V. Lima 1977:72), the Jeje nation is the one that is losing its distinct existence, just as the Congo nation has virtually disappeared into the Angola nation.

It is difficult to ascertain the age of non–Quêto/Nagô resistance to the Quêto/Nagô metalanguage, but such resistance, whatever its failures, is central to Pai Francisco's religious life. For example, Pai Francisco frequently declared that his *mãe-de-santo* had required her initiates to learn the words for all sacred objects and actions in Jeje, but that, much to his regret, he had forgotten many of them. Therefore, one of his greatest dreams is to visit the land of his nation's origin and thereby renew what he and his diaspora nation have forgotten. As I learned my way around Bahian Candomblé, I often reported my discoveries to Pai Francisco in order to seek verification and clarification. Two of our common friends, who perhaps had reason to envy him, told me that his house is "mixed" (*misturada*), which I innocently reported back to him. At first he reacted with anger, since he regarded the description as slanderous. Then he implicitly confessed the truth of their words, poignantly explaining the inevitability of his nation's fate: "Jeje is a dying nation. If you just sing Jeje songs at your festival, you'll end up singing alone. The songs are more complex [than Quêto's], and people [in the audience] don't know the reponses. So, we have to sing Quêto songs too." Thus, Pai Francisco must compromise the purity of his nation by singing Nagô songs in honor of his Jeje gods. Later, he added an explanation which implied that it was indeed Jeje's *superiority* that accelerated its extinction: unlike Angola and Quêto priests, Jeje elders refused to pass on much of what they knew— evidence of both their admirable commitment to secrecy and their fear of the power that their juniors might, once fully instructed, acquire over them (see also Larose 1977). Recognizing the tragedy of the consequences, Pai Francisco concluded, "The African is stupid!"—*Africano é burro!* Yet, just as Martiniano and Aninha's "return to Africa" in the second decade of the 20th century superseded the hierarchies and protocols of the Bahian Nagô nation, one male Jeje priest returned to Africa in the 1970s and reversed the course of Candomblé's incorporation into the national projects of the Nagô and of Brazil as well. Through him, Jeje's influence is at the root of an African renaissance.

In Bahia, I had never heard the name of Jamil Rachid, the Jeje priest perhaps most responsible for "re-Africanizing" the Afro-Brazilian religious community in distant São Paulo. From the 1930s onward, as Ortiz (1991[1978]) and D. D. Brown and Bick (1987) have shown, middle-class whites in the Center-South endeavored to legitimize Afro-Brazilian-inspired possession religions and their participation in them by arguing that these religions had ultimately originated outside of sub-Saharan Africa, in the "higher civilizations" of India and Egypt, and by stripping the existing religion of ritual elements considered too "African" and therefore "low" (*baixo*) (such as animal sacrifice, malevolent magic, and cigar smoking). The new religion that resulted—Umbanda—grew rapidly and opposed itself, above all, to the "African" forms that Brazilians associate paradigmatically with the Bahian Candomblé. D. D. Brown and Bick (1987) show that much of the impetus to "purify" Umbanda of what looked African came from the ideological preferences of the white middle-class leaders of the federations that could secure the greatest protection from police persecution.

Reginaldo Prandi (1991) has documented the reverse phenomenon since the 1950s, when vast numbers of Northeasterners in search of work in the expanding Center-South economy introduced Candomblé into São Paulo. Under their influence and amid Mãe Menininha's nationwide renown, the subsequent decades witnessed an equally vast "re-Africanization" of Umbanda. Since then, hundreds of white, middle-class Umbanda centers have adopted Candomblé practices long anathema to Umbanda (such as sacrificing animals to Exú) or wholly converted to Candomblé.[5] Transmigrant Nigerian students and traders in São Paulo command a lucrative market for African clothes unsustainable in Bahia, since many Candomblé members in São Paulo now wear them as liturgical garb.

It should also be noted that Mãe Stella's efforts in the 1980s to eliminate syncretism from Candomblé were not simply a continuation of the uninterrupted efforts or priorities of her nation. Indeed, the fact that she had lived in São Paulo before inheriting the leadership of Opô Afonjá in 1976 suggests that she had observed the early efforts of Jeje priest Jamil Rachid to "re-Africanize" Umbanda in the 1970s.[6] From the 1970s until Mãe Stella of Bahia began her campaign in 1983, the advocate of re-Africanization most widely quoted in the Brazilian national press, as well as its foremost advocate in the Center-South, was Jamil Rachid. Jamil, aka "Dofonitinho de Baluaê," was a son of Omolú, the leader of his own Umbanda temple consecrated to Omolú (Templo Espiritualista de Umbanda e

Candomblé São Benedito in Pinheiros, São Paulo), and, around 1976, the leader of the 350,000 members of the state's largest Afro-Brazilian religious federation, the Union of Spiritist Temples of Umbanda and Candomblé of the State of São Paulo (União de Tendas Espíritas de Umbanda e Candomblé do Estado de São Paulo). According to newspaper reports, he had taken his religious training "at the 'Roça da Ventura,' in the Bahian city of Cachoeira de São Felix, the greatest and strongest redoubt of Candomblé in Brazil [sic]. His grandmother-in-saint [i.e., the person who had initiated the person who initiated him] was a Guayacu [a female priest consecrated to a female god, also spelled guaiacú] of the 'Gêge Marrim' nation, whose ritual is most practiced in Dahomey."[7] On the occasion of the report, he was embarking on a voyage to Dahomey (now called Bénin) and southern Nigeria to research his religion. The trip would also take him to Egypt and Kano, Nigeria, to visit his Lebanese relatives. At the height of Brazil's military dictatorship, the report mentions that Jamil's trip had the "collaboration and total assistance of General Nelson Braga Moreira, President of the Superior Organ of Umbanda of the State of São Paulo," an organization of which Jamil himself would serve as the coordinator in 1983.[8] I cannot say whether Jamil ever converted to Candomblé. What is clear is that, in the fastest-growing and most prosperous Candomblé community in the country—São Paulo—Jamil was the most famous advocate of re-Africanization and that his literal voyage to Africa attracted major attention. Moreover, he selected priests of the Bahian Jeje Candomblé as his guides to re-Africanization. He both proclaimed publicly and demonstrated, for all of Brazil to see, that the traditions of the small Jeje nation belonged in the most exalted position among the Afro-Brazilian religions and that Africa stands at the center of them all. Despite the material wealth of the Center-South, its influence on the Bahian Candomblé has been limited. In Bahia, both anagonization and the feminization of priestly power have remained a juggernaut.

GENDER AND THE JEJE STRUGGLE FOR SURVIVAL

Since Landes's transnational conversation with Arthur Ramos and Édison Carneiro in the 1930s, Brazilian scholars and Northeastern journalists have given a bad name to male priests and to any hint of commerce in Candomblé. Male priests have not been able to demand the same resources from the state and the state-subsidized tourist industry that some priestesses receive. And priestesses who wish for access to those resources have had to renounce the benefits of doing business. Consequently, since the 1960s, Northeastern and female-led temples have enjoyed disproportionate access to outside sponsorship, while male priests in Bahia and the Center-South are forced to rely on commercial strategies. Priests of all genders and regions have learned to tap the benefits of projecting their images and interests through the Brazilian mass media and transnational scholarly networks.

Nonetheless, as in the entire Candomblé priesthood, the position of male priests was declining not only statistically but also symbolically in the Jeje nation. And the best-known chief priestess in the Jeje nation, Mãe Nicinha, articulated positions in Bahian newspapers that grew more and more exclusionary toward men over time, her close collaboration with several male Jeje priests notwithstanding. On the one hand, in 1975, she was quoted explaining the ambiguous gendered logic of her succession as follows: "In our nation, the only person who can occupy this post [chief priest] is a woman who has a male saint [i.e., who is consecrated to and possessed by a male god]."[9] By 1986, the survival of the Jeje nation was already in danger, and the Bogum temple itself was in a serious state of disrepair. In pursuit of an opportunity for state sponsorship of temple repairs, Mãe Nicinha asserted her temple's prestige and worthiness in terms even more expressly attuned to post-Landean standards of "African purity" and yet perfectly inconsistent with what she knew empirically of the Jeje nation and with what scholars have known about its West African counterparts: "Men do not get possessed by the *voduns* and are just *ogãs* [nonpossession priests or sponsors]. The matriarchate is total in the Jeje nation.... the same custom is observed in the regions where the Jejes live in Africa."[10] The prestige of Mãe Nicinha's Bogum temple has been enhanced further by repeatedly published claims that it is the *only* Jeje Marrim or the only Jeje temple in Bahia (her sacred kinship with Pai Vicente de Ogum and collaboration with Pai Francisco notwithstanding), and by claims that the Jeje Marrim *terra*, or subnation, and the Jeje nation in general are "totally" matriarchal—that is, even more female-dominated than the Quêto nation.[11] In fact, during the 1930s and 1940s, some of the most prominent Jeje priests were men—for example, Pai Manuel Falefá and Pai Manuel Menêz. And according to a 1983 survey of Candomblé temples in Salvador, the percentage of male chief priests in the Jeje nation (40 percent) is significantly higher than in the Angola nation (26 percent) and the Quêto nation (30 percent; see dos Santos 1995:19). The statistical decline and ideological marginalization of male priests are the backdrop of Pai Francisco's struggle for respect on behalf of himself and a dying nation.

A MAN IN THE "CITY OF WOMEN": HOW PAI FRANCISCO GOT HERE

While Jamil Rachid found Bahia an opportune place to seek authorization for his leadership in São Paulo, Pai Francisco lived in the wrong place to find an easy road. He was born into the "City of Women." I am not alone in pointing out the novelty and weight of this "cult matriarchate." Indeed, numerous male priests I have met in Salvador, Rio, and São Paulo have described *female* leadership in Bahia as the consequence of a "revolution."[12] Pai Procópio was descended from Jejes, while Joãozinho was of the Angola nation, as well as a mulatto, a *caboclo* worshiper, and a reputed lover of men. Pai Joãozinho fled the City of Women and built a dynasty in Rio. Here, I narrate the struggle of one man who remained behind in the City of Women. I do so using terms, tone, and causal logic as close as possible to how Pai Francisco related it to me during two recorded interviews in 1987. Supplementary information is drawn from numerous casual conversations between us and from Regina Pires (1992).

What follows is a human portrait of what might otherwise seem seismic or monumental forces. Transnationalism, the conflicting options of diasporism and indigenism, the degradation of male priests, and the ambiguous consequences of state and bourgeois sponsorship all have material and emotional consequences for flesh-and-blood people. Above all, this story shows that enormous cross-cultural flows and social transformations are often embedded in interpersonal dialogues. Indeed, "dialogue" is a useful model of the intercultural transformations that created Candomblé.

Pai Francisco lives close to the transnational routes of the black Atlantic. Since adolescence, he has long been close friends with a Quêto/Nagô priest descended from one of the great families of trans-Atlantic travelers. And the scion of another such family, Dr. George Alakija, worked in the same hospital as Pai Francisco for several years. At the time, according to Pai Francisco, all of Bahia knew of this family's illustrious trans-Atlantic history. Dr. George remains a leader in the invocation of Candomblé as an emblem both of Regionalist-nationalist identity and of transnational Black identity. Both of these influential associates of Pai Francisco continually travel the circum-Atlantic, as Pai Francisco himself now does. Yet he has spent decades negotiating an ambiguous relationship with the spirits of 19th-century Indianist nationalism (i.e., the *caboclos*), the gods of early 20th-century African purism, and the flow of the North American–inspired Black nationalism into Brazil during the 1970s. In the 1990s, though, transnational environmentalism appears to have opened up opportunities that could elevate Pai Francisco to even loftier heights of prestige and access to state resources.

As a male priest, Pai Francisco began his life struggling, and he continues to struggle against the now-entrenched elite view that the Candomblé possession priesthood, and therefore the office of chief priest, is the preserve of women. His maleness appears to have incapacitated him in the pursuit of state largesse as much as it has capacitated his less lucrative pursuit of commercial opportunities. What has remained evident throughout Pai Francisco's rise in the world of Candomblé is that the preponderance of the financial resources that made it possible came primarily not from the sort of Euro-Brazilian bourgeois sponsors discussed in the recent antiessentialist ethnographies of Brazil but from Afro-Brazilians of his sacred family and from Pai Francisco's own earnings as a state employee and businessman.

On his mother's side, Pai Francisco was born to a long line of Candomblé initiates. His great-great-grandmother was an African captive, a daughter of the rainbow and snake god Oxumarê, and, by Pai Francisco's inference, probably a Jeje.[13] She was the one who reared his grandmother, Vovó Crispina; thus, he knows little about his great-grandmother, except that she too was a slave, and a daughter Obaluaiê—an avatar of the god of pestilence, Omolú. His grandmother Crispina was a knowledgeable and initiated women—an *ekedi*, or nonpossession priestess, affiliated with the house of Pai Procópio. Though Pai Francisco's maternal grandfather was a "white man with kinky hair" (*branco de cabelo duro*), it was, according to Pai Francisco, because of his graham-colored mother's "black blood" (*sangue negro*) that Pai Francisco could not get away from the Candomblé. Pai Francisco's biological father was also a white man, but as far as Pai Francisco is concerned, the thunder god Xangô is his real father.[14] A daughter of the sweetwater goddess Oxum, Pai Francisco's mother had been initiated in an Angola house led by a woman who received a *marujo*—the hard-drinking, womanizing spirit of a sailor who had died in war. A *marujo* is, in equal parts, a *caboclo* spirit, an Exú (the Nagô messenger god), and an *egum* (or spirit of the dead; see Santos 1995:126–31). The *marujo* of that Angola priestess guided Pai Francisco through much of his youth.

Though the young Pai Francisco was devoutly Roman Catholic—frequenting the church, praying daily, and helping to celebrate the Mass—he suffered ominous misfortunes. He almost died when he was born, and he swelled up at seven months. After every recovery, he suffered a relapse. At age seven he started kicking the walls; he infers in hindsight that he had probably seen an Exú—lord of communication and mischief—and become so energized with fear that he managed to kick down the wall of his bedroom. At fourteen he almost died in a car accident. The car was bearing down on him and braked virtually on top of him. Good Catholic that he was, he screamed, "Save me, Mother Mary!" When the car stopped, Pai Francisco felt the presence of death. That night, when he went to sleep, he dreamed of a man telling him, "You have some nerve calling for the help of that woman [Mary]. You're lucky I didn't take you away [i.e., kill you]." The man Pai Francisco saw speaking in his dream was the Exú that is seated (*assentado*) on his altar to this very day, as a "slave" (*escravo*) of Pai Francisco's Xangô. Now that this Exú has been properly "treated" (*tratado*), "seated," and "cleansed" (*limpado*), he serves Pai Francisco, too. Beforehand, however, he used to get next to the young Francisco (*encostar nele*) and make him misbehave. When his mother then beat him, Francisco would run out and take revenge by throwing a stone at the first person he saw or hitting the person with a stick. As Pai Francisco recounted these stories in November 1987, I observed aloud that his age on these occasions of affliction regularly corresponded to numbers sacred in the Candomblé. For example, the ideal intervals of key initiatic events, or *obrigações*, are, from the time of ritual birth itself, three months, one year, seven years, fourteen years, and so forth. He denied, however, that this parallel was meaningful.

He told me of other afflictions but could not remember his age at those times. In the worst case, the soles of his feet were covered with watery blisters. He could not even put his feet on the floor and so had to be carried everyplace. One day, the doctor would remove the blisters and medicate the foot. On the next, the blisters would return. His mother took him to Saint Isabel Hospital every other day for six months, until she finally gave up and said, "If those feet are going to rot, they'll rot; if they're going to get better, they'll get better," whereupon his feet healed and never blistered again.

Another time, on the Day of Saint John, the thorns of a plant pierced both of the young Francisco's knees, which then swelled to the size and shape of a hammerhead. After several ineffective treatments, the doctor prescribed surgery. Instead, the *marujo* spirit of his mother's *mãe-de-santo*, or chief priestess, ordered that red palm oil be applied, and the knees healed. Pai Francisco says he always had a lot of faith in that *marujo*. Red palm oil is both the foremost condiment in African-Bahian cuisine and a metonym of all Candomblé practice.[15] On another occasion, Francisco's abdomen started contracting, as though someone were squeezing it. When he shouted, "Help! Help! Save me, Marujo!" the contractions stopped. Another time, Francisco ate manioc flour along with two different types of banana—*banana de àgua* and *banana de prata*—and then swelled up. The doctor diagnosed the problem as something called *hidrofésia* and concluded that Francisco would soon die. Despite many such predictions of death, Francisco would always get out of bed, completely healed, on the very day when death was expected.

Francisco's mother frequently took him to centers of Kardecist spiritism and to Candomblé houses in pursuit of treatment, but never during a festival—evidently for fear that he would get possessed and have no alternative but to undergo initiation. So she would take him only to the temples of people she knew, where there was no danger that they would initiate him forcibly, and she never let him go alone. Though his father was an *ogã* (a sponsor or non-possession priest) of a temple, he shared the mother's preference that the boy never be initiated into the possession priesthood. According to Pai Francisco, his father was ethnically

"Portuguese," though of Brazilian birth and citizenship. He cared little for his numerous and scattered children. But he did care enough to threaten to beat Francisco if he joined the Candomblé. He said he did not want his son in the Candomblé because "every man who enters the Candomblé turns into a faggot [*dá para viado*]. . . . every man who gets possessed by the saint is a faggot." It hurt Pai Francisco's feelings that his father so misunderstood. In fact, Pai Francisco is not a *viado*. It is just that, according to Pai Francisco, few men got possessed in those days; men were more often *ogãs*, or non-possession priests, who assisted in animal sacrifices, kept order in the temple, or sponsored its activities.[16]

Pai Francisco reflected further that maybe his family's resistance arose from their wish to protect him. He was a very attractive young man with "fine" features, which he himself used to admire in the mirror. Some of our common friends have told me that Pai Francisco's African ancestry is evident in his swarthy skin and strong wavy hair. However, I could not easily have distinguished him from some Iberians, had I not already known his medium-brown mother. Though now he is, in his own words, "ugly and fallen-apart [*caído*]," people found him pretty. Otherwise, his family might have allowed him to be initiated earlier.

When it came to Candomblé festivals, curiosity got the best of the young Francisco, despite his family's trepidations. He had long planned to attend the festival at Pai Paulo do Brongo's temple, which Pai Francisco could hear from his house. He went and positioned himself inconspicuously, lest someone report to his mother and get him spanked. At Paulo's, when the drum rhythms for Xangô known as *ilú* and *alujá* started playing, Pai Francisco felt something odd, as though something were lifting him off the ground, and he got scared. He screamed, "God, help me!" He called for the help of God, the *marujo* spirit, and all the *santos* (i.e., gods) connected to his mother; those were the only ones he knew. Once he recovered his senses, he bolted. Fortunately, his mother never heard of the episode. She would have beaten him for sure.

Pai Francisco attended the Maria Quitéria School and the Anísio Teixeira School. When he was about thirteen, during the entrance exams for Brazil *Gymnasium*, (Ginásio Brasil) that Francisco met his friend Oscar, grandson of a *babaláwo* diviner who repeatedly crossed the Atlantic from the mid–19th century to the mid–20th century and profoundly influenced Candomblé practice. After the exam, Oscar spoke to him. They agreed that they had both done poorly. Oscar offered reassurance, "If you did badly, you'll try it again," thus launching many years of counsel to Francisco. The next day, by coincidence or by divine will, they showed up at the exam center at the same time and ended up talking at length. For years, Oscar was Francisco's only friend. Throughout the subsequent three decades, they taught each other and arranged innumerable opportunities for each other to grow in the priesthood. Yes, Pai Francisco does believe the saints, or *voduns*, give friendships to people. Despite the sort of tensions that trouble any friendship of this length and depth, Pai Francisco and Pai Oscar still support each other at critical moments.[17]

Though he had long suffered, the afflictions apparently troubled Francisco ceaselessly after age fourteen, prompting his mother and grandmother finally to consider initiation for him. He was taken to the Angola temple of his mother's *mãe-de-santo*, where the priestesses performed various cleansing rites (*ebós*) to tide him over, but they refused to initiate males in that house. His grandmother then checked out two houses run by men, neither of which apparently satisfied her. She then took him to a house run by a woman she knew named Jerônima, who, it turned out, was not initiating anyone in those days. People thought that she had initiated her own biological daughters, but it turned out that they had really been initiated at the Terreiro do Omolú temple. Thus, in 1960, Pai Francisco came to meet Mãe Vitória and her people. Francisco's grandmother, Vovó Crispina, was satisfied with the Terreiro de Omolú.

Why Mãe Vitória's house? Because his family "wanted the house of a person of respect, that did not have a bunch of men, a house of great respect." Pai Francisco says that his grandmother did not fear homosexuals as much as she feared his entering a house where they did not know much about *santo*. Perhaps some explanation of Pai Francisco's admission to the Terreiro de Omolú, a Jeje temple, lies in the 1983 statistics: among self-described Quêto, Angola, Jeje, and Caboclo houses in Salvador, the Jeje houses were the most likely to feature male chief priests (Santos 1995:19). Bahian Jeje houses were at one point far more receptive to men in the priesthood than is suggested by Mãe Nicinha's 1986 testimony in the newspaper.

The Terreiro de Omolú was then located in Sussuarana, which is now in the center of Salvador, but with the state of public transportation in those days, Francisco and his family had to walk a long way to get there. After they had sat and rested, the elderly priestesses performed divination. They "looked to see what it was and what it was not." Meanwhile, Francisco's family waited tensely. Would the Terreiro de Omolú take him? Mãe Vitória and her people hemmed and hawed. They were clearly ambivalent, which Pai Francisco found humiliating. Would they initiate him, or would they not? They rushed to perform some *ebós*, cleansing sacrifices to remove evil or unwanted influences. With these various *ebós*, they managed to delay the pending crisis, and it took nearly three years after their initial meeting for Mãe Vitória and her people to take any major step on Pai Francisco's behalf.

Though his family resisted, Francisco's youthful indiscretion forced them to act. When Pai Francisco was fifteen, his curiosity drew him to a

Candomblé festival, which he attended without his family's knowledge. There the *santo* unexpectedly grabbed Francisco, who then "flipped" (*bolou*). His midair somersault was an unmistakeable sign of the god's calling. Following established practice, the priest claimed Francisco, by straddling his supine body, and sequestered (*recolheu*) him for initiation. His grandmother Vovó Crispina—in Pai Francisco's words, "the father of the family"—despised the idea of such a man's claiming her grandson. Since Francisco was a minor, his grandmother simply called the juvenile court to get him out. State officials could be useful in projects beyond their understanding. Pai Francisco has never been back to that temple, but, he confesses at times, what happened was good for him. He had to be initiated, or he would have gone crazy. If it had been left to his family, the conclusion to this story would have been much worse.

However, at age sixteen, Francisco was still uninitiated and troubled. His grandmother considered sending him to Rio de Janeiro to stay with an aunt, in the hope that the problem would go away. But on the day of his planned departure, Francisco started trembling and lost consciousness. Vovó Crispina's plan was foiled, but something still had to be done (R. Pires 1992:57). So Francisco's grandmother finally persuaded Mãe Vitória to rescue the boy. At age sixteen, Francisco "made the saint" at the Terreiro de Omolú, after which he spent practically a year on the temple grounds, or *roça*. For a time, Francisco's afflictions ended. In the years following the initiation, his elders at the Terreiro de Omolú discovered his virtues and made a point of his continuing to progress in the religion. He was always the first to arrive and the last to leave on the occasions of sacred obligations.

In the meantime, Francisco completed courses in general pathology and accounting at Brazil *Gymnasium*, but he discovered that nursing was the profession that paid in those days. So, at age twenty-one, he went to study at an important local hospital, receiving the second-highest score on the final exam in the nursing course. For fourteen difficult and ungratifying years, he practiced nursing at local hospitals, at times in the shadow of Dr. George Alakija, psychiatrist and son of African-Bahian traveler Maxwell Alakija.[18]

Francisco made two important friends during his nursing career. One was his psychiatrist friend Jaime, who now practices in the town of Ilhéus, Bahia. Because Jaime is not only a friend but also an *ogã*, or non-possession priest and sponsor, of Pai Francisco's house, they frequently exchange visits, referrals, and services. The other is Dr. Rubem Amoroso, another psychiatrist. Dr. Rubem is an affiliate of several African purist temples in Bahia and an *ogã* in Pai Francisco's house, as well as a generous, but occasional, financial contributor there. Both Dr. Jaime and Dr. Rubem are white.

It was when Pai Francisco reached twenty-three years of natural age and seven years in the saint that the next struggle began. As was his habit, he again resisted the calling, and it almost killed him. Every time Pai Francisco's Xangô arrived during a festival, he would say that Pai Francisco must become the head of his own temple. But Pai Francisco himself did not want to do so. Moreover, the officials of the Terreiro de Omolú temple were afraid of such a young man's assuming office and not taking it seriously. So Pai Francisco's *mãe-de-santo* had him buy some okra, and she prepared a sacred dish called *beguiri* (from the Yorùbá *gbègìrì*, a type of bean-based stew), which she then set way up in a sacred Lôco tree. So Xangô stopped his demands for a year and a few months. But the demands started again, and Francisco fell ill. He suffered from repeated five- or ten-minute-long fevers for two whole days. His psychiatrist friend Dr. Jaime was then in training at a hospital for contagious diseases. Suspecting typhoid fever, Dr. Jaime took Francisco to that hospital, where he stayed for ten days without showing any further symptoms. Nonetheless, the doctors refused to discharge him.

People were very worried at home. Francisco's family went to Mãe Vitória's biological daughter Clarice, who divined for them the source of the problem. She said, "I'm not seeing any sickness. What I'm seeing here he already knows very well. It's his father [Xangô] who's doing this. He [Francisco] doesn't want to be a *pai-de-santo*, right? We can't do anything. He doesn't want to take Xangô home [i.e., take his personal Xangô altar home from his *mãe-de-santo*'s house]." Clarice continued, "We don't want to grant [him independence either], but Xangô wants it. We can't do anything."

Dona Joana, who is now an *equedi* (nonpossession priestess) in Pai Francisco's own temple, had helped Pai Francisco's mother to rear him. One day she bribed a guard at the hospital to breach the quarantine and let her in. Thus, she delivered the news that Xangô was causing his affliction. Or, as Pai Francisco usually explained the mechanism of such callings, Xangô withdrew his protection and gave Exú free reign to upset the victim's—that is, Pai Francisco's—life. So when Pai Francisco finally left the hospital, his first stop was at Mãe Vitória's kiosk, where she sold medicinal herbs and her own home-roasted manioc flour daily.[19] She, in turn, sent him to her biological son Augustinho and daughter Clarice for them to fix the date of his ceremony of independence, or *decá*. At the time, Mãe Vitória was old and so did nothing without contacting her children. Pai Francisco paused to remember her:

She was a very pretty black woman [*preta muito bonita*]. Really very pretty. She had features like a—as you don't like for me to say—she had white features. Those very fine features, that sort of thing, that cute little tiny nose. She had pretty lips. A clean, clean, clean skin. She didn't have a blemish, that skin. You would touch her and feel that softness of skin. She was a very pretty black woman [*negra muito bonita*]. She died [extremely old]. Until [old age] she didn't have a single wrinkle. She had perfect teeth. She read the newspaper without eyeglasses.[20] She died [extremely old] with complete lucidity.

When Pai Francisco turned ten years old in the saint, the saint again revealed that he wanted Francisco to be a *pai-de-santo*, or chief priest. "Everybody knew that I had a road to follow, everybody knew . . . that I would become a *pai-de-santo*." But for a long time, Francisco did not want to. Nonetheless, everyone who divined for him said the same thing—that he was destined for great things.

Pai Francisco often remembers something that the *marujo* spirit of his mother's *mãe-de-santo*, or chief priestess, told him. Once, when Francisco had already been initiated, he fell down on the way to the *mãe-de-santo*'s house. When he arrived, the *marujo* said clairvoyantly, "Don't fall down again, because the sons of Xangô do not fall. Your road is long, and there are a lot of people for you to take care of." From then on, Francisco never fell down again. Sometimes he loses his balance, but he never falls. He remembers the *marujo*'s words and secures himself.

Francisco underwent two rites in preparation for his role as a chief priest. The first was called the *obrigação de sete anos* (seventh-year ritual), in which an initiate ceases to be an *iaô*, or junior initiate (i.e., a "wife," "child," or "slave" in Bahian Nagô, depending on the translator), of the god and becomes a senior initiate—*ebomim* (in Nagô) or *deré* (in Jeje). In truth, few initiates perform this ceremony on time. For one thing, it takes time to assemble the costly, and often imported, materials involved. If one was initiated in a group, or *barco* (boat), of several people, one is expected to wait for them as well. In any case, it is, in some houses, considered rude to rush one's own promotion. Pai Francisco underwent his seventh-year ritual when he was eleven years old in the saint.

The second rite is called the *decá*—the conferral of an eponymous "calabash" containing the instruments and substances with which the celebrant will then be able to initiate his or her own *filhos-de-santo*.[21] Pai Francisco received his *decá* at twenty-eight years of natural age and twelve years of age in the saint. The *decá* is the equivalent of a diploma, formally ending the apprenticeship to one's initiator, such that the independent priest can now initiate followers of his or her own. On this occasion, the newly independent priest receives a calabash full of the items needed in the conduct of initiations. The *decá* give its name, *pars pro toto*, to the entire rite of passage.

Pai Francisco's seventh-year ritual and *decá* were performed at his own new temple. That is how Xangô wanted it. Pai Francisco's baptismal godfather had given him the land (Pires 1992:63).[22] What is remarkable is that, in his mid-twenties, Pai Francisco managed to erect the temple building itself, almost to completion, with his own resources. Pai Francisco estimates that the cost of the seventh-year ritual—which required three ceremonial outfits, a chair, and the animals to feed all his saints—was about US$3,000. The members of Mãe Vitória's house contributed some clothes, sacred metalwork for the altars, and sacrificial animals. Mãe Vitória's family continued to assist Pai Francisco's temple with advice and labor and to donate the money for a four-legged sacrificial animal virtually every year. Pai Francisco testified that it was not until his twenty-fifth anniversary of initiation that he himself began to buy the four-legged offering himself. For his part, Dr. Rubem contributed enough money to buy "food, beans, meat, sugar, flour, coffee, light, the bulbs that were missing, [and] the paint" to prepare the house for its inauguration. Pai Francisco's relatives contributed, too (R. Pires 1992:63–65).

With all the remaining expenses, Pai Francisco had no money left to add windows and doors to the temple. Nonetheless, in obedience to Xangô's command, Pai Francisco moved his saints to the new house. Xangô told him to put a banana leaf on the door, and no one would enter. Potential tresspassers would see a door where there was none. Pai Francisco never followed the directive, but no one bothered his altars during the year and several months that the house remained open and unoccupied. Ultimately, Pai Francisco's boyhood friend and transnational traveler Pai Oscar gave him a front door, a rear door, and a front window. Oscar also contributed some of the most important contents of the *decá* calabash. For his part, Pai Francisco did no less than hunt down the plot of land on which his friend Oscar's Quêto/Nagô temple is now located. Their friendship is long and deep, dense with many more acts of goodwill than I could enumerate here.

Two and a half months after his *decá*, Pai Francisco initiated his own first *barco*, or "cohort of initiates," which included his blood sister Maria and one other woman. As is customary, he did so under the supervision of Mãe Vitória. Since then, he has initiated dozens of women and a handful of men. He says he dislikes initiating male *filhos-de-santo* because they are less responsible, less respectful, and far harder to get along with than are the females. Pai Francisco has also performed the seventh-year ritual and conferred *decá* on another dozen or so people who have thus placed themselves under his authority. Half a dozen *equedis* have been nominated by the manifest gods and/or been ritually confirmed in Pai Francisco's house, as have a half dozen *ogãs*, including psychiatrists Dr. Rubem and Dr. Jaime and, most important, José.

José is an accountant and had attended a number of Candomblé festivals elsewhere before a friend took him to one at Pai Francisco's. There, the Oxum manifest in Pai Francisco's mother nominated José as an *ogã*. Some time passed before he confirmed this nomination ritually by having his guardian *voduns* seated in altars within the house, but his emotional attachment to Pai Francisco, Pai Francisco's mother, and Pai Francisco's eldest niece has made him into another family member. He has served as a godfather, or *pai pequeno* (little father), in a number of initiations in the house and has therefore contributed financially to those initiations (R. Pires 1992:92–93). This shining-black son of the warrior god Ogum is now a central figure in both the affections and the finances of the entire house. I am flattered that few of our common acquaintances fail to note the similarity of our appearance. Some, like Pai Francisco, observe similarities of character as well. José's earnings and checking account are constantly involved in everything from the financing of festivals to the purchase of clothing for Pai Francisco's three nieces. During the many years when Pai Francisco owned no car, José's car was needed to bring animals from the market, to convey ritual offerings (*ebós*) to distant highways and crossroads, and to transport Pai Francisco and others to late-night festivals at other temples. Such a close friendship inevitably knows antagonisms, but even at the height of interpersonal conflict between Pai Francisco and José, José bows lower than any other *ogã* to Pai Francisco's manifest Xangô.

As a nurse, Pai Francisco continued working with mentally and physically ill patients for five years after opening his temple. But by 1979, he could not tolerate working with sick people anymore. Over the objections of Drs. Rubem and Jaime, he quit and spent three years with religious services as his only source of income. Then he decided it was wrong to "wait at home for clients to give him pittances," so he took a job monitoring arrivals and departures for the municipal bus company.

In time, the sacred community under Pai Francisco's leadership outgrew his small temple. On 7 January 1986, Pai Francisco purchased a huge plot of land on which another minor temple for Xangô had already sat. He did so with major contributions from an *ogã* and from the oldest biological daughter of Mãe Vitória, Dona Xica. The property cost an enormous sum of money.[23] With the proceeds of the sale of the original property and contributions from the Terreiro de Omolú, Pai Francisco built a new temple (see also R. Pires 1992:66). He, his biological family, and his sacred family moved in on the Day of Corpus Christi, 6 June 1986.

Since the mid-1980s, Pai Francisco has held a job with the municipal electoral commission, in voter registration. However, in recent years, state austerity measures intent on ending inflation so reduced his rate of salary increase that Pai Francisco had to seek other means of earning a living. Since the late 1990s, he has run a series of shops vending Candomblé supplies and tourist souvenirs. Thanks to Dr. Rubem, Pai Francisco's sister Maria obtained training and a job as a nursing assistant. Thus, she is able to rear her three daughters. Two of her daughters are now teenagers and have been assisting Pai Francisco in his shops. Far and away, Pai Francisco's earnings are the most important source of support for the sacred and secular functions of the house.

Dantas (1988) and others have represented implicitly white and culturally alien *ogãs* as indispensable sources of sponsorship and as the force behind major changes in Candomblé practice. On the contrary, this highly successful young temple, its finances, and its ritual standards have been shaped foremost by a largely black network of Candomblé insiders, including financially modest black priestesses and prosperous black professionals and property holders. In Pai Francisco's evaluation of any given festival or rite of passage in his house, and thus of his progress as a leader, his measure is not the attendance or approval of an ill-defined crowd of *Brazilian* elites. Rather, it is the attendance of *os mais velhos*, the oldest initiates of his temple of origin, the representatives of the oldest priestly lineages, and of priests closely linked to the transnational ideology of "purism" in the early 20th century—that is, the lines of Bogum, Casa Branca, Gantois, and Pai Oscar's temple. The generosity of Dr. Rubem has undoubtedly been great, but he attends only intermittently. Furthermore, according to Pai Francisco, that generosity has never required him to adulterate the temple's ritual standards and could never do so.

On the other hand, Pai Francisco's black female elders in the 1950s and 1960s were clearly imbued with a sense that female priests embodied the ideals of ritual competency far more than their male counterparts. His white father was the most preoccupied with the nearly universal but untrue conviction that all male Candomblé priests are homosexual, but the women of Pai Francisco's family may have shared the concern that their beautiful son and grandson might have been vulnerable to unscrupulous male priests. These views of the gendered nature of competency and the unrespectability of male priests appear to have arisen only a few decades before Pai Francisco's elders began to propagate and act upon them

Euro-Brazilian society undeniably shapes the language and ideological expectations of Candomblé participants. If anagonization and African purism lie well beyond the determination of the Euro-Brazilian bourgeoisie, the matriarchalization of the Candomblé leadership surely bears traces of the Regionalist dialogue with nationalism and feminism abroad. But a sense of proportion urges the observer to recognize first and foremost the overwhelming complexity, gravity, articulateness, and power of the forms of black agency involved. And these are irreducible to collective folk memory. Nor can we overlook the ways in which the so-called elite patrons of the Candomblé, the bourgeois organs of communication, and the very imagination of the Brazilian nation have been

transformed by the priests of the Candomblé. The Euro-Brazilian bourgeoisie is itself encompassed by a broader Atlantic, Anglophone-centered,
and Black nationalist dialogue in which the Candomblé priesthood is a
major participant. Only by recognizing that dialogue can we understand
why prosperous Euro-Brazilians—as well as Europeans, West Africans,
white North Americans, and Black ones—choose to visit and support
one temple rather than another. This is where I enter Pai Francisco's
biography.

An Afro-Atlantic Dialogue

From the first day that he met me, in August 1987, Pai Francisco thought
I (like many visitors and students in Bahia in the 1970s and 1980s) had
come from Nigeria, which is at least one reason he immediately invited
me into the shrine of the messenger god Exú to witness the annual sacrifice. I am certain, however, that some measure of spontaneous affection
was also among his motives. We have been friends ever since. To most
people in Brazil—not to mention West and West-Central Africa—I look
African, but Pai Francisco was none too disappointed to discover that I
had simply spent a great deal of time in Africa, mostly in Nigeria, where
I had studied the gods of the Nagô-affiliated Yorùbá people.

Pai Francisco has always been disinterestedly kind toward me, but
soon after we met, he asked me to teach him some Yorùbá. Some of it
rang a bell for him, and some did not, though he had spent much time in
the great Nagô temples of Bahia (such as Casa Branca and the temple of
his friend Pai Oscar), where the priests and congregants are reputed to
speak and, above all, to sing in "perfect Yorùbá." Pai Francisco soon
tired of our lessons and set his goals higher. He wanted to go to Savalou,
the African capital of his Jeje-Savalú subnation, which I had told him
was located in the People's Republic of Bénin. He was anxious to visit
there and to recover all that his Brazilian Jeje-Savalú forebears had forgotten or had, out of spite, selfishness, and stupidity, refused to divulge
to their successors before they died.

I could not imagine the immediate means of getting my friend to
Savalou, but I promised to visit that West African town on his behalf
and to find out as much as I could. I also agreed to bring him some
African wild buffalo horns and instructions on how to make African
black soap, of which the normal, imported variety simply costs too
much in Brazil. I quietly strategized on how I would get him plenty of
these and other ritual items, which I knew mainly by their Yorùbá
names—efun (white river lime), osùn (camwood powder), and atare
(alligator pepper). A priestess friend—or actually the lively caboclo

Indian spirit who spoke to me through her body at the priestess's fort-nightly Candomblé "sessions"—sent me off with an order for a whole suit of *pano da Costa*, the costly handwoven Yorùbá cloth worn around the waist, typically in more affordable quantities, by female Candomblé elites. Other priests and priestesses simply sent me off with so much love and generosity that I was anxious to bring them whatever ritual items, photographs, and information I could find. Indeed, their demands, cu-riosity, concerns, and preoccupations profoundly shaped my questions and activities during the month that I spent in Savalou and the twelve additional months I spent in Ìgbòho, Nigeria, in 1988 and 1989.

From one travel fellowship to the next (with more or less extended stopovers in London, Paris, Chicago, Washington, Cambridge, and Williamstown in between), I squirreled away ritual items, photocopies, books, song texts, liturgical poetry (*oríkì*), newspapers, sculptures, vo-cabulary lists, and other information whose value would multiply when I took items to friends in places where they were more sacred and less easily available. I have not yet been able to get Pai Francisco to Savalou, but his several university-sponsored trips to lecture in the United States have, besides teaching my students what I could not teach them and bringing greater depth to our friendship, added some small height to his already enormous stature within the Candomblé priesthood.

Pai Francisco has thus extended a dialogue that began in the mid–19th century, with Brazilian Jeje returnee Joaquim d'Almeida's commerce among Brazil, Cuba, and the Coast, as well as numerous 19th-century Afro-Cubans' passage through Brazil en route to Lagos. In the 1980s, this triangular dialogue blossomed in the form of multiple in-ternational conferences of òrìṣà and *vodun* worshipers in Ilé-Ifẹ̀, New York, Bahia, and São Paulo. On his sojourns in the United States, Pai Francisco spent time with *santeros* and received gifts of ritual objects from this Cuban tradition, which he admires for its visual beauty despite unfamiliar aspects of its ritual protocol. He especially admires the dimi-nuitive, concrete-and-cowry-shell Elegguás—the Ocha counterpart of the Brazilian Exús. In turn, commercially imported Brazilian dolls and statues representing the Candomblé gods are now increasingly available in U.S. *botánicas*, the shops where North American practitioners of Ocha purchase their ritual supplies.

As of 2003, I am not the only friend or follower to have facilitated Pai Francisco's travels abroad. His itinerant children-in-saint (*filhos de santo*) have hosted him in Switzerland, France, Italy, and Spain. Nor is Pai Francisco the only person who has benefited in some minor way from my transnational gratitude. Nor am I the only Afro-Atlantic trav-eler to maintain such a network of friends who have some material, in-tellectual, or religious investment in the traveler; I am the successor or the predecessor to millions of such travelers. Indeed, as we have seen, there have been times in Brazilian and West African history when such investments were the basis of many livelihoods and were the fundamental conditions of major religious trends. Since Martiniano do Bonfim, numerous scholars—including Melville J. and Frances S. Herskovits (1934; also Price and Price 1999:168–71); William Bascom (1969, 1972), Pierre Verger (1981), Mikelle Smith Omari (1984), Robert Farris Thompson (1983), and myself (1994b)—have conveyed from contemporary Africa forms of information and persuasion potentially transformative to New World beliefs and practices.[24]

My friendship with Pai Francisco is thus but a minor illustration of a major dialogue that has continually shaped the cultures and politics of the Afro-Atlantic world. Far from being the scattered and fading (or, in a few cases, heroically preserved) remains of "African culture," the cultures of the Afro-Atlantic have continually been refashioned through the voluntary exchange of people, objects, images, and ideas. The modes of musical creativity, the religious practices, the motor habits, and the linguistic dispositions that are cited as evidence of African-Americans' collective "memory" or the "survival" of the African past have continually been reinscribed not simply with meanings specific to their local American or African contexts but also with meanings generated specifically by *dialogue* among Afro-Atlantic locales.

It has long been conventional for authors to preface books with a summary of the circumstances of their research and thanks to its facilitators, but since the reflexive turn in anthropology in the 1980s, it has become obligatory for ethnographers to detail the autobiographical terms of their analytic focus, as well as the daily, concrete conditions of their observations. Such a preface allows the reader to assess the empirical claims of the text in light of the biases that any author brings to the act of observation and in light of how the locally recognized social position of the observer might have shaped what he or she was able (or unable) to see in the field. The disadvantage of presenting such information primarily in the preface is that it allows some readers to believe that the author's biases, positionality, and circumstances of research are somehow prior to and outside of the material observed. I have chosen to describe the cultural processes I study as a "dialogue" precisely because cultural interlocutors are constantly in the process of transforming each other through their exchange of glances, ideas, services, and materials. The processes by which my social role, scholarly reasoning, and personal imagination have been changed over the course of my investigation both illustrate and result from the circum-Atlantic "dialogue" that I now describe.

Thus, the story that I am about to tell does not simply reveal the antecedent circumstances of my research or thank those who contributed it, though it also does these things. Nor does it simply reveal the antecedent biases that I brought to this investigation, though it may do that as well. For example, being a Black person in the Americas, I am more attentive to the contradictions in American nationalisms than to what is consistent and homogeneous in them. And, since my very existence and birth in the Americas depended on translocal processes ranging from five hundred to forty-three years old, I am ill inclined to emphasize the newness of transnationalism; I am congenitally biased, as it were, in favor of recognizing what is old about transnationalism. The following tale illustrates for the reader the historical path of my own anagonization, which in some ways resembles and in other ways differs from the anagonization that Pai Francisco and others have experienced in their lives. One difference is that my U.S. national citizenship has never been sufficiently unambiguous, emotionally affirmative, or central to my social status for my anagonization to cause me much pain.

The conditions of my entry into this dialogue with Brazil are in many ways the story of the anagonization of an entire class of Black American intellectuals who, since the Cubist appropriation of African artistic forms and the Harlem Renaissance, have regarded African art as an appropriate emblem of Black racial dignity and pan-African unity. Since the 1960s, Yorùbá art in particular has been the most extensively published and praised of sub-Saharan aesthetic traditions. And, on account of the 1959 exile of numerous Cuban practitioners of Ocha from their island home, the Yorùbá-related òrìṣà/orichas have also been the African gods most available for ritual incorporation into Black North American cultural nationalism.

My personal anagonization began before I was born and accelerated inexorably. I was born, literally and metaphorically, at Howard University, in Washington, D.C. (no less a capital of international Black convergence than Harlem and Lagos in their respective epochs). Founded in 1867 to educate those recently liberated from U.S. slavery, it became in the 1950s, the "Capstone of Negro Education" not only in the United States but also internationally. Howard has hosted West African and Afro-Caribbean students since the 19th century, but its internationally *central* role first emerged alongside that of the Harlem Renaissance, when Howard University philosopher Alain Locke anthologized the major writings of that literary movement. Like the Harlem Renaissance itself, Locke's anthology included authors who had converged on Harlem from Africa, the Caribbean, and culturally diverse regions of the United States. Locke's collection of African art has, through its generational cycles of storage and reappearance in the university's public space, influenced generations of Black artists and thinkers. Since World War II, Howard has hosted tens of thousands of students from the

Caribbean and Africa, including the late Dr. Olubadejǫ Adebǫnǫjǫ, the Ìjèbú-Yorùbá man who introduced my parents to each other when they were all undergraduates at Howard.

My sister Yvedt, who ultimately became a surgeon like my father, took Robert Farris Thompson's African art history class at Yale University in the 1970s, and I attended with her on more than one occasion. Her further studies and artistic endeavors took her to Ghana and Nigeria on multiple occasions, following a path well traveled by African Americans since the 1960s. I, too, would follow that path after college, on a Rotary Fellowship, and in my graduate study of anthropology. Yorùbáland in particular became the holy of holies in my family's post-1960s, bourgeois nationalist African-American pilgrimage route, much as postgraduate pilgrimages to Paris and London had long been and remain rituals of racial and class self-affirmation for our white class peers. On most of our journeys, our parents' fellow Howard alumni have hosted and oriented us. Ultimately, Dr. Adebǫnǫjǫ stood in for my father during my wedding, when I married my wife, Bunmi, who comes from the Èkìtì-Yorùbá town of Ìgèdè.

My anagonization, however, took place as much at the hands of Cubans, Puerto Ricans, and Brazilians as of Nigerians. Dra. Aleyda Portuondo, an Afro-Cuban professor of Spanish language and literature at Howard University, counseled me in times of trouble and taught me about her religion during my undergraduate vacations from Harvard College. After college, when I taught at Phillips Academy, Andover, I found my best friends among the *santeros* of the neighboring, working-class town of Lawrence, Massachusetts. Since then, I have come to understand much of my daily experience, in religious and nonreligious settings, in terms of the vocabulary of human character and causation elaborated in the Yorùbá-Atlantic family of religions. That vocabulary is particularly important to understand in connection with my incorporation into the transnational networks that unite these religions.

I am a son of Ogum—Ogum Onirê, to be specific.[25] He rules my head and molds my personality. He makes me strong like steel. Because of him, I am a pathbreaker, ever ready to invent or organize something new, to look at things in a new way. I detest administrative duties and the excessive quotation of classics. Let me see and judge with my own eyes. Ogum is the god of war and iron. Elsewhere around the Atlantic, his avatars Gu, Ogún, Ogoun, and Oggún (or Ogún) are therefore credited with command over technology and revolution. Like my *orixá* father, I like the color blue and the number seven, but my Ogum is closely connected to Oxalá, elderly god of purity, peace, and wisdom. That is why my Ogum bears the surname "Onirê"—because Oxalá is the second in a team of gods who rule and protect me. His color is white, which often conveys the idea of spiritual essence when worn by black Atlantic people. Beyond that, it looks electric next to our skin. Brazilians often associate Oxalá with Jesus Christ.

Thanks to Ogum, I am a warrior. It is not my first instinct to hold my tongue in the face of untruth or injustice, great or small. However, the Oxalá behind my Ogum blunts my spirit, cools my tongue, and sometimes manages to keep me out of a disagreement altogether, though not as often as I would like. I do not like to fight, but as long as I stay black in the Americas, I am destined to do so regularly. And, indeed, the sons and daughters of Ogum are usually not only feisty but also, at least in Brazil, luminously black like anthracite. Wherever we go, we are known by a further virtue and flaw—we are workaholics.

On account of Sobô, the god who rules his head, Pai Francisco is generous, self-indulgent, epicurean, portly, domineering, and, by his own description, lazy. If this last adjective is valid, it applies only outside the sphere of his priestly duties, which are onerous and gladly assumed. They range from leading the numerous all-night rituals of sacrifice and initiation that characterize Candomblé, to an insistent, hands-on involvement in the lives of dozens of family members, scores of initiated "sons-" and "daughters-in-the-saint" (*filhos-* and *filhas-de-santo*), and hundreds of spiritual advisees. Annually, Pai Francisco organizes a half dozen lavish festivals (*festas*), where the public is invited to honor and appreciate the beauty of the *vodum* gods manifest in their human horses. The word in Salvador is that Pai Francisco's own manifest god is "the most beautiful Xangô in Brazil."

I met Pai Francisco at the time when his fame was taking off. He had just moved into an enormous temple compound in a neighborhood famous for its history of black self-assertion, organization, and cultural innovation. In the 1980s, the redemocratization of Brazil and the Brazil-wide media coverage given to the funeral of priestess Mãe Menininha of Gantois established Candomblé as an object of deference by the public and the state. Since then, the architecture of Pai Francisco's temple—like his priestly reputation—has undergone yearly expansion and renovation.

International networks, divine lucre, and a *pedido de Exú* (a request to the messenger god Exú) were what first brought me to Pai Francisco's house in 1987. Yemoja was the goddess who first called me to the study of Yorùbá-Atlantic religion. Though in Nigeria she rules a river, Cubans (who call her "Yemayá") and Brazilians (who call her "Iemanjá") credit her with mastery over the sea. News of the enormous New Year's Eve festivals for Iemanjá on the beaches of Rio de Janeiro had persuaded me to seek her first when I began my Nigerian researches. Moreover, as a child of the Atlantic slave trade, I could not but cross the sea on my journey home. Yet it has been the daughters of Ọ̀ṣun (in Nigeria), Oxum (in Brazil), and Ochún (in the Cuban diaspora) who have first met me on every continent and assisted my comings and goings. The year was 1981, and the first òrìṣà worshiper I knowingly met was Howard University's Dra. Portuondo, a daughter of Ochún—lovable and prodigious goddess of honey, sweet water, love, amber, coral, and gold. The Cuban legend has it that she smeared her body with honey to lure Oggún out of the forest and into the service of civilization. Where Dra. Portuondo led, I could not but follow. I would not have found my way without access to some small part of the wealth her *oricha* patroness commands. International travel is costly.

In October 1986, I met Fernanda in New York City, at the Third International Congress of Orisha Tradition and Culture. Like other Brazilian daughters and sons of Oxum, she is beautiful and honey-sweet, as well as a self-possessed connoisseur of male beauty, with primordial ties to the people of Oxum's ex-husband, Ogum. From then until nearly a year later, when the National Science Foundation enabled me to see her in Brazil, Fernanda kept me boyishly dumbstruck. Undeterred, she took me by the hand and led me where I most needed to go. Her friend Lúcia, daughter of the stormy goddess Iansã, is perhaps Bahia's most outspoken advocate of the *caboclos*, or Indian spirits. Though they possess people like the *orixás* and enjoy high esteem in the larger, whiter Umbanda religion of São Paulo and Rio, the *caboclos* reputedly draw contempt from the most Afrophile and most quoted priests of the Candomblé.[26] In New York, I had also met another daughter of Oxum, the beautiful and serious Teresinha, who, upon my arrival in Bahia, initiated my rapport with the world of the "great" and well-established Nagô houses.

Yet it was Fernanda and Lúcia who led me into the intimate world of the small but "serious" houses founded in the present generation, houses where the gods of Africa defined absolute value, but the *caboclos* serve a survival function that not even the most ambivalent priest could deny. The *caboclos* require little paraphernalia or personnel to put on a good show or to render healing, advice, and service on short notice. In September 1987, Lúcia had been invited to attend Pai Francisco's annual sacrifice to the mischievous lord of communication and sexuality, Exú—or, in Jeje, Leba. In turn, she broadened the invitation to include Fernanda and me. Never one to favor politeness over adventure (son of Ogum that I am), I forged ahead, undeterred by the laconic and somber expression with which Pai Francisco inspected me upon my arrival. For reasons then unknown to me, he invited me into the inner sanctum to witness the sacrifices up close, leaving Lúcia and Fernanda outside. Only hubris and much craning of the neck had previously allowed me to witness sacrifices at the "great" multigenerational

Nagô houses, where the principle of initiatic seniority ideally places noninitiates around the corner and at the back of the crowd during such high ceremonies.

In fact, despite the appearance of unambiguous hierarchy in this sacred world, the cognoscenti know that a little hubris is necessary even for insiders to get ahead (see Rodrigué 1991:219). Pai Francisco later told me that he suspected guile when, during the sacrifices, Lúcia and Fernanda suddenly became possessed by their female Exús, or Pombagiras, and these manifest goddesses demanded entry into the shrine. Never having entered Pai Francisco's new Exú shrine building, Lúcia and Fernanda had finally engineered a way to see it from the inside. The visual beauty and power of Candomblé shrines is an object of great pride and interest, but the appearance of unlimited access would besmirch the reputation of the entire temple community. To call a temple exclusive, or "closed" (*fechado*), is the highest encomium in Bahia.

Pai Francisco also told me later that, in exchange for that year's enormous sacrifice of roosters, guinea fowls, and he-goats, he had asked his Exú to bring him a companion. That day, two people showed up, following Exú's usual modus operandi, uninvited and unexpected. Pai Francisco was not sure which was meant to fulfill his prayer, but, as is also normal for the proceeds of deals with Exú, both of these companions ended up causing the supplicant a lot of trouble.

After a long absence, one person who showed up that September day was Clara of Iansã, an attractively thirtyish white widow. In the subsequent months, she visited often. Her pubescent son—the mischievous and affectionate Adilson—became a permanent denizen of the house. Like me as a child and like my own son, he delighted in teasing the people he loved. Also, like most sons of the hunter god Oxossi, he was light skinned and strikingly beautiful. His spindly forms were not destined to fill out into the handsomeness of my friend Chris Dunn—who, though born in Mount Kisco, New York, and reared in Evanston, Illinois, was also destined to join lives with Pai Francisco and me in Brazil. He, too, is a son of Oxossi, and I sometimes see Adilson when I look at him.

My son, Adu, also shares in the unmistakable beauty of Oxossi's sons. But he is doubly beautiful, exceptional in the glistening chocolate blackness of his skin. By 1995, I was no longer a layman to this genre of divinatory knowledge. When I saw Adu as an infant, I knew that he belonged to Oxossi. Pai Francisco independently confirmed my judgment upon consulting with the cowry-shell oracle, of which he is widely reputed to be a master. My Adu is, exponentially, a son of Oxossi, born a magical three years after Adilson's death in 1992, and three years after our familial sojourn in the city where Adilson was killed. My wife, my daughter, Ayǫ, and I had spent February, March, April, and much of May

1992 in São Paulo. We had not known that Adilson was there with us, but weeks before setting off for Bahia, we received a call from Pai Francisco telling us that he had died near us, felled by a policeman's bullet. Even when his body falls, the spirit of the hunter remains to protect us.

Besides Adilson's mother, Clara, I was the other unexpected guest on the day of Pai Francisco's offerings to Exú in 1987. Like the close companions of many Candomblé priests, I have, over the years, received much generosity, assistance, and affection, but I have also been a lightning rod of hostility from Pai Francisco's followers and detractors alike. Some opined that he shared too much information with me and not enough with them. Some resented the affection I received and they did not, or Pai Francisco's use of their resources to make me comfortable. Some envied his trips to the United States as my guest, as well as many more important forms of personal devotion that I reserved for him. Our friendship has at times cost Pai Francisco dearly. The first hint of trouble actually took place in another temple but involved both Adilson's mother, Clara, and me.

One day in October 1987, I had set out for Mãe Stella's esteemed Ilê Axé Opô Afonjá temple, but a wrong bus dropped me on the doorstep of Pai Frederico's Ilê Iansã temple, in the Largo do Retiro neighborhood of Salvador, where I subsequently became a frequent guest. It turns out that Clara was also a frequent guest there and was romantically involved with the chief priest. One afternoon, when Pai Frederico's household, including me, was setting out for a festival at another priest's house, Clara showed up late. Annoyed, Frederico proceeded to berate Clara, blame her lateness on a dalliance with Pai Francisco, and impugn Pai Francisco's masculinity with every venomous word in the Bahian lexicon. Pai Frederico knew that both Clara and I were close to Pai Francisco as well, so it was difficult not to suspect that Frederico's tirade was at least partly directed toward me. Either fear or good judgment silenced Clara, but fifteen minutes of forbearance was as much as I could manage. Both Ogum and the child of privilege in me spoke up—unstrategically in this extremely hierarchical setting. I informed Frederico that Clara had the right to go where she pleased; she was neither his daughter nor his daughter-in-the-saint (I was also indirectly clarifying the terms of my own independence). I added that Pai Francisco had done nothing to deserve Pai Frederico's opprobrium. At that point, I liked Pai Francisco and Pai Frederico just about equally, but Frederico's talk had so enraged me that I could not hold my tongue. Though irate, I thought I had disciplined myself into a few reasonably respectful words and a calm tone of voice. So when Pai Frederico stopped shouting, I thought the matter was over.

Upon arrival, Frederico, as he had promised, asked permission for me to film this festival of Oxalá (the god of purity) and conferral of the *decá* (i.e., the calabash diploma of ritual independence) upon the celebrant. When I changed videocassettes, Pai Frederico's son-in-saint Raimundo volunteered to hold the used one for me. Before I had finished the next one, some alleged breach of ritual protocol at the festival caused Pai Frederico to fly into a rage and storm out of the festival with his entire entourage in tow. I was so perplexed by Pai Frederico's performance that I forgot to recover the used cassette from Raimundo before we parted company that evening. When I went to his temple to collect it a few days later, Pai Frederico shrugged his shoulders and declared through a disingenuous grin that he had no idea where it was. As far as he knew, Raimundo had never taken it. Raimundo cast his eyes to the floor. I was so disturbed that Frederico could lie so casually and think so little of destroying our new friendship, despite whatever might have happened the night of the festival, that I left weeping over our great mutual loss.

Unfortunately, the cassette that Pai Frederico had stolen included a lengthy interview with Pai Francisco, and the label specified its contents. The interview was wholly biographical in substance and in a new 8mm video format that would have made it impossible for Pai Frederico to view it. However, I learned days later from Clara that Frederico had begun telling everyone in his limited circle of contacts that he had confiscated the tape because Pai Francisco had been "revealing all the secrets of the religion." Fortunately, Pai Francisco gladly conceded me a new and longer interview on the same matters. Now comrades-in-arms, we began a friendship that has now lasted for eighteen years and will undoubtedly last forever.

Whatever further trouble I might have caused Pai Francisco, I cannot apologize for my presence in his life or his presence in mine; neither of us would have had it any other way. But the least I can do here is to vow not to cause any further injury by betraying the secrets of his religion. Pai Francisco consistently refused to divulge religious secrets. Of course, his was not the only temple where I observed and learned much about Candomblé ritual and lore, but I am so closely identified with him in that world that he might be blamed for any lapses on my part. I have therefore chosen not to divulge any information that would be considered secret according to Pai Francisco's high standards. Fortunately, a thoroughgoing analysis of the transnational and gendered qualities of this dynamic tradition remains entirely possible without the divulgence of such secrets. I can only assure the reader that my words are in every way consistent with the secrets I know or have inferred.

Since September 1987, Pai Francisco has been my friend and guide, helping me to see myself and experience the world in a new light. Until we met, most Brazilians in the know had identified me as a son of Oxalá, with Ogum second-in-command of my head. I would thus be a son of Oxaguiã. Since then, however, Pai Francisco and every diviner trained by him has independently given the same reading as he: Ogum Onirê. My Ogum stands in front. Over the years, Pai Francisco has remained certain that I was incapable of being possessed by any god (although I have often wished to know such loss of self-awareness and self-control), but he has vacillated in his will to initiate me into other official duties in his house. The plan of Pai Francisco's gods to nominate me as an *ogã*, or quasi-priestly sponsor and official of the house, was preempted by my nomination at the temple of his friend Pai Oscar. Yet Pai Francisco remains my primary spiritual adviser, partly because his *vodum* father and his office give him the very princely social skills, political judgment, and enjoyment of administration that I aspire to possess.

I am guileless and often, when it comes to judging other people's complex motives, clueless. In the 1970s and early 1980s, the Maret School and Harvard College weaned me on the "Forrest Gump," white-liberal doctrine that sustained goodwill, simplicity, and honesty will prevail in the end. An articulate and skilled heir to both Mediterranean and Afro-Atlantic political legacies, Pai Francisco came ready to teach me a different and more Machiavellian lesson about how the world works for those not born to privilege. Through both words and ritual deeds, he has protected me from others' malice and repeatedly cleansed me of dangerous "influences" (*influências*)—both other people's envy and the discarded substance of others' misfortune, which is so easily and accidentally picked up on the street.

With the aid of his cowry-shell oracle, Pai Francisco has counseled me on job prospects and on the physical illnesses of my loved ones. With both his hands and his heart, he has even nursed me through a several illnesses. Pai Francisco had never met my mother when she fell ill in 1987, but his divination revealed her to be a daughter of Iansã—the goddess of wind and storm and the wife of the royal thunder god Xangô. Knowing my mother's warrior-like character, her taste in jewelry, and the autumn tones of her favorite clothes, I would, in hindsight, have surmised the same. When my daughter, Ayọ, was an infant, Pai Francisco divined that she belongs to Xangô. Indeed, ever since she learned to speak, even American adults with no knowledge of Candomblé have, on account of her dignified bearing and self-direction, spontaneously described her as a "queen." When Pai Francisco divined that my wife, Bunmi, belongs to Oxum, it came as no surprise because we had informed him that the Nigerian Yorùbá river goddess Ọṣun is said to have originated from Bunmi's hometown of Ìgèdè-Èkìtì, Nigeria, and is worshiped there during times of crisis. However, Pai Francisco's divination did cause me to notice for the first time that most people who meet Bunmi immediately describe her as "sweet," much as they did the late Mãe Menininha of Gantois, another daughter of Oxum and the most famous Candomblé priestess in Brazilian history. Menininha is also undoubtedly the most famous òrìṣà priestess in the history of the world. News of her funeral in 1986 commanded two columns of print in the *New York Times*.

Cuban myths have it that Oggún and Changó are rivals for the affection of Ochún; she left the one for the other. That is why, say the Cubans, Oggún and Changó tend to fight when they possess people during the same ceremony. Pai Francisco does not necessarily believe this Cuban story, but it helps me to understand why he and I sometimes argue more than we should. I have also wondered aloud if we fight because of that Exú that he once said a jealous priest had sent to attack me because of my greater loyalty to Pai Francisco. Through a dramatic sacrifice called an *ebó*, Pai Francisco cleansed me of that Exú back in late 1987, but we still had regular arguments during the first half decade of our friendship. We argued about whether Brazil is racist (he usually avows that it is not). We argued about whether Candomblé can be "purely African" (he is insulted when I question or deconstruct the claim). He used to get angry that I am not sure that I believe in invisible spirits, and all the more so when I would answer the queries of other priests honestly. It also took him some time to stop bringing up the crabbiness to which I subjected him during his 1990 visit to me in the United States; perhaps I should not have invited him while I was writing my dissertation. It's just that I didn't want to miss the opportunity that became available then.

The irony is that, while I seek mythical explanations for our fighting, he proposes a more mundane one. At times, he has pointed out that he takes all my questions and challenges as indications of a single unresolved question. He wonders whether I care for him. If I do not befriend him *in order* to seek mystical protection, *in order* to get rich, *because* I believe, *because* his religion is "purely African," or *because* Brazil is a racial paradise, then why *would* I seek his friendship, visit his country, or study his religion? The preceding words must give some hint of the answer to all of these questions, ambiguous and insufficient as any answer must be. My love for Pai Francisco, for the Candomblé, and for Brazil are bigger than any particular purpose, cause, or virtue, and it endures beyond the lapse of any particular purpose, cause, or virtue.

Our friendship demonstrates, microscopically, certain broad truths about group loyalties and imagined communities that social scientists must leave for poets to explain. We anthropologists content ourselves to illuminate the historical, material, and cultural conditions that make relationships possible, confer strategic advantage, and favor certain feelings. Our university-style analytic models are not altogether different in purpose from Candomblé's sacred personality theory, which constitutes a "model of and model for" (Geertz 1973) social relations in the world (indeed, as should now be obvious, I find the latter quite persuasive). But the structural conditions highlighted by such models pale before the affective substance of sacred brotherhood, parenthood, and wifeliness. Such affect regularly outlives the structural conditions that made it possible and the material interests that might once have made it profitable.

PAI FRANCISCO'S *CABOCLO*: AMBIGUITIES OF RACIAL AND NATIONAL IDENTITY

The last time I met Pai Francisco's *caboclo* spirit—Pena Grande, or Great Feather—was at Pai Oscar's 1992 *caboclo* festival, where Pai Francisco had not planned on getting possessed. When Pena Grande called me for a chat in the back room, he told me without malice that he does not like black people (*pretos*), just Indians (*vermelhos*). Instead of taking it personally, I have processed it primarily as data for my hypothesis that Indians and Africans remain rival symbols of national identity in both the Candomblé nations and the Brazilian one. My impression remained unchanged when, during his visit to the United States in April 2003, Pai Francisco clarified that Pena Grande meant to say not that he dislikes black *people* but that he dislikes the *orixás*. Moreover, I've convinced myself that friendship with Brazilians and with most non-*estadounidenses* requires a North American to relativize his own etiquette of talk about race.

For me, Pai Francisco's *caboclo*, sometimes celebrated and sometimes hidden, also speaks to the ambiguity of Pai Francisco's own racial identity and strategy for priestly success. Indeed, Pai Francisco's *caboclo* and his "purely" African gods embody the racialized tension between Brazilian indigenism and a black diasporism that now has as much to do with the North American Black Power movement and the belated independence of the Lusophone African nation-states as with Jeje and Yorùbá trans-Atlantic commerce. Deftly handled, the symbols of both Brazilian nationalism and those of black transnationalism yield potentially valuable but different sorts of symbolic and material affirmation to a man of Pai Francisco's complexion.

From World War II until the 1970s, black transnationalism ebbed as *mestiço* nationalism flowed in Brazil. But the 1970s and 1980s witnessed renewed Black assaults on the nationalist aesthetics of *mestiçagem* and Indianness in Bahia. For example, from 1975 to 1978, the Ilê Aiyê Carnaval club made a splash by declaring publicly that no whites would be allowed to join the club, whose name in the sacred Yorùbá language means "life" or "the world." Because racial classification in Brazil is subject to a range of context-specific phenotypical and social indices, the organization devised its own objective admissions test: if the candidate's skin did not turn ashy white when scratched, he or she was pronounced officially white and therefore ineligible for membership (Antônio Carlos dos Santos [Vovô], personal communication, 13 July 1992). Bahia had long been accustomed to segregated *white* Carnaval clubs, since the social logic of *mestiçagem* had always prioritized the dissolution of black distinction over the disappearance of whites and their supremacy. Thus, the avowed exclusivism of an *Afro*-Brazilian club marked a new social and psychological phase in the history of Bahia and of Brazil generally.[27] It reflected the fast-spreading transnational influence of the North American Black Power movement and black Lusophone pride in the late victory of the Angolan and Mozambican movements of independence from Portugal.

Since 1983, the antisyncretic and anti-"immediatist," or antimagical, discourses with which Opô Afonjá's Mãe Stella dominated the news coverage coincided with her private but widely rumored discourses of contempt toward *caboclo* worship.[28] Unlike the similar sentiments of the 1930s and 1940s, the anti-*caboclo* movement of the 1980s and early 1990s was so hegemonic that it emerged in print primarily in indirect forms. That is, virtually every anthropologist who wrote about Candomblé during this period felt obligated to note that, despite well-known rumors to the contrary, even the famous Quêto houses worship the *caboclos* a bit. They might not do so publicly, or they might forbid members of the temple from doing so on temple grounds, but those members probably visit smaller houses to carry out their obligations to the *caboclo* spirits (e.g., Henfrey 1981; Santos 1995:23). In other words, the acknowledgment or praise of the *caboclo* in the 1980s and 1990s seemed to scholars and journalists like a subversion of established order.

Pai Francisco and his friends have told me many stories of a time when he served the *caboclo* Indian spirits far more than he has done during our acquaintance since 1987. His *caboclo* regularly held sessions in his first temple and danced at the temples of other priests. Such devotion is characteristic of small houses, which naturally have much to gain by attracting numerous clients in need of the immediate help that *caboclos* give. The year that I first met Pai Francisco, however, he had recently moved into his new temple. There, he had built the outdoor *caboclo* shrine at the back of the house. And he held no *caboclo* festival that year. Since I have known him, Pai Francisco has devoted little attention to his *caboclo* spirit and has all but disavowed him verbally, periodically denying any knowledge of how to worship or take care of a *caboclo* at all.

Ilê Aiyê sings of transnational Black pride and privileges people with dark black skin. Such a policy takes a sword to the *mestiço*-nationalist

conviction that Brazil is a "racial democracy." It also implicitly under-
cuts the credibility of Pai Francisco's claim to African purity. The following discussion of the contrast between these two great male-headed
Bahian institutions is based entirely on my own observations. Indeed,
what Pai Francisco most emphasizes is the importance of love among
people of all colors, and so he refuses to criticize even the racial sepa-
ratism of the institutions that practice it.

By the time I met him, Pai Francisco exemplified the value of "African
purity" and had entered the pantheon of Salvador's dozen most celebrated temples. He remains among the few males in that company, and
they all lag far behind the female chiefs of Casa Branca, Gantois, Opô
Afonjá, Bogum, and Alaketu in both their national name recognition
and their access to the state's largesse.

Moreover, Pai Francisco is the lightest-skinned chief priest among
them all. Though most Brazilians would recognize him as white, he occasionally avows that he is black (*negro*) and prominently displays
photos of his black grandmother and black *mãe-de-santo* in his public
ceremonial space, or *barracão*. Notwithstanding his color, or perhaps in
part because of it, African purism on religious matters was an important
dimension of Pai Francisco's self-presentation in the late 1980s and early
1990s. He wanted to learn Yorùbá, to travel to Savalou, and to convince
me and anyone who might doubt it that his Candomblé is purely
African. While he is willing to admit that the dresses of the dancing
priestesses and gods do not look very African, he insists that he and like-
minded priests are purely African in the shrine room (*peji* or *quarto de
santo*) and the initiation chamber (*roncor*). Every icon, gesture, and
ritual in the temple's holy of holies is, he swears, African. The rest is
decoration (such as the beautiful festival costumes) or mementos of
beloved forebears (such as the chromolithographs of Saint John and
Saint Sebastian that Pai Francisco inherited from his grandmother and
displays in the private living quarters of the house). However, even in
the public festivals, the songs performed by Jejes and Quêtos are to Pai
Francisco proof positive of these nations' Africanness.

He vacillates, however, on the issue of whether he is Roman Catholic.
He enjoys attending the Mass, and in fact sponsors Masses in honor of deceased loved ones, but no more than at the Afrocentric Opô Afonjá temple
do Pai Francisco's initiations require the subsequent visits to Catholic
churches that many Candomblé temples require or once required.

While Pai Francisco shares Gilberto Freyre's reservations about Black
racial nationalism, he is not precisely a Brazilian nationalist. Like most
Brazilians I know, he detests São Paulo and its residents in principle,
though he does not go as far as those who regard *paulistas* as "foreigners." Pai Francisco's friends in Rio tell me that he complains endlessly

about the people and the food of that city, too. Pai Francisco himself verifies that, for him, there is no place like Bahia. Whenever he is away, he wants to go back. Far more than Brazil itself, Bahia is Pai Francisco's country. He has said so both inside Brazil and on visits to me in the United States.

Whatever his ambivalence toward Africa, Pai Francisco has long wished to travel there—and hopefully first to Savalou, his ancestral *terra*. So he abandoned the ineffective Bahian Federation of the Afro-Brazilian Cult (successor to the union established under Édison Carneiro's inspiration in the 1930s) and once affiliated himself with the most Africa-centered cultural organization in Bahia, with the possible exception of Stella's Opô Afonjá—that is, Argentine Juana Elbein dos Santos's Society for the Study of Black Culture in Brazil (SECNEB), which hosted the organizing sessions for the National Institute of Afro-Brazilian Culture and was reputed to have secured airfare for priests and other Bahians to attend international conferences of òrìṣà worshipers, such as the 1986 International Congress of Orisha Tradition and Culture in New York City, where I first met Fernanda and Teresinha of Oxum. In 1987, Pai Francisco became actively affiliated with the National Institute of Afro-Brazilian Tradition and Culture (INTECAB), which in turn came to rely on him to enlist new members. The fact that he is personally charismatic and leads a small but respectable house within a small nation makes him an attractive spokesman to the "great" Quêto/Nagô houses that INTECAB most wants to attract. The choice of Pai Francisco as the Bahian state coordinator of the international conference committee avoided giving the appearance that any *particular* "great" house or large nation dominates the organization. Pai Francisco's help notwithstanding, the international coordinating committee of these conferences is now terribly divided, and SECNEB's international sponsors are less willing and able to fund travel than perhaps they once were.

Pai Francisco's temporary engagement with SECNEB and the range of other institutions and personalities to whom Pai Francisco matters help me to understand why, in the mid-1990s, much to my surprise, Pai Francisco resumed public festivals for his *caboclo* and denies ever having neglected them. It seems that trial and error taught him not to keep all his eggs in SECNEB's re-Africanizing, transnational basket. Now Pena Grande's shrine sits in *front* of Pai Francisco's house—not as elaborate as the shrine building of the wind and storm goddess Iansã, not as central as Iemanjá's new shrine or the new fountain featuring a mermaid-like representation of the sweet-water goddess Oxum. These shrines, too, dot Pai Francisco's front yard. But the *caboclo* shrine has now been removed from its backyard obscurity and promoted into their company. Its white walls, decorative wrought iron gate, and sacred tree are visible from the front gate and all the way down the cascading entry stairs of the compound.

Pai Francisco's changing degree of attention to his *caboclo* is, to me, a metaphor of the priest's own socially ambiguous position. He is a man whose income, complexion, and features qualify him to be white in Brazil. North American–inspired Black nationalist discourses, which privilege dark and self-described "Black" people in the membership and leadership of Afro-Brazilian cultural institutions, threaten to turn Pai Francisco's racial advantage in Brazilian national society into a disadvantage within the transnational religion to which he has with great sacrifice committed his entire adult life. Yet his dark family background, his hard-won priestly pedigree, and the spirits that guide him also keep him in the running for the leadership of the struggling Jeje nation. Thus, the greatest obstacle to Pai Francisco's success is not his color but his gender.

Entrepreneurialism versus State Largesse: The Gendered Structure of Opportunity in Pai Francisco's World

Contemporary Regionalists (and others concerned to represent Candomblé as a pristine and premodern root of their own imagined communities) take great offense when a great Black Mother is caught engaging in commerce. For example, when a São Paulo advertising firm paid the Gantois temple's Mãe Menininha thousands of dollars to advertise Olivetti typewriters as Mother's Day presents, a rival advertising company and several scholars who study Candomblé pitched a fit, declaring that the innocent Mãe Menininha and the culture of the "folk" had been exploited. Thereafter, an Umbandista federal legislator threw fists and feet into the attack on the guilty firm.

Meanwhile, the male-run Bahian Federation of the Afro-Brazilian Cult laconically avowed the dignity of Candomblé, observed the irrelevance of typewriters to it, and said it would not punish Mãe Menininha—as though it had any power to do so![29] One male priest from Rio simply added, "If at least it had been a bar of soap, an incense burner, an herbal bath, perfume or any product linked to the religion, then it would be understandable."[30] Thus, the male priests who run the federation (and the priests leading small houses) are far less determined to regard this religion as a Romantic idyll—as innocent, as an inversion of everything adultlike in their own lives—than are the bourgeois Euro-Brazilian advocates of the Candomblé. Priests who admit that they practice magic or demand payment for their professional services are spared no invective by Bahian journalists.[31] Those journalists seem more comfortable with government donations to the "great" houses, as we saw in chapter 4. On the other hand, it is not surprising that newspapers in highly industrialized and business-oriented São Paulo (which exports vast quantities of its manufactures to other states) would show a lively but respectful interest in the range and cost of Afro-Brazilian religious supplies, as well as in the demand for and cost of Candomblé's magical services.[32] With some logical consistency, Candomblé in São Paulo and nearby Rio is more lavish and more male-dominated than in Bahia. At least in the 1930s, the national journalistic community thought it was more ridden with "black magic" as well (see Landes 1947:183).

Menininha's ultimate silence on the consenting nature of her business deal with the São Paulo advertising company kept her house eligible, following her death, for far greater dividends—costly renovations financed by the municipal government and the state petrochemical industry—than Olivetti was offering. The "great" Quêto mothers have made a decision, conditioned by both their intelligence and their limited opportunities. They have gained financial and symbolic capital for their institutions by downplaying their own entrepreneurialism. This strategy has worked significantly for the mothers of Casa Branca, Gantois, Opô Afonjá, and Bogum, perhaps more than it could have or did for any male priest.

Only Pai José of Bate-Folha might be considered a runner-up to these women's success in the arena of attracting state largesse.[33] Equally exceptional among the Northeastern priestesses, Mãe Stella of Opô Afonjá has felt free to engage in small commerce. During the late 1980s, she had two shops for the sale of Candomblé supplies, one of them on the grounds of the temple. Mãe Stella stands atop so prestigious a legacy of "African purity" that, as we saw in chapter 4, she feels free to be selective in her acceptance of state and bourgeois largesse.

The male priests most successful at attracting a following and enriching their houses generally did so in Rio or São Paulo. Pai Caio in São Paulo built his Candomblé temple after prospering in his own meatpacking business and converting from Umbanda. Having worked a "white-collar job . . . in the schools" (Landes 1940:397), Joãozinho da Goméia moved from Bahia in 1948, with his *caboclo* Indian spirit Pedra Preta, to Duque de Caxias, Rio de Janeiro, where he presided over the rapid expansion of Candomblé there and in São Paulo up until his death in 1971. Then there is Waldemiro Costa Pinto—"O Baiano" or "Waldemiro de Xangô"—of the small Efã nation.[34] He has long traveled, conducting business among Bahia, Rio, and São Paulo, and maintains temples in the first two cities. This sort of mobile commerce is especially common among male priests.[35] Some men—such as Pai Joãozinho da Goméia, Pai Luiz da Muriçoca, and Mestre Didi dos Santos—added to their fame by cutting phonograph records of sacred music.[36]

Thus, with little chance of attracting largesse from the state or major corporations, the male priests made virtue of necessity. Their commercial wealth simply enhanced their reputation among their Afro-Brazilian followers and initiates. Moreover, they remained free to seek secondary authorization from the "great" female-headed *axés* as well. For example, Pais Caio, Joãozinho, and Waldemiro have increased their symbolic capital by undergoing post-initiatic rites at the hand of, or claiming other sorts of affiliation with, Mãe Menininha of Gantois, as did the Bahian Jeje priest Vicente de Ogum.[37] Similarly, in the 1930s, Pai Bernardino of Bate-Folha and Pai Procópio of Ogunjá enhanced their prestige by seeking the frequent company of the "great" Quêto mothers (Landes 1940:396–97).

Other male priests, such as Luiz da Muriçoca, have entered the limelight by providing articulate commentary to numerous journalists.[38] And, with the assistance of French photographer and ethnographer Pierre Verger, the Bahians Didi dos Santos, of the Egungun priesthood, and Balbino of Xangô were able to undertake symbolically invaluable and career-enhancing trips to Nigeria and the People's Republic of Bénin (Santos 1988[1962]:34–35). Some have undergone initiations or coronations by Nigerians visiting or living in Brazil. For example, a newspaper reported the double-barreled exaggeration that, in 1975, "the king of Candomblé in Nigeria" had crowned the president of the Candomblé Federation of São Paulo State as the "king of Brazilian Candomblé."[39]

In this generation, the male priest most successful at combining material prosperity with the emergent media standards of respectability among the "great" Quêto houses is the grandson of transnational merchant and *babaláwo* Felisberto Sowzer and great-grandson of Báŋ́gbóṣé— Pai Francisco's friend Oscar of Oxalá, whose prestige owes as much to his impeccable pedigree, ritual competency, and dignified personal bearing as to his business sense.

PRIESTLY CORPORATISM: MEN'S RESIDUAL DOMAIN

While the matriarchs of the "great" Nagô houses benefited individually from state largesse, most women's temples trudged along in obscurity and in submission to the patriarchs of the Bahian Federation of the Afro-Brazilian Cult (Federação Baiana do Culto Afro-Brasileiro). Founded in 1946, the federation replaced the earlier Union of the Afro-Brazilian Sects of Bahia (União das Seitas Afro-Brasileiras da Bahia) and shared the union's mission to guarantee rectitude in priestly practice and to protect its members from police persecution.

Winning the approval of the Bahian state, the federation became a public utility in 1960. However, in 1967, the state governor declared the Candomblé a form of folklore rather than a religion, thus exempting it from certain constitutional protections. With that justification, he placed the religion under police supervision. The priests organized no official protest until 1973, when the Casa Branca temple founded the Confederation of Candomblé for that purpose. This move reinvigorated the federation and persuaded Bahia's Governor Santos to liberate Candomblé from police regulation in 1976 (Barbosa 1984:70–71). Some priests, however, announced their preference for police regulation, for fear that the federation would attempt to dominate previously independent temples.[40]

With the lapse of police regulation, the federation indeed vastly expanded its authority.[41] In 1979 it was made responsible for the inscription of Candomblé priests in the social security program (Barbosa 1984:71). Though the federation was not a regulatory arm of the state, it used the newspapers to convey the impression that it was, publishing every new federation decree and suggesting that the police would shut down violating temples. Like federations and unions of Afro-Brazilian religions in other states, the Bahian Federation forbade the deposition of *ebós*, or "sacrifices," in the street.[42] The federation also began to register and tax the *baianas de acarajé*, or street vendors of a traditional bean-fritter snack, and regulated other sales of Candomblé-identified foods.[43]

Federation activities and pronouncements harkened back to the 1930s and 1940s, when President Getúlio Vargas incorporated civil institutions into the government as instruments of regulation, representation, and patronage. Yet, in the spirit of the post-1964 Brazilian dictatorship, the federation conveyed to its members much more regulation, taxation, and threats of violence than representation or patronage. It is not clear how much power the federation ever had to back up its threats to close down temples or have priests arrested. What is clear is that the center did not hold. Once the federation attempted to bar children's participation in nighttime rituals, Mãe Stella of the Opô Afonjá temple simply withdrew her temple from the federation, and did so with impunity. Between 1977 and 1979, various other "great" Quêto/Nagô houses followed her lead.[44]

Between the bean-fritter sellers and the priests of the smaller temples, the vast majority of the federation's constituency is female, and many are terrified of the powers they attribute to the federation. Yet, the offices of the federation have remained almost entirely in male hands. Amid Menininha's extraordinary fame, the male officials of her Gantois temple and other temples of the same *axé* monopolized the leadership of the federation from at least the early 1970s until the beginning of 1983.[45] In sequence, an *ogã* (or male nonpossession priest) from the Gantois temple and then one from the Casa Branca temple, both "great" Quêto/Nagô houses, served as presidents of the federation in the 1970s. In the mid-1980s, the presidency passed from a succession of

Gantois's *ogãs* to an Angola father-in-saint, who reigned until 1994. Though the president in 1995 was a father-in-saint from a small Nagô house, almost all the other board members were Angola men, and much of the federation's activity from the mid-1980s to the 1990s was aimed at resisting the monopoly of the "great" female-headed Nagô temples over official and public recognition.[46]

The federation still avows authority over a wide range of issues. Officials report that the federation has even patented the term *axé* (meaning "amen" or "ritual power"), which years ago became generalized as a hip way of saying "Right on!" in urban Brazil. Now, say federation officials, no one outside the religion can use this term without official federation permission.[47] If only it were so easy!

Though many smaller houses fear the federation, the ferocity of its roar has grown with the weakness of its bite. The authority of the federation is, with some consistency, not only male and largely Angola but also marginal. Nonetheless, it benefited from the significant largesse of the Bahian state in 1994. By the authority of Governor Antônio Carlos Magalhães, who of course took personal credit, the federation received the gift of a beautiful building for its new headquarters in the historic Pelourinho neighborhood and a loan of more than $127,000, only 40 percent of which has to be repaid.[48]

Pai Francisco briefly served on the executive board during the 1980s but withdrew because of the absence of any evident material benefits. The federation did not channel revenue, facilitate travel, or enhance access to what he needed. Pai Francisco has ultimately found greater opportunities in commerce and, potentially, in the transnational environmentalist movement.

A RISING STAR?

The gendered minefield of post-Landean Bahia presents many constraints and a few opportunities to a priest like Pai Francisco. He tried the federation, a zone of male authority, but ultimately saw no material benefit. Nor did the African purist Society for the Study of Black Culture in Brazil facilitate his transnational travel as he had hoped. And, although his initiatic pedigree and ritual competency guaranteed his fame and stature among African purist priests, the Ilê Aiyê Carnaval club and a wave of its prosperous, young sympathizers resurrected transnationally inspired forms of *racial* purism that, in principle, delegitimized people of Pai Francisco's complexion. Thus, even as Mãe Stella resurrects the anti-*caboclo* discourse of the 1930s black elite, Pai Francisco has rehabilitated his *caboclo* spirit. In Pai Francisco's small temple, there remains the possibility that the immediacy of the priest's intervention in people's lives will yield greater benefits than does Pai Francisco's now-contested claim to purely African origins.

The only attention he receives from politicians, he says, has so far come in the form of unfulfilled promises. His international contacts have afforded him greater knowledge of Europe and the United States than of the land that would matter most in the rescue of his struggling nation— Savalou in the Bénin Republic. Access to public funding long depended on the temple's having a female leader, and such leaders were forbidden to engage in commerce. As a male priest, however, Pai Francisco has had nothing to lose symbolically by setting up shop in the new, government-subsidized shopping district in the Pelourinho neighborhood. Moreover, his articulacy and membership in the phytolatric Jeje nation offered him an opportunity to tap into the power of the transnational environmentalist discourses of the 1990s.

During the last four years of her life, Mãe Nicinha, the chief priestess of the "great" Bogum temple, had fought to save the life of the fire-damaged Lôco tree near the Saint Lazarus Highway.[49] Soon after it succumbed, so did she, on 6 October 1994.[50] In 1994, an *ogã* (non-possession priest and sponsor) of the temple named Jaime Sodré proposed that the city replace the tree, a suggestion that the municipal secretary of the environment, Juca Ferreira, gladly incorporated into his beloved Project Gardens of the Sacred Leaves. As Jaime's mother-in-saint and the most respected Jeje priestess in the state, Mãe Nicinha had been expected to conduct the rituals that would make the new tree the abode of the god Lôco. However, with her unexpected death, my friend Pai Francisco was called upon not only to conduct the ritual but also to address the press about the meaning of the event. More than at any point in the previous years of our friendship, he appears to have invoked the syncretic talk of environmentalism and Candomblé with feeling and expertise. The journalist quotes him indirectly:

> The replanting ceremony was performed by *doté* [senior *vodum* priest] Francisco Agnaldo Santos, . . . who, filled with emotion as he placed the sapling in the ground, said that this was an event of great importance for the followers of Candomblé, since the Iroko [the West African Yorùbá spelling] is a sacred tree. *Doté* Francisco Santos emphasized, still, the ecological importance of the replanting, since nature, some years from now will give the population a leafy tree, which will fill the air with more oxygen and serve the practice of works specific to the Afro religions.[51]

Moreover, when quoted, he asked to be identified by the Jeje name for his temple. Among the specific rituals performed at the foot of this sacred tree is the placement of candles and sacrifices. Once resented as a source of pollution, these works were anticipated and accommodated by the new city administration. City engineers built a protective wall and railing to protect the tree from the candle flames, simultaneously, in the words of the secretary of the environment, "guaranteeing space for the arrangement of sacrifices."[52]

Dantas, Henfrey, and others have narrated the conversation between the Brazilian and the African-Brazilian nations as a history of "ideological manipulation" by the "dominant classes," as Dantas puts it (1988: 160–61, 164, 191, 215–16, for example). Yet it is more evident to me that, like the Anglo-Yorùbá purism invoked by priests since the last century, today's priestly environmentalism actually invokes international discourses and transnational networks of power to transform the conduct and historical identities of "elite" Brazil. While the priests of Candomblé are Brazilians and act regularly in a Brazilian political arena, they are also elites in a set of nations that subdivide and crosscut Brazil. The symbolic, material, and emotional power of *those* "imagined communities" clearly supersedes the capricious romanticism and administrative projects of Brazilian nationalist elites. The Bahian Candomblé is a major center of religious and ideological gravity not only within the territorial nation but all across the Atlantic world. Perhaps that is what Mãe Aninha meant when she said, "Bahia is the black Rome."

But there is division within the empire. My friend, a male Jeje priest, had finally gotten his moment in the sun of the Brazilian national media. He displayed himself, his nation, and his religion with grace, just as he had done during his lectures in the United States in 1990, 1996, and 2003. Ferreira's Project Gardens of the Sacred Leaves went further than any previous government project in validating male priestly legitimacy. In the form of commemorative plazas, the project also built monuments to the founders of several "great" temples. The two plazas commemorated Mãe Ruinhó of Bogum and Mãe Pulchéria of Gantois. But the third and apparently best-publicized plaza honored a male priest, whom Ruth Landes's *City of Women* (1947) had recognized in only the most grudging terms—Pai Manoel Bernardino da Paixão of the Congo-Angola nation. Also known as Tata Umbamandezo, he founded the famous Bate-Folha temple in 1916. Naturally, journalists asked the opinion and repeatedly noted the approval of Mãe Stella, supreme matriarch of the equally supreme Nagô nation. She, too, has helped to expand the place of man in the City of Women.[53] Only the mighty can afford largesse. If the next publicly funded Candomblé project broadens its target still further, Pai Francisco's beautiful and highly respected temple near the black Pelourinho neighborhood will be an attractive candidate.

Pai Francisco stands on the brink of fame and media-propagated respect like no other male priest in recent times. However, one fears that

Mãe Nicinha prepared the media well enough for "total matriarchy" to return Pai Francisco, in the end, to the background of his depopulated but highly respected sacred nation.

CONCLUSION

Ethnic and religious phenomena called "Djedji" and "Jeje" have emerged on two sides of the Atlantic during times recent enough for us to document their renaissance and decline. Despite internal fears of this nation's death, Bahia hosts at least two strong scions of this dynamic nation, which necessarily negotiate their fate with reference not only to the resources of the Brazilian bourgeoisie and state but also to the power and prestige of a trans-Atlantic Yorùbá identity, transnational feminism, Black nationalism, and transnational environmentalism. Historically conditioned claims of racial and cultural "purity" are among the most powerful emblems of authority in this transnational dialogue.

By relating the story of one rising Jeje star in the Bahian Candomblé, I have tried to illustrate two other historical phenomena reshaping the Candomblé in the 20th century—the struggle against *caboclo* worship and the symbolically linked marginalization of men in the priestly leadership. In the 1930s, ethnographers reproduced and amplified a novel Bahian discourse scandalously conjoining male priests, homosexuality, mixed race, *caboclo* worship, the Angola nation, newness or lack of tradition, and "black magic"—all in a fictive antitype to Nagô and "purely African" religion. Yet all these symbols of otherness within the African purist nations that inspired Regionalist pride created a "loophole" for the commercial endeavors of male priests, who also stood the least chance of securing state largesse.

By relating the story of my friendship with Pai Francisco, I have tried to illustrate the ongoing role of transnational movements (such as Black nationalism and environmentalism), transnational travelers (such as Pai Oscar, Dr. Alakija, and numerous scholars), and their often multigenerationally transnational families in consequential local dialogues about African "tradition," Brazilian national identity, gender, and race. We have seen that communities are imagined and reimagined and traditions invented and reinvented (Hobsbawm and Ranger 1983) in ways that are not only situational and strategic but also cosmopolitan. These imaginations and inventions in Candomblé have long occurred in a transnational context that shapes local values, constraints, and opportunities. In the nature of the Afro-Atlantic dialogue, there remains an inequality of opportunity for mobility, profit, knowledge, and prestige among the interlocutors, but the wealthier, the more literate, and the more mobile parties are not the only ones with the power to shape the terms of the dialogue. This long-term encounter has deeply anagonized the transnational academy itself.

In the next chapter I will propose a new analytic vocabulary for the study of local lives in such cosmopolitan contexts and will reconsider those analytic terms that have allowed so many scholars to overlook at least two centuries of transnationalism and many more centuries of its translocal precedents.

CHAPTER SEVEN

Conclusion
THE AFRO-ATLANTIC DIALOGUE

CANDOMBLÉ IS A FIELD site defined not by its closure or its internal homo-geneity but by the diversity of its connections to classes and places that are often far away. At the root of this diversity is the universally human capacity of worshipers to imagine their belonging in multiple communi-ties that crosscut any given community, encompass it, or even distinguish themselves from it. Hence, the "survival" of African practices, or "cogni-tive orientations," and the adaptive requirements of white-dominated American institutions are entirely inadequate to explain the shape of Can-domblé. This Afro-Brazilian religion thrives and grows because of strate-gic verbal and ritual performances of assimilation and differentiation, of hybridization and purification, that draw on a circum-Atlantic repertoire of references and are incomprehensible outside the context of the priest-hood's centuries-old and ongoing participation in a circum-Atlantic move-ment of people, commodities, texts, recordings, photographs, and ideas.

Many cultural historians of the African diaspora have assumed that African culture endures in the Americas, if it endures at all, only among the poorest and most isolated of black populations. What I have seen is quite the opposite—that African culture has most flourished in urban areas and among prosperous populations that, through travel, com-merce, and literacy, were well exposed to cultural Others. And they chose to embrace African culture in full knowledge of numerous alterna-tives. Because the most purist and self-consciously traditional "folk" cultures are often taken as proof of local cultural closure and isolation, Candomblé and other similarly African-inspired African-American prac-tices provide perhaps the best proof that broadly translocal and cos-mopolitan fields of migration, commerce, and communication are the normal conditions of human culture and its reproduction.

Candomblé is unmistakably African in its practices, its ritual logic, and the ethnic identities that divide it into denominations. However, Candomblé and its constituent practices are not African "survivals" (Herskovits 1958[1941]), simply enduring in the Americas beyond the death of any obvious social function. Nor is Candomblé, as Bastide (1967) would have it, a "preserved" or "canned" (*en conserve*) African religion, in contrast to "living," changing, and adaptive counterparts like Vodou.

Candomblé certainly results partly from the selective reproduction of African "cognitive orientations" that could usefully be institutionalized in the spaces, within these racially hierarchical societies, where African-American creativity was allowed (see Mintz and Price 1992[1976]). But if the term "creolization" is taken to imply that—following one understanding of creole languages—each local African-American religion quickly formed a bounded, internally integrated, and enduring "system" highly resistant to exogenous change, then a different term must be sought to describe the history of Candomblé, where exogenous transformations have been as central as any quick-forming foundations. The local institutions of the racist American republics have never been big enough to contain the communal imaginations and aspirations of African-American peoples.

This case recommends several lessons to anthropology and African-American studies. The first is that it is less places than translocal dialogues that produce cultural self-representations, which, throughout history, have been occasions of collective self-making. This process is a critical source of situational and longitudinal dynamism in human culture. Second, this case illustrates the fact that oceans, seas, and deserts have long been foci and conduits of such translocal dialogues, and that maritime and desert perimeters deserve as much attention from ethnographers as they have received from historians, art historians, and literary critics. Third, this case calls attention to the analytic metaphors that, unless carefully considered and revised, stand to hide some cultural dynamics of critical concern to both anthropology and African-American studies.

CIRCUM-OCEANIC AND CIRCUM-DESERT FIELDS

Transnationalism is not new. Nor, as Sassen (1998) and Cooper (2001) point out, is it shapeless, embracing all regions equally or simultaneously. For several millennia, desert and ocean perimeters have structured dense webs and literal flows of interconnection among phenotypically, culturally, and ecologically diverse locales. Because they are relatively unpopulated and ungoverned, oceans and deserts facilitate long-distance movement between points of radical difference.[1]

Several historians offer models of economic and ideational flows between pairs of regions, such as Eric Williams's (1993[1944]) pathbreaking work on the role of the Anglo-Caribbean colonies in the making of English capitalism; C.L.R. James (2001[1938]) and Gaspar and Geggus (1997) on the flow of ideas and resources between the Caribbean and France; Walter Rodney (1982[1972]) on Europe's active role in "underdeveloping" Africa; Amitav Ghosh (1992) on the ancient movement of people and goods between Egypt and India; and Verger (1976[1968]) on the commercial and cultural links between Bahia and West Africa. One might add anthropologist Paulla Ebron's (2002) work on the flow of people and music between the Gambia and the United States.

Other scholars have examined the pathways of particular products (such as Mintz [1985] on sugar), of particular multinational corporations (such as Watson [1997] on McDonald's), and of particular peoples (such as Ong [1999] in the Chinese diaspora and Gilroy [1993] on the black English speakers of the United States, the Caribbean, and England) across multiple locales. Long before Harvey (1989), Jameson (1991), and Sassen (1998) sought to outline the processes by which global capitalism is similarly reshaping life in a range of industrial and postindustrial countries and cities, Wolf (1982) outlined how a range of indigenous Asian, African, and American societies were transformed and interconnected by the expansion of European-centered capitalism.

My aims are both more modest and more ambitious. A truly global ethnography is beyond the scope of my competency and, in my view, of the ethnographic genre. Yet the success of a number of historians at documenting the interactive character of multiethnic, transgeographical regions—as Braudel (1992[1949]) described the Mediterranean perimeter; Chaudhuri (1990) the Indian Ocean perimeter; Bovill (1999[1958]) the circum-Saharan world; Bourgeois, Pelos, and Davidson (1996[1983]) the perimeter of the larger Afro-Asiatic "desert crescent"; and Thompson (1983), Gilroy (1993), Thornton (1992), Linebaugh and Rediker (2000), Bailyn (1979, 1996) and Roach (1996) the Atlantic perimeter—poses a major challenge to the ethnographer. How does one render the changing cultural realities and human predicaments of transregions like these, which possess such disproportionately weighty significance in world history? For enormous stretches of world history, flows across such trans- or circumgeographical regions have been as important to world commerce, to the long-distance spread of ideas, and to the rise and fall of local polities as they have been resistant to representation in the discursive conventions of ethnography.

Thus, the setting of this study has been a new kind of ethnographic field site and cultural field, one of the sort structured by the Atlantic slave trade or by the spread of Islam across the great Afro-Asiatic desert, to name just the two most obvious among some very old cases. In such fields, local ethnic identities and cultural canons have almost invariably taken shape in direct dialogue with multiple ethnic Others sharing the same transgeographical field.

If this book has proven anything, it has demonstrated by exhaustive example that Candomblé and the late 19th- and early-20th-century Afro-Atlantic world constituted just as massive a flow of people, in proportion to the contemporaneous population of the world—and just as massive a disjuncture among the constituting flows of "people, machinery, money, images, and ideas"—as does today's "late capitalism." The sometimes hybrid, sometimes reactively purist choices allowed by these disjunctures are the essence of the story I have told about Candomblé.

Major segments of the population of the Cape Verde Islands (Halter 1993) and the West Indies (Reid 1969[1939]) have engaged in fast-moving transnational and transimperial labor migration since the mid–19th century, and not only sent enormous numbers of U.S. dollars back to western Europe's black colonies but also variously invoked their nonnative cultures, languages, status as French or British subjects, and loyalties to European royals to protect themselves from racist assaults in the United States. During the second and third decades of the 20th century, the Jamaican Marcus Garvey enrolled approximately eight million people from the United States, the Caribbean, and Latin America in his transnational Universal Negro Improvement Association, whose ultimate goal was the return of New World blacks to Africa and the establishment of an independent state. He founded a steamship line intent on easing transportation and communication among black Atlantic locales. Long after the collapse of his organization, Garvey's ideas and the symbols of the movement he founded continue to influence Black nationalist movements around the Atlantic perimeter.

In the early and mid–20th century, we also witnessed the similarly massive disjuncture in sub-Saharan Africa entailed by the often-simultaneous influx of western European colonialism and nationalist ideology, Islam, Soviet and American military funding, Chinese capital, movies from India, merchants from India and Lebanon, African students returning from Black North American universities, and African-American music and television programming—all with politically momentous and culturally central consequences for African societies.

And even if one insists on emphasizing that today's flows are faster and bigger than their predecessors, it should be remembered that both the Atlantic and the intra-American slave trade were increasing in speed and volume over the entire course of their existence, and, after the mid–19th century, they were even exceeded by the migration of Europeans to the Americas. Gross quantitative increase does not a qualitative difference make, and there is no reason to discard the lessons of today's transnationalism in our analysis of the past. Nor should it be forgotten that warfare, the rise and fall of economies and polities, the construction and decay of transportation infrastructures, changing technologies, epidemics, and famines have always made the volume and speed of migration and communication rise and fall. Immigration and travel restrictions following 9/11, George W. Bush's protection of the U.S. steel and farming industries, and America's territorial occupation of Afghanistan and Iraq should impress us that teleology has little predictive value even in the present era. "Deterritorialization" and the demise of the nation-state are options, rather than the universal and inevitable trajectories that the theories of transnationalism have tended to argue.

In fact, once anthropologists are as attentive as historians, literary critics, and art historians to the long-running centrality of ocean and desert peripheries to the cultural history of the distant past, we are likely to learn a few further lessons about the present, including the motives behind our own willingness at times to silence the past. For example, scholarly assertions that the current social conditions of the author's home culture are utterly new have been a persistent theme in Western social science since the 19th century, when such assertions served at least two rival functions. Such assertions could either naturalize the hierarchical difference between colonizer and colonized, or be invoked to criticize the current social realities in the analyst's home society.

Today's scholarly prophecies about the recency and inevitability of transnationalism serve similar critical functions: they endeavor to undo the enormous power of certain territorial nations over their residents' lives by pointing out clearly demonstrable gaps in *any* territorial nation's ability to monopolize the loyalty of and control over its residents. Yet these scholarly prophecies also set up an a priori and arguably demeaning contrast between, on the one hand, those who, by reason of class, race, or religion, have only recently come to doubt the virtues of the nation-state and lately pursued options beyond it and, on the other hand, those who, by reason of class, race, or religion, had never been persuaded to invest their entire imagination of community in the nation-state or any other territorial entity. The "newness" of transnationalism is less a careful observation of history than a hierarchical proposition, a temporal trope of present-day social inequality.

I employ the terms "black Atlantic" and "Afro-Atlantic" heuristically, with no intention to exaggerate this super-region's geographical isolation from other ocean and desert perimeters, or from the other "races" of the Atlantic perimeter. Nor is the black Atlantic isolated from the Pacific Rim or the Indian Ocean perimeter, which has supplied numerous important goods, ideas, and icons to the Afro-Atlantic religions—such as the Kwan Yin and Buddha statues with which Cuban *santeros* syncretize the orichas, and the Hindu lithographs that have become common in Beninese Vodun (on the latter case, see Rush 1997, 1999), not to mention the omnipresent Maldivean cowry. Yes, this geographical area is old, dense, and intense with political, commercial, and discursive interconnections within its borders, but it is not hermetic.

Europeans, Euro-Americans, and Native Americans are centrally involved in the Afro-Atlantic dialogue. Their role could not have been neglected in any of the preceding chapters. The Native American presence and image have also been important. For example, one of the most important slave-trading officials of early-19th-century Dahomey was Francisco Felix de Souza, the Bahian-born son of a Portuguese father and a Brazilian Indian mother. His marriages with multiple West African women produced one of the most important commercial dynasties of the 19th-century West African Coast (J. M. Turner 1975[1974]:88–93).

In Brazil, the Caribbean, and the eastern United States, disease, starvation, and murder have reduced socially recognized Native Americans to a small proportion of most national populations. Their demographic presence in the Afro-Atlantic world is further diminished by the norms of social recognition. In these three places, African ancestry regularly outweighs Native American ancestry as a determinant of social status and eligibility for ethnic group membership. That is to say, the millions of people with both African and Native American ancestors are more likely to be classified as Black than as Indian.

Nonetheless, the symbolic presence of Native Americans in the Afro-Atlantic cosmos is powerful. In Bahian Candomblé, where the spirits of the deceased Native Brazilians are described as the "owners of the land," many Candomblé devotees feel that failing to worship such *caboclo* spirits is not an option. Priests specify that the Angola nation worships the *caboclos* most actively because, due to the concentration of both groups in rural Bahia and their cooperation in the flight from slavery, the Angola slaves developed the closest alliances with the Natives. Black North Americans in New Orleans express a similar sense of alliance with Native Americans and so celebrate them with elaborate Mardi Gras festivities. However, Bahian *caboclo* worshipers have also welcomed into their practices 19th-century Euro-Brazilian uses of the Native Brazilian as an emblem of Brazilian patriotism. The space of *caboclo* worship is usually festooned with Brazilian national flags during *caboclo* festivals, and the manifest *caboclos* are feted with beer and a Native American cultigen—tobacco—which has been thoroughly integrated into a range of other Afro-Brazilian and Afro-Caribbean religions as well.

Joseph Roach (1996) cites numerous traditions in which Europeans, Africans, and Native Americans have mimed each other in theatrical performances and thus hybridized, reinvented, and clarified their own ethnic identities in a circum-Atlantic "interculture." I have argued that much American nationalism, even in regions denuded of one Atlantic "race" or another, relies on the semantic presence of all three. For example, the white rebels of the 1773 Boston Tea Party dressed up as Indians as they destroyed an unwelcome shipment of Crown-taxed British tea in Boston Harbor. Mid-19th-century Brazilian nationalists proposed that Euro-Brazilians adopt Tupi as their national language, and, in the mid–20th century, the Brazilian Indian "cannibal" became a major metaphor of the nation in the cultural nationalism of urban Brazil.

The late-19th-century to mid-20th-century tradition of black minstrelsy in the white-dominated nations of the Atlantic perimeter suggests a much more oppositional relationship between blackness and the free, independent white self. In the Dominican Republic, the official Indianness of the nation's dark citizens seems intent on denying the cultural worth and demographic importance of the nation's African ancestry. At least as late as the 1930s, as we saw in chapter 4, some Afro-Brazilians perceived the Romantic Indianism of Euro-Brazilians as an indirect putdown of the nation's accomplished Black population.

In these and other ways, Atlantic racial ideologies have created especially dense webs of transnational interconnection among people of African, European, and Native American descent in this geographical field, the likes of which cannot be ignored in our explanation of the shared religious network that has produced and is being produced by Candomblé, Santería, and Nigerian òrìṣà worship; the network producing and produced by black Protestantism in Africa and the Americas; by Black American missionaries and historically Black American universities and colleges; by Black cultural and political nationalism; by global Rastafarianism; and by the dialogical evolution of black and black-influenced music around the Atlantic perimeter (Matory 1999a; in preparation).

The black Atlantic is a commercial, migratory, and discursive field. It is, simultaneously or in variable, situational alternation, an ethnoscape, an ideoscape, a mediascape, a technoscape, a finanscape (Appadurai 1990), and, to coin a term, a lawscape—that is, a unit defined by the distinctive operation of laws ranging from the maritime conventions of the slave trade, state-sponsored abolitionism, segregation, and exclusion from full national citizenship to passport regulations, the differential regard for different ethnic or religious groups' demands for reparations, affirmative action, and the international copyright laws governing unwritten "folk" and popular music, dance, and ritual objects. And "disjuncture" has always characterized the relationship between the black Atlantic and the nation-state.

However, it should be remembered that the Afro-Atlantic is but one of *many* historically constituted migratory, commercial, and discursive fields in which black Atlantic people stake their claims of belonging. We people of African descent sometimes "imagine" ourselves primarily as members of a specific territorial nation, sometimes as subjects of a particular empire, sometimes as Europeans, sometimes as Jejes, sometimes as Native Americans, sometimes as white, sometimes as middle-class people, sometimes as young people, sometimes as women, sometimes as environmentalists, and so forth. The black Atlantic is thus a geographical focus, an identity option, and a context of meaning-making, rather than a uniquely bounded, impenetrable, or overdetermining thing. The Afro-Atlantic field is one of many historically constituted fields that overlap in Brazil, but it is one that remains indispensable to the understanding of social history and cultural meaning among the African-descended populations of Europe, Africa, and the Americas. And it has been indispensable for five hundred years. Yet we must not mistake the density of the black Atlantic field for any type of boundedness, since the Atlantic has long been a conduit to the densely networked Indian and Pacific Oceans as well. Long before there were telegraphs, telephones, automobiles, radios, televisions, jets, and Internet terminals, oceans and deserts were major media of "time-space compression" (Harvey 1989).

ANTHROPOLOGY'S FOUNDATIONAL METAPHORS

Some analytic metaphors are unsuited to the realities of the black Atlantic and of translocalism generally. This study suggests the greater utility of "rhizomes" and, more important, "dialogue" than "roots," "survivals," and "diaspora" to explain the Candomblé and the Afro-Atlantic cultural transformations that gave Yorùbá culture its preeminence among the West African models of New World black religiosity and artistry. The "rhizome" and "dialogue" metaphors illuminate the similar historical dynamics of music, dance, politics, education, and so forth all around the Atlantic perimeter and other heavily traveled transgeographical complexes, such as the Silk Road and the circum-Mediterranean, circum–Indian Ocean, circum-Saharan, and medieval Islamic worlds.

This scholarly project employs the insight of Lakoff and Johnson (1980) that all perception and thought are mediated by metaphors, which involve the comparison of one domain of experience (the metaphoric "target") to another (the metaphoric "source"). Due to its experiential concreteness and clarity, the *source* helps to illuminate the nature or structure of a *target* that is less concrete and less clear to those who understand it. Any given metaphor, continue Lakoff and Johnson, tends to highlight particular aspects of the target in ways that another metaphor would not, while hiding other aspects of the target that another metaphor would not. For example, representing an argument as a battle (as in the expressions "Your claims are indefensible" and "He shot down all of my arguments") highlights the competitive and mutually destructive aspects of a debate while hiding the cooperative aspects of argument (such as the shared effort toward mutual understanding; Lakoff and Johnson 1980:4, 10).

While Lakoff and Johnson are concerned chiefly with the metaphors that structure daily language and experience, I am keen to address the specialized set of metaphors that structure our *scholarly* language about and perception of reality—"*analytic* metaphors." Like theories, they should be judged not as true or false but as more or less useful in highlighting important dimensions of the realities we seek to explain and more or less parsimonious in the explanations they allow. The empirical phenomenon that I reconsider in this volume is the cultural history of the African diaspora. We are heirs to a family of well-established and productive analytic metaphors in our area of study. While they usefully highlight much that had previously been ignored about that cultural history, they also hide dimensions of that cultural history that are long overdue for recognition. At times, I will argue, they even patently distort the phenomena that they purport to explain. For example, the term "culture" itself represents intergenerationally learned aspects of collective lifeways as something like *cultivation*—the agricultural selection, training, and reshaping of nature for the fulfillment of human biosocial needs. Even diasporas are compared, in the Greek etymology of the term, to *sowing*, or the dispersal of seeds.[2] Anthropologists have also represented human lifeways metaphorically as organisms (as in A. R. Radcliffe-Brown's organic metaphor) and as symbolic systems akin to languages or texts (as in Claude Lévi-Strauss's structuralism and Clifford Geertz's interpretive anthropology).

Such analytic metaphors have been convenient tools in highlighting the *synchronic* functioning of any given society or culture, and they emerge from an early-20th-century anthropology that was generally averse to the speculative histories of armchair ethnologists, evolutionists, and diffusionists. These metaphors propagated the image of the self-contained, isolated, bounded, and internally integrated society or culture. Yet sustaining that image in an era of worldwide colonialism required the erasure of history and the fiction of the "ethnographic present." Whereas British social anthropology typically endeavored to capture precolonial sociopolitical orders (often complementing the colonialist objective of resurrecting them as administrative units), the German-American Franz Boas and the French Claude Lévi-Strauss attempted to archive precolonial lifeways before they were swept away by Western influence and expansion.

Yet the proximity of Columbia's Boas and his students to the Black intellectuals and political leaders of New York City—including W.E.B. Du Bois, Marcus Garvey, and the literati of the Harlem Renaissance—made it inevitable that American cultural anthropology would turn its analytical eye toward the politically central theme of black difference. Yet African-Americanist anthropology remained marginal to the broader discipline. Two reasons stand out. First, from the standpoint of Boas's antiracist anthropology, it was not obvious that African-Americans were culturally different from white Americans. Thus, Boas's influence lent itself equally to (1) Melville J. Herskovits's pursuit of what was historically and therefore remained subtly distinctive about African-American life-ways, and to (2) the argument of Howard University's influential sociolo-gists and jurists that any difference of conduct or values among the races in the United States was but circumstantial, quantitative, and largely the result of unequal structures of opportunity (see Baker 1998). Neither model, however, suited African Americans to the image of absolute Oth-erness that had long dominated anthropologists' choice of their objects of study. Second, both the cultures and the very presence of people of African descent in the Americas were so obviously the consequences of a recent, highly transformative, and ongoing history that an analysis de-void of history seemed not only fictional but also false. African Ameri-cans are collectively the product and agent of one of the oldest, largest, and most revolutionary colonizations in the world. Indeed, it bears re-peating that the triangular trade, on which these colonizations depended, renders absurd the fashionable conviction that transnationalism is new.

Before it was obvious about any other group that anthropologists have studied, it was obvious about African Americans that they could not be an-alyzed in isolation from global political, economic, and cultural forces. Thus, notwithstanding the apparent marginality of African-Americanist anthropology, scholarship on this region has arguably been the sand in the oyster of the discipline, foreshadowing late-20th-century efforts to restore global political, economic, and cultural forces to the portraits of local soci-eties. Early African-Americanists articulated, under these and other names, theoretical themes that have only recently become standards across the discipline—such as "cultural resistance," "alterity" and the "subaltern," "hybridity," "reflexivity," "collective memory," "embodiment" and "habitus," "diaspora" and "transnationalism," and the more general re-thinking of continent-based area studies divisions.[3]

African-Americanist anthropology stands to make a greater contribution still to the discipline, in a way anticipated by Herskovits's "work of comparing cultures in a single historic stream" (e.g., 1958[1941]:15–17; 1966[1956]: esp. 76; 1966[1930]; 1966[1945]). That is, the African re-gions that supplied the slave trade are culturally varied but not infinitely so, and the peoples of African descent in the Americas are culturally varied but not infinitely so. Moreover, many of the political and economic forces that have shaped the difference between African and the African-American cultures, as well as those that have shaped the differences among African-American cultures, are well documented. Therefore, it is difficult to imag-ine a finer terrain for the comparative study of how geography and history shape culture, and of how culture has shaped local imaginations of geogra-phy and history. The dividends of comparativist Afro-Americanist anthro-pology are potentially both empirical and methodological. Indeed, on the basis of a comparison among African and African-American cases, the present study is intended to revise the range of analytic metaphors devoted to the anthropological study of all dispersed populations.

A History of Afro-Americanist Metaphors

When Africa is regarded as part of the cultural and political history of the African diaspora, it is usually recognized only as an *origin*—as a "past" to the African-American "present," as a "source" of cultural "survivals" and "retentions" in the Americas, as an essence "preserved" in collective "memory," or as the "roots" of African-American branches and leaves. Neither Herskovits nor many of his numerous followers have ignored the transformative adaptations, "syncretisms," or "creolizations" that have reshaped African culture in the Americas, but most have focused on the one-way transmission of culture from a past Africa to a present America. Yet this representation of the cultural history of the Afro-Atlantic world is itself culturally, politically, and historically conditioned in ways worthy of study in themselves. What that representation highlights is valuable, but what it hides is often equally important.

From the time of Herskovits, the central analytic metaphors of Afro-Americanist anthropology could not but invoke history and imply the need for its reconstruction and preservation. The conjunction of tropes in which that history was narrated—not only the recently popular trope of "diaspora" but also the older "survival" metaphor—bear the signs of a uniquely Black-Jewish dialogue. In itself, the survival metaphor in Herskovitsian anthropology has roots in E. B. Tylor's cultural evolutionism and suggests that certain practices outlast the evolutionary stage during which they had served some material purpose. The metaphor therefore brackets the question of what present-day social purposes and symbolic meanings such "survivals" embody. It thus suggests that human cultural practices have a life of their own and, like organisms, possess their own will to live on, independently of any particular people's volition to reproduce them or to direct them toward a present-day purpose. Yet in an age and in a nation where collective political identity was increasingly associated with territory and with the new, republican principles defining the role of all citizens, it seems no surprise that an intellectual project involving chiefly Jews and Blacks in New York City would generate models of corporate "survival" that rely on continuities of ritual, musical, and sartorial form.

European and immigrant American Jewish identities had had an ambivalent and, more often, contrary relationship to territory, defined alternately by the group's distance from its ancestral home territory, its alienness within the land of residence, and its exclusion, in many places, from the right of landownership. Rituals commemorating the past became much of the substance of Jewish collective identity and corporate survival. African Americans, too, could be imagined as a landless people, defined alternately by their distance from an ancestral home territory, their alienness within the land of residence, and their exclusion from numerous rights of citizenship within the American territorial nation-state. To an even greater than usual extent, the Blacks of New York City from the second decade of the 20th century onward were refugees from places where they had worked the land and where there were more recent memories of being owned *like* the land than of being landowners. For similar reasons, neither group had reasons for full confidence in the nation-state's much-vaunted promises of "modernity," as Gilroy (1993) summarizes the fundamental dilemma of the African diasporic experience. Yet for both of these groups the option of abandoning one's past, renouncing pre-American identities, and melting into the mainstream seemed to require more struggle, reflection, and debate than it did for *other* nonindigenous ethnic groups in the United States.

Thus, in dialogue with the Black intellectuals of the Harlem Renaissance, Herskovits generated a series of metaphors highlighting continuity against the odds and implicitly acknowledging subaltern agency in the transmission of culture over time: "survivals," "retentions," and "preservations" of not only practices but also "deep-seated cultural orientations" that guided the selection and "reinterpretation" of cultural precedents in the American context. Herskovits borrowed the term "syncretism" from Afro-Brazilianist Arthur Ramos to describe the processes of selection and reinterpretation that allowed African practices and creative principles to "survive" in apparently Western milieus. For Herskovits, syncretism is "the tendency to identify those elements in the new culture with similar elements in the old one, enabling the persons experiencing the contact to move from one to the other, and back again, with psychological ease" (1966[1945]:57). In the classic example, Herskovits employed the term to describe Haitians', Brazilians', and Cubans' use of the images of Roman Catholic saints in the worship of iconographically similar African gods.[4]

In their day, Herskovits's "survival" and "reinterpretation" metaphors helped to highlight what had been overlooked—the reality of cultural continuity and transformation in African-American history. However, one might justly be dissatisfied with Herskovits's and Herskovitsians' relative inattention to diverse contextual meanings of apparently similar signifiers. For example, highlighting the formal similarities among possession by the Holy Spirit in North America, the river goddess Yemọja in Nigeria, and the eponymous sea goddess Iemanjá in Brazil might lead the analyst to overlook the radically different theologies and ritual complexes that buoy them.

Herskovits is far more attentive to psychological and unconscious "dispositions" than to agency and strategy in the reproduction of cultural forms. What, for example, might be any given actor's *motive* (beyond inertia) to reproduce an African cultural form, or to identify such a form as "African"? Such an actor is likely to have alternatives to the African-looking form and might risk reprisal or disapproval for adopting even the most camouflaged of non-Western forms. And he or she might choose to interpret that form as *non*-African in origin. Are antecedent and intergenerational "dispositions" or the desire to hold on to the past (trendily called "cultural resistance" during the 1970s and 1980s) sufficient explanations for the genealogy of African-American cultural choices?

Herskovits's methods have invited the same criticisms that diffusionism generally has attracted—that it renders culture as a "thing of shred and patches" rather than as an integrated, synchronic whole. Indeed, Herskovits deserves more credit for his contributions to the study of cultural *process* than to the study of synchronic issues of meaning, function, and structure. Moreover, as later generations of anthropologists have come to doubt the integrity of any given culture and to recognize the diversity and contradictory relationship among local discourses, it is not clear that Herskovits's critics are more correct than he.

In African-Americanist popular discourse, the most direct successor to the "survival" metaphor is "roots," which implies that social units are like trees, bounded and enduring through time just as the roots of a tree share an ideally uninterrupted boundary and selfhood with the trunk, branches, and leaves that emerge from them over time. The metaphor reifies bounded social units, minimizing the heterogeneity of the historical sources of any given local culture or population and minimizing the degree to which the people who identify with that culture and live within that local population can, when they wish, claim membership in crosscutting cultural and social units.

The roots metaphor is as consistent with the white nationalist vision of "Western" history as it is with the black nationalist vision of "Black" history. "The West" is said to have sprouted from Greece, unfolded in Rome, Germany, France, England, and, ultimately, the United States. This vision ignores the ancient Greek dialogue with the Afro-Asiatic world (Bernal 1996, 1997), medieval Europe's costly and transformative pursuit of Chinese goods and ideas (e.g., porcelain, paper, pasta, the compass, and the printing press), Europe's use of African and then American precious metals in that pursuit, Europe's adoption of Indian numerals as taught to them by Arabs, and Europe's mutually transformative, oppositional dialogue with the Islamic world. No less important is "modern" Europe's use of African labor and technology to cultivate the Americas (Wood 1975; Carney 2001) and to subsidize its own industrialization (Williams 1993[1944]).

The conventional narration of "Western" history also ignores the transformative effects of the Enlightenment-era and 19th-century evolutionist self-construction of the West in contrast to the non-Western Other (whether as noble savage or as primitive inferior). Scholars such as Orlando Patterson (1991) and Susan Buck-Morss (2000) have shown how important the image of the African slave was in white Atlantic articulations of the culturally and politically central definition of "freedom." And William Pietz (e.g., 1993, 1995) traces the history of a discourse first generated in the 16th-century commerce between Europeans and West Africans. He shows that the concept of the "fetish" originated in an intercultural misunderstanding over the nature of commercial debt and, thereafter, became an important trope in a series of influential Europeans' efforts—such as those of Marx and Freud—to understand their own societies in the 19th century.

It was a global dialogue that produced the lexical riches of English as we know the language today, and an enormous economic and psychological subsidy from overseas colonies financed the formation of European and Euro-American nation-states. The narrative of a hermetically bounded Western culture has been useful in excluding nonwhites from the full rights of citizenship and in a range of military projects, including the justification of U.S. involvement in World War I (Levine 1977), the joint Western European and Euro-American resistance to Soviet socialism, the alliance among the settler colonies of apartheid South Africa, Israel, and the United States, and the pursuit of a Euro-American alliance to control Middle Eastern oil. This narrative (like the tree-roots-and-branches metaphor on which it is based) is, on the other hand, relatively useless in the balanced scholarly documentation of the past.

Alternatively, Deleuze and Guattari's "rhizome" metaphor recommends ways of writing cultural history that mimic the network-like roots of grasses, which unite multiple roots with multiple shoots. Deleuze and Guattari (1987[1980]) thus reject monocausal narratives, single-source constructions of group history, and inattention to the multidirectional ramifications of any genealogy of events. Deleuze and Guattari do not so much deny the existence or legitimacy of linear narratives and the "arborescent" (i.e., tree-roots-and-branches) metaphors that constitute them, as mark them out for special explanation: they are the products of hierarchical power and hegemonic intentions. Gilroy (1993) puts the rhizome metaphor to excellent use in charting the cultural dynamics of the Anglophone black Atlantic musical and literary culture.

"Diaspora" shares some of the misleading implications of arborescent metaphors, insofar as the "diaspora" concept suggests that homelands are to their diasporas as the past is to the present. Candomblé and many other African-diasporic phenomena are not simply outgrowths of their homelands but also, and just as important, outcomes of an ongoing dialogue with a coeval homeland. African homelands and diasporas—much like Europe and its American, African, Middle Eastern, and Australian settler colonies—have engaged in a long and influential dialogue of mutual transformation.

Afro-Americanist anthropology is indebted to Sidney Mintz and Richard Price (1992[1976]) for their "creolization" model. Mintz and Price focus on the New World social institutions, organizations, and sanctioned spaces that made the practice of African creative models possible in the Americas, and the situational constraints that transformed them into distinctly *African-American* logics of creative production.

Mintz and Price argue that such newly African-American logics took shape quickly and enduringly—on the slave ships and in the early years of any given American slave society. "The rapidity with which a complex, integrated and unique Afro-American religious system [or, presumably, other cultural systems] developed" and stabilized in a particular American locale is shown to be analogous to the rapidity with which the local creole *language* variety developed and stabilized (Mintz and Price 1992[1976]:22–26, esp. 25). Yet Mintz and Price are ultimately less concerned to argue that all African-American cultures formed rapidly or remained fixed than to propose a search for the precise local *institutional history* by which each one took shape and changed over time (Richard Price, personal communication, 12 April 1999).

What I add to these insights is that the political and demographic contexts shaping African-American cultures are seldom produced through a once-and-for-all departure from Africa and are *seldom* isolated from a broader, circum-Atlantic context. I submit that, unlike languages (as they are conventionally understood), African-American cultures should not be considered integrated, internally systematic, and bounded in discrete units; they are crosscut (à la Bakhtin) by multiple transnational languages, discourses, dialogues, and imagined communities.

The most recent analytic metaphor to take anthropology (including its Afro-Americanist varieties) and Afro-American literature by storm is the conception of "collective memory" and "forgetting" as the chief mechanism of sociocultural reproduction and (in the cases that most concern us here) the chief mechanism of diasporas' relationship to their homelands. The rhetorical effect of the "memory" metaphor in African diaspora scholarship is to highlight the implicit expectation that African Americans have (amid the ravages of enslavement, oppression, and poverty) lost contact with their past but retain at least mnemonic traces of that past.

Following Maurice Halbwachs, Bastide and others have argued that the recollection of myths, for example, can occur only when the social relationships or places to which they refer or the institutions to which they are relevant remain intact. For example, Bastide argues that the Yorùbá goddess Yemoja continues to be thought of as a mother in Brazil because (according to Bastide) lower-class Afro-Brazilian families tend to be "matriarchal." In general, asserts Bastide, the preservation of ancestral images and practices depends on the social relationships and landmarks in which they are "preserved" (Bastide 1978[1960]:243, 247–48).

Bastide has contributed much more to the study of social structural issues in the practice of Afro-Brazilian religions than to the study of "memory" per se. Yet the failings of his "collective memory" metaphor are instructive. Bastide equates the collective preservation of ancestral images and practices with *memory* because he understands *personal* memory as a paradigmatic manner of preserving information and technical competency. Yet there are ways in which the reproduction of an image and the teaching of a technique in society are *not* self-evidently forms of "preservation"; they are as likely to be forms of appropriation, quoting, mockery, propagandistic nostalgia, and so forth. The selective and strategic interpretation and invocation of the past are not the same as the "preservation," "retention," or "memory" thereof.

Bastide's conception of "collective memory" focuses on what "memories" are structurally possible or conditioned by the circumstances rather than on social actors' *choice* among possible practices and images to reproduce or on the purposes and motives behind their selective reproduction. Indeed, the "memory" metaphor seems semantically inconsistent with such agency.

In fact, Connerton (1995[1989]) identifies bodily practices as a privileged form of collective memory on the grounds of his view that they are exempt from controversy and therefore from the influence of ongoing reinterpretations of the past. On the contrary, I argue that bodily practices—such as ways of walking, dancing, or hand gesturing—become the subject of controversy *as soon as* they are recognized as signs of a collective past and, therefore, of a uniquely legitimate present. Cross-culturally, styles of sacred dance and postures of prayer, for example, often become signs of rival factions and charters for new cult groups. Even scholars who declare the historical significance of a bodily practice are moved by particular socially conditioned motives when they choose *which* bodily practice to highlight. Furthermore, these scholarly choices often enter into the public debate over which versions of verbalized history are right. For example, Herskovits's argument in *The Myth of the Negro Past* (1958[1941]) that "shouting" demonstrates Black North Americans' African cultural roots is neither politically disinterested nor uncontroversial. The debate over this matter between Herskovits and E. Franklin Frazier is significant less for the *scholarly* correctness of one or the other argument than for how the argument framed a debate that would continue in the general African-American population throughout the century. Diverse programs for the uplift of African Americans would follow from each position. In other words, scholarly debates over bodily practice ultimately affect political conduct and are, therefore, part of the object of study.

Invariably, the involvement of multiple parties, multiple classes, and multiple interest groups makes the intergenerational reproduction of culture a dialogue and a "struggle for the possession of the sign" (Hebdige 1979) rather than a form of "preservation." If it is obvious that the intergenerational transmission of culture resembles and entails a dialogue, it should be almost as obvious that personal memory, too, is shaped by dialogue and controversy.

However, the contributions of the literature on "collective memory" are not to be overlooked. In this genealogy of thought appear K. M. Brown (1989), Gilroy (1993), Dayan (1995), and Shaw (2002) who have shown that present-day ritual and verbal forms *indirectly* reveal past cultural dispositions, events, and conditions that have otherwise been overlooked (see also Scott 1991). Brown argues that the prominence of warrior gods in Haitian Vodou reveals the Haitian experience of military dictatorship, and Dayan shows that the coquettish goddesses of Haitian Vodou reveal the experience of mulatto mistresses under the regime of slavery. Gilroy (1993) argues that memories of slavery are manifest through their metaphoric displacement by lyrics of lost love in contemporary black music. To Gilroy, lost love is itself the key metaphor in African diasporic popular culture's commentary on the failed promises of "modernity." Shaw (2002) nimbly demonstrates how the ritual forms of Temne divination interpret present-day suffering through imagery of kidnapping and death reminiscent of the trans-Atlantic slave trade. The "collective memory"–based literature on the black Atlantic thus contains several rich discussions of the ways in which changing social conditions and political needs shape the selective reproduction, meaningful transformation, and meaningful reinterpretation of past cultural forms. Diaspora scholars would do equally well to emphasize that *commemoration* is always *strategic* in its selections, exclusions, and interpretations.[5]

But why call such cultural reproduction "memory," a term that hides rather than highlights the unending struggle over the meaning and usage of gestures, monuments, words, and memories? It implies the organic unity of the collective "rememberer," anthropomorphizing society rather than highlighting the heterogeneity, strategic conflicts, and unequal resources of the rival agents who make up social life and negotiate its selective reproduction. Of course, experts on the psychology of personal memory know how complex, variable, and socially conditioned memory is. For example, personal memories are reshaped by conversations and conflicting collective interests. However, if we are seeking a metaphoric *source* that makes the process of cultural reproduction *easier* to understand, is it wise to choose an analytic metaphor whose implications and entailments are so unsettled and so unclear? Metaphors are usually chosen for the concreteness and clarity of the *source*, which enables them to clarify what is otherwise unclear or inchoate about the *target* (Fernandez 1986). As the scholarly literature indicates, the source itself (i.e., personal memory) needs just as much clarifying as the target (i.e., the collective interpretation, material commemoration, and bodily reenactment of the past).

Invoked casually, the comparison of collective cultural practices to the recollections of an individual mind suggests a certain passivity, involuntariness, absence of strategy, and political guilelessness and neutrality that seem quite foreign to the processes that have in fact shaped African and African-American cultures over time. In general, the "memory" metaphor complements the nationalist personification of the nation and the fiction of innocent, pristine, and primordial folkways, which give proof that the nation has always been one and is rooted in God and Nature, rather than in human strategy or history.

Among the unavoidable failings of the "memory" metaphor is that, by making a figurative *person* of the collective rememberer, it risks concealing the fundamental pluralness of the agents of this "collective memory." Any memory invoked by a *group* of people is ready ground for sharp disagreements and interested rivalries, and any form of *personal* memory describable in terms of "sharp disagreement" and "interested rivalries" within itself would be considered nonnormative—or sick.[6] In fact, as richly suggestive as all of them have been, the "survival," "reinterpretation," and "creolization" metaphors also share the potential for similar underestimations of heterogeneity and conflicting strategy. They lend themselves to the premise that, like the survivor of a disaster, like the interpreter of a text, or like a creole language, any given Afro-Atlantic culture is self-existent, internally integrated, bounded, and possessed of its own agency and autonomous authorship; that it is not rent by multiple and contradictory discourses, languages, perspectives, and interests; and that each such Afro-Atlantic culture has evolved in organic isolation or discreteness from the others.

"Dialogue" and Its Precedents

An alternative metaphor might represent these cultures not as self-existent but as organically part of a *dialogue*—less as evolving *langues* (Saussure's term for languages as internally systematic codes) or as isolated collective remembers of self-contained national pasts than as interacting and changing sets of participants in a conversation. The metaphor of dialogue places traditions—or strategically constructed genealogies of cultural reproduction—into a context beyond nation and region, a larger, transoceanic context from which the nation and region themselves are rarely extricable.

Bakhtin's "dialogic" employs a similar analytic metaphor to the explanation of the novel, which embodies the speech of such diverse social classes that Bakhtin doubts the authorship, or agency, of the novelist (Bakhtin 1981). Applied to cultural reproduction generally, Bakhtin's analytic metaphor might rightly suggest that *multiple* agents are powerfully involved in the production of any given culture and that multiple cultures are "quoted" in the production of any *one* culture.

The Comaroffs (1991, 1997) represent the unequal dialectic of mutual transformation between the southern African Tswana people and European society, represented by the Nonconformist missionaries, as a "long conversation," entailing words and other "signifying practice," a conversation both verbal and nonverbal (1997:428n73). As in the case of Candomblé, that dialogue lies at the root of emergent ethnic identities and syncretic practices on both sides of the African-European divide that might otherwise be mistaken for primordial or antecedent to the intercultural encounter itself.

Sex and the Empire That Is No More (1994b), the prequel to this volume, represents culture as a "debate," "cosmopolitan dialogue," and a "field of intertextual contradiction," in which diverse classes of actors—including rival religions, rival priesthoods, and rival genders—struggle to impose different meanings, and therefore different material outcomes, upon the shared signs of social life (esp. 1994b:129; also 1994a:497). Ọ̀yọ́-Yorùbá social life has long included insiders and outsiders to this ethnic group, multiple languages, and a transoceanic array of hybrid cultural forms, which in turn facilitate the "intercultural dialogue" that is ethnography (Matory 1994b:216–30).

Tedlock and Mannheim (1995) employ the metonym of dialogue to illustrate their view that neither languages nor cultures are preexistent systems. Rather, systems are the descriptions of the ever-evolving patterns of verbal and bodily interaction among people (consider also Bourdieu 1985[1972]). Ethnography, they argue, is a moment of such a dialogue that includes the anthropologist him- or herself. In their view, expository ethnographies are an exercise in hegemony, and the authors recommend instead that ethnography become the less autocratic record of a dialogue.

However, as Clifford (1988) points out, the direct presentation of a dialogue is no guarantee that the ethnographer has conceded equal power to the interlocutor. Moreover, problems of the historical depth and geographical scope that I examine here are ill suited to such literal forms of dialogue transcription. The text would be inaccessibly long. Nonetheless, I recognize Tedlock and Mannheim's proposal as, in some ways, ideal. In this project, therefore, dialogue serves as a metaphor of the processes by which interlocutors not only produce their own local culture but also, through their acts of quotation and collective self-objectification, produce their local cultures *in systematic dialogue with faraway cultures*. As Tedlock and Mannheim have urged, people's representation of their conduct as the product of an objectified cultural tradition (which requires and justifies what they do) should not be mistaken for a merely objective report, then to be reproduced without attribution by the ethnographer. The style, media, and motives of such self-objectifications must be analyzed (1995:18).

In the same year, art historians Michael D. Harris (1999) and Moyọ Okediji (1999) and I (Matory 1999a) simultaneously and independently published essays likening the transformative exchange of people and ideas between Africa and the Americas to a "dialogue." There must have been something in the tides that year, pooling together a range of conceptual and empirical precursors around this term.[7]

Actors produce their cultures in dialogue and confrontation with cultural others near and far, and with groups more powerful and less powerful than they. Bakhtin's analytic metaphor might, however, *incorrectly* imply that all these agents and quoted cultures are equally powerful in the production of local or regional culture. I cannot deny the forms of power inequality that imbue this ethnography any more than I can deny that it is the product of a dialogue. Indeed, the forms of dialogue and inequality in this text mime, and are part of, the very unequal dialogues that I have attempted to document in the broader Afro-Atlantic world. And these are inequalities in which the richer, the whiter, the better-armed, and the more masculine do not always exercise the greater degree of control. For example, literacy and the speaking of stronger languages (Asad 1986) imbued the African-Brazilian circular migrants with the power to define important cultural standards and priorities for a white-dominated Northeastern region of Brazil. And in every chapter of this analysis, I have been constrained by my Portuguese-speaking interlocutors' willingness to speak. There were matters that, for example, Pai Francisco refused to discuss and certain stories that he refused to repeat, on the grounds that my request implied, to him, doubts about his prior account. Pai Francisco's decision was final and, by his own reasoning, deprived me of an opportunity for verification. I also translated the entire text for him and, by agreement, deleted or modified all those portions that he thought inappropriate. This narrative remains a powerful and, in part, strategic presentation of one extended moment in Pai Francisco's voluntary dialogue with me.

A major service of Gilroy's *Black Atlantic* (1993) is that it demonstrates that Blacks in the United States are not the lone creators of black Atlantic culture. Gilroy usefully illustrates the role of Anglophone West Indians and Black Britons in the process. However, this insurgent and democratizing move also recapitulates a form of Afro-Atlantic inequality that becomes more obvious when one broadens the scope of analysis. The Afro-Atlantic dialogue encompasses the speakers of not only English but also a half dozen other European languages, a dozen creole languages, and hundreds of African languages. Moreover, literacy, economic power, commercial contacts, and command over the English or French language have made certain Afro-Atlantic voices louder than others. The speakers of these Afro-Atlantic languages far outnumber the approximately forty million black or mulatto native speakers of English in the United States, the West Indies, and the United Kingdom. However, Anglophone blacks have exercised a disproportionate influence on the contemporary making of all Afro-Atlantic cultures.[8] It is no less true—notwithstanding Gilroy's effort to question the centrality of Black North Americans—that the Blacks of the United States have been more influential than any other Anglophone group in the 20th-century reproduction and reshaping of Afro-Atlantic culture.

The empirical fact that people of the African diaspora have traveled, carrying goods and ideas among various locales on the Atlantic perimeter, has long been known. The intercontinental movement of corn, cassava, cowpeas, peanuts, tobacco, palm oil, and cowries has, over the past five hundred years, wrought incalculable demographic and political changes everywhere on the Atlantic perimeter—changes dwarfing the oft-cited consequences of Europe's importation of the potato. Afro-Atlantic peoples were not only victims but also major agents of these seismic shifts, during and long after the trans-Atlantic slave trade. Focusing on the migration and commerce between Bahia and West Africa, Pierre Verger described this phenomenon as *flux et reflux* (flow and counter-flow; Verger 1976[1968]). Yet similar ongoing transoceanic exchanges of people and goods also gave birth to the terrestrially bounded social spaces of Liberia, Sierra Leone, Angola, and South Africa. Nearly two decades ago, Elliot Skinner (1982), too, theorized a "dialectic between diasporas and homelands," showing that the African, Jewish, Irish, Indian, and Chinese diasporas have often aided in the formation of independent nations in their homelands, and that such independent homelands often enhance the political stature of their diasporic kin, despite the ambivalence that homelanders and diasporans often feel toward each other. Many apparently local, continent-bound or language-bound cultural formations—such as the Harlem Renaissance, Senegalese and Martinican Négritude poetry, Ghanaian pan-Africanism, Afro-Cubanismo, and Congolese Soukous music—in fact have origins and reciprocal outcomes in geographically and linguistically distant regions of the African diaspora (Matory 1999a). Thus, the Afro-Atlantic dialogue is not only transnationally religious but also transnationally economic, political, literary, and musical.

Just as Africa and its diaspora are linked, diverse African *diaspora* locales are linked to each other by migration, commerce, and the mutual gaze among them, which is the subject of *The Black Atlantic*. With this example, Gilroy retheorizes the scope and mechanisms of cultural reproduction that he sees posited in the nationalist and racist cultural histories of blacks in the West. For him, the cultural exchanges *between* diasporic locales undermine and falsify the boundaries that nationalists imagine around the races, nations, and cultures of the Atlantic perimeter. Thus, for Gilroy, the ships that carry ideas and cultural artifacts *between* locales are more emblematic of black Atlantic culture than are the national boundaries and watery divides that separate one locale from another. In the place of continuous forms of "memory" constituting geographically bounded cultural units, Gilroy prioritizes the "discontinuous" forms of cultural reproduction by which ideas and images from one place have constantly amplified and modified the cultural genealogies of other places.

Gilroy prioritizes cultural exchange across territorial boundaries over the divergent but uninterrupted "memory" of Africa posited by Bastide and many Herskovitsians in the genesis of the Anglophone black Atlantic culture. Yet Gilroy remains curiously comfortable with the notion of the African diaspora's *continuous* "memory" of slavery. For example, Gilroy believes that black Atlantic ballads about lost love symbolically commemorate slavery. But when genres inaugurated by one ethnic group, nationality, or race are adopted by another, whose collective past is being "remembered," and who is the rememberer? Weren't the regimes of slavery in the different "remembering" communities, nations, and regions different? Wouldn't the diversity of musical genres suggest that the rememberers are diverse, and at least *partially* constituted by nationality? Yet Gilroy, like Appadurai (1990), represents the territorial nation and the local identities it generates as contradictions to (rather than agents or optional objectives of) these shipborne cultural crossings.

And if, for Gilroy, "collective memory" usefully highlights something about Afro-Atlantic people's relationship to their past, why is the "memory" of Africa selected for denial in Gilroy's analytic model? Perhaps Gilroy is simply much more *interested* in the cultural exchange between blacks and whites, as well as among English-speaking black diaspora locales, than in Africa. Gilroy's representation of "black Atlantic" culture as an essentially "modern" (read "Western") response to blacks' exclusion from the promises of the European Enlightenment renders Africa irrelevant. Yet the credibility of this representation of black Atlantic cultural history relies less on a persuasive critique of the existing contrary positions than on a silence about them.

The main fields of African-diaspora culture that Herskovitsians and Bastideans have represented as the products of "continuous" reproductions of African culture have been the lexically, structurally, and texturally African-looking forms—or "Africanisms"—in American religion and music. Yet Gilroy studiously ignores the aspects of African-diaspora music that have been cited as proof of "continuity," while inexplicably dismissing African-American religion as "the central sign for the folk-cultural, narrowly ethnic definition of racial authenticity" (1993:28). On the contrary, African-diaspora folk culture and religion are usually enlisted as evidence of transethnic and transnational cultural commonalities (and not national closure) among peoples of African descent, and *whites'* adoption of African-inspired folk culture and religion from their African-diaspora neighbors is a frequent theme in the writings of Herskovits, Bastide, and their successors (e.g., Herskovits 1966[1935]; Bastide, 1978[1960]; P. Wood 1974, 1975; Philips 1990).

Hence, my criticism of Gilroy does not result from any belief on my part that cultural reproduction is "continuous" or nationally bounded. In fact, Gilroy is correct in emphasizing the "discontinuity" and cross-territorial nature of cultural reproduction. Rather, I question Gilroy's neglect of Africans' participation in this cross-territorial phenomenon and his premise that African-diaspora cultural history *began*—temporally and conceptually—at the moment that blacks encountered the ideas of the European Enlightenment. Such a premise smacks of the same anti-cosmopolitan boundary-making that Gilroy's entire oeuvre is arrayed against.

Perhaps, however, there is a single oversight that at once shapes Gilroy's historical periodizations and his failure to cite Herskovits and his scions. At first sight, the type of "culture" documented in the ethnohistorical literature appears distinct from the type treated in the cultural studies literature. On the one hand, Herskovits, his avowed successors, and his critics focus on the *anthropological* stuff of culture—the everyday practices and beliefs of the every(wo)man, including the religions and musical genres of what appear to be common people. On the other hand, Gilroy and other cultural studies specialists focus on the writings and political gestures of the highly educated, the transformative efforts of cultural reform movements, and the sort of art forms diffused by professional publishers, recording industries, and museums. To my mind, this division of intellectual labor results less from the empirical accuracy of the "low culture"/"high culture" distinction in the Afro-Atlantic cultures than from many of these authors' trained *inattention to the intense dialogue that unites these cultural spheres.*

This book has concerned the empirical and theoretical utility of attending to this interclass dialogue. Artists, scholars, and travelers—including those with trans-Atlantic social connections, commercial interests, and political aspirations—have played a major role in shaping the local religions and other "folk" cultures of the Afro-Atlantic. Moreover, both Africa and non-English-speaking parts of the Americas were deeply involved in this dialogue.

Joseph Roach has given a further treatment to the cross-cultural exchanges shaping Afro-Atlantic culture, highlighting their symbolic mediation and the rhetorical uses of these exchanges. Through his notion of "circum-Atlantic performance," Roach (1996) observes that diverse peoples on that ocean perimeter have mimicked each other for five hundred years as a fundamental condition of each such people's formation of its own cultural identity. Roach, too, largely ignores the active role of continental Africans in this dialogue. However, unlike Gilroy, Roach calls attention to the frequently ironic and fictional quality of one people's mimicry of another—such as African-American and European theatrical performances of Africanness and of Native Americanness. Such mimicry, often based on literary and celluloid fabrications—and the prejudices they propagate—are a further important element of the Afro-Atlantic dialogue.[9] Nigerians, Jamaicans, Brazilians, and U.S. African Americans often perceive each other through books and films, many of which are produced or distributed by whites. Such perceptions at times inspire mimicry and, at other times, oppositional self-definition and its attendant cultural self-transformations.

The "dialogue" metaphor does not posit that every interlocutor is an atom, autonomous or equal in his or her power to act. Relative wealth, linguistic proficiency, nationality, and access to the means of communication distinguish *groups* of actors from each other. On the other hand, the "dialogue" metaphor does posit one type of equality: temporal equality. It posits the radical "coevalness" (Fabian 1983) of Africa, Latin America, and the United States in a dialogue that, even following the conclusion of the slave trade, has continually reshaped them all. In other words, Africa is not, as Herskovits's ethnohistorical "laboratory" method would suggest, the past of Afro–Latin America (Herskovits 1958[1941]:15–17; 1966[1956]:76). Nor does either of these sets of cultures illustrate the past trajectory of Black North American assimilation.

The dialogue metaphor also poses a radical challenge to conventional models of resistance. The enslaved and the oppressed do not simply react to their local oppressors in the spirit of their inherent desire for freedom or of their self-existent love for the ways of their ancestors. Nor is their response a culture-free pursuit of class interest. But their desiderata are not all culture-specific, and their means of pursuing such desiderata are often consciously shaped by symbols, ideas, and social forces from far away. The embrace and manipulation of the supralocal often far exceed the referential power of the term "resistance"; they redefine local orders of power and knowledge.

If my suggestion is taken seriously, African-Americanist cultural historians will no longer begin books with the conventional chapter about Africa (usually based on colonial era accounts of African culture and history) and then demonstrate the stages by which an African-American culture evolved over time *out* of that original, African one. The "dialogue" metaphor would instead highlight the ways in which the mutual gaze between Africans and African Americans, multidirectional travel and migration between the two hemispheres, the movement of publications, commerce, and so forth have shaped African and African-American cultures in tandem, over time, and at the same time. It highlights the ways in which cultural artifacts, images, and practices do not simply "survive" or endure through "memory"; they are, rather, interpreted and reproduced for diverse contemporary purposes by actors with culturally diverse repertoires, diverse interests, and diverse degrees of power to assert them. As if in a literal dialogue, such interpretations and reproductions can also be silenced, articulated obliquely, paraphrased, exaggerated, quoted mockingly, or treated as antitypes of the legitimate self.

So What?

My graduate students who read the manuscript of this book posed two key questions. The first: Isn't transnationalism qualitatively new because of the enormous speed of transportation and communication enabled by jets and the Internet? To which I replied that such changes are indeed momentous but quantitative, nonlinear, and unevenly spread, and that the consequences of these changes might be illuminated better by the circum-oceanic and circum-desert social fields of the distant past than by the fashionable vocabulary of historical "rupture" and "break," which literally forbids the transfer of lessons from one period to the other. The recent changes that have been described so perspicaciously under the rubrics of "transnationalism" and "globalization" highlight precisely those aspects of the distant past that have until recently been hidden by the agrarian and arborescent metaphors of sociocultural anthropology. This is the greatest unsung gift of Appadurai, Hannerz, Harvey, and Sassen to cultural history.

My students' second question mobilized my own argument against me. If the African-American experience has always troubled the dominant narrative of pure, bounded, and fixed identities, couldn't one easily conclude that the black Atlantic experience is unique, an exception to the general rule that human cultures were or are normally bounded and that citizenship in the nation-state was once the universal standard of human identity? By insisting on the universality of translocalism and transnationalism, am I not denying African Americans due credit for forcing scholars to recognize the value of hybridity and additivity? My students could be right. I do believe that the African diaspora, along with the Jewish diaspora, has been disproportionately accountable for the willingness of the West to imagine parts of itself beyond the boundaries of race and national territory. However, the fact that African Americans have so often inspired this willingness does not prove that African Americans are the only peoples who resisted territorially bounded identities and hybridized the cultures of their host nation-states. It proves that African Americans had the incentive, the genius, and the power to *advertise* that fact. The African Americans who dwell in the largest settler colonies of the Americas—Brazil and the United States—have long sat close enough to the heart of the West to describe and contradict the West's mythologies about itself in ways that the inhabitants of faraway colonies could not or would not do because traveling colonial subjects and postcolonial immigrants often have a stake in the mythology of citizenship that the victims of slavery, settler colonialism, and intranational racial hierarchy lack.

Whether the postulate of transnationalism's antiquity illuminates as much in the history of other peoples, places, and times as it does in Islam, Western Judaism, and the African diaspora is a purely empirical question. I am confident of the answer. However, I will consider this book a success if the reader is now seriously in doubt about the assumption that, before some late-20th-century sea change, monolithic territorially bounded local identities and cultures were the lone or even preeminent context of human social life. By the African-American example, I offer the hypothesis that cultural identities, as well as the lifeways that their reification modifies, have everywhere and always been created in translocal contexts that host multiple alternative and crosscutting collective identities. The nation-state never possessed a monopoly on people's collective identities or allegiance, such that latter-day transnationalism could have suddenly deprived it of such a monopoly. Nation-states, like religions and empires, are both the products and the producers of transnational forces. Indeed, most of the world's nation-states were founded by people who looked across an ocean and saw a nation-state that they wanted to be like but no longer wanted to be an extension of.

Particular places and epochs are not incidental to the lessons we learn there about all times and all places. The great lesson of sociocultural anthropology to those who have sought to construct universal histories and sociologies—such as Maine, Morgan, Marx, Durkheim, Weber, and Toynbee—is that we must be no less careful in the study of there than of here. As anthropology embraces history and turns back to the search for universal trajectories, we must take seriously the lessons of non-Western and subaltern histories: we must be no less careful in the study of then than of now.

Geechees and Gullahs

THE LOCUS CLASSICUS OF AFRICAN "SURVIVALS"
IN THE UNITED STATES

CLOSER TO HOME, the "dialogue" model requires us to reassess the locus classicus of "African survivals" in the United States, embodied, as it were, in the Gullah people of the Georgia–South Carolina Sea Coast Islands (Vass 1979; Creel 1987; Stuckey 1987; L. D. Turner 1949; Holloway 1990). The Gullah people of the Georgia–South Carolina Sea Coast Islands have long been the focus of scholarly investigation into what remains culturally African about Black North Americans, despite their generally high degree of acculturation in Western ways. Relative geographical isolation long kept the speech and lifeway of the Sea Coast Islanders distinctive. Various scholars have sought to explain that distinctiveness as a debt to the cultures of what are now the settler colonies of Sierra Leone and Liberia.

In the second half of the 18th century, the rice farmers of the Sea Coast Islands drew many of their workers from the rice-growing regions of Sierra Leone and Liberia, and the term "Gullah" might derive from the ethnonym of Sierra Leone's "Gola" people. On the other hand, African captives had come to these islands from many other regions as well, and students of the local creole language, known as "Gullah" or "Geechee," have identified an extremely diverse set of African origins in its lexicon and in its justly famous basket-making tradition. Indeed, some identify the term "Angola" as the more likely source of the term "Gullah" (Rosengarten 1997; L. D. Turner 1949).

As Black North Americans have grown more willing to embrace Africa as a cultural model and emblem of collective identity, the decline of Gullah language and crafts has been reversed. Indeed, the "Africanness" of Gullah basketry has become its major selling point and a means of livelihood for many craftswomen in coastal South Carolina. However, it was Joseph Opala (a Euro-American anthropologist and former member of the Peace Corps in Sierra Leone) who recently established the local conviction that Sierra Leone in particular was the source of the islanders' Africanness and the appropriate target of their "return" to the motherland.

Indeed, the interest in this ahistorically specific tie was reciprocal. President Joseph Momo of Sierra Leone paid a highly public visit to the

Sea Coast Islands in 1986 and encouraged the islanders to visit their "ancestral homeland," which a party of them did in 1989. President Momo continued the American tradition of attributing the islanders' linguistic distinctiveness to their African roots and identified the similarity of Gullah language to Sierra Leonean Krio, or Creole, as proof.[1]

In fact, both language varieties are predominantly English in their lexicons, since Krio initially came about due to the interaction of African-American returnees to West Africa, diverse British-educated recaptives, British administrators, and Anglophone missionaries in Freetown. Thus, these Sierra Leonean and South Carolinian creole language varieties share features primarily on account of *the parallel circumstances of their genesis*, not on account of some shared, primordially African roots or of Gullah's having originated in Krio. In sum, the shared features of Gullah and Krio are highly ambiguous evidence of the Gullah people's Sierra Leonean or African roots.

Nonetheless (and this is what makes the "dialogue" metaphor useful) a complex, politically, economically, and academically shaped dialogue has made the highly creolized Gullah dialect into emically persuasive grounds for a powerful new kinship—a web of living, albeit recent, social connections between Sierra Leoneans and South Carolina Geechees.

The Afro-Atlantic dialogue also encompasses the white neighbors of the Gullahs. Paralleling the Afro-Brazilianist emphasis of Regionalism in Northeastern Brazil is an important strand in the regionalism of the U.S. South from 1889 onward, during a period strikingly parallel to the emergence of Afro-Brazilian studies. As in Northeastern Brazil, many of the elite southern whites beleaguered by economic and political decline, and domination by the elites of a whiter region, generated a sizable literature demonstrating their knowledge of and investment in African-American culture. Like so many anti-colonialist "*negrigenismos*," this southern regionalism was intent on demonstrating the legitimacy and beneficence of southern whites' sovereignty over local Blacks and denouncing Yankee "interference"—such as the Civil War, the Emancipation Proclamation, the Thirteenth, Fourteenth, and Fifteenth Amendments to the Constitution, Reconstruction, and the overall destruction of a noble planter "civilization." While the elites of Northeastern Brazil had faced no such major obstacles to oppressing their darker neighbors, the recent neo-Marxist ethnographers seem to suspect similar motives in their modes of power/knowledge.

And there are still greater parallels. Many southern white elites—like Georgian Joel Chandler Harris (1848–1908), Mississippian Newbell Niles Puckett (1897–1969), South Carolinian DuBose Heyward (1885–1940), and Charleston's all-white Society for the Preservation of Spirituals—linked their nostalgia for a past regime to this careful documentation of a distinctly Black regional culture.[2] These southern regionalists constructed this Black culture as perhaps inferior to the European or metropolitan white culture but praised it as traditional, uniquely authentic, sincere, and, best of all, *characteristic* of the region.

Moreover, these white writers and performers claimed to have grown up with this Black culture, to have loved it, and to have embodied it—claims that appear to be as truthful as (albeit less forthright about the violence of slavery than) Freyre's similar representation of his Brazilian self and nation. The most dramatic example is the Society for the Preservation of Spirituals. Made up exclusively of the children and grandchildren of slaveholding Carolinian planters, it has, since around 1923, endeavored to preserve the 19th-century, pre–"concert hall" versions of the Negro spirituals that African-Americans themselves have progressively abandoned. The group not only studies and transcribes these "authentic" forms of the spiritual but *performs* and *records* them in carefully reproduced Gullah dialect. Whatever last-ditch efforts at paternalism one might suspect, the musical results and the sincerity of the performers are sometimes moving and beautiful. Moreover, the organization's documentation of this tradition will forevermore be available for uses other than the construction of postslavocratic white identity (Howe 1930; Smythe et al. 1931; Society for the Preservation of Spirituals [recording] 1955; Puckett 1969[1926]).[3]

Equally parallel to the Northeastern Brazilian phenomenon, many of these white southern elites have long read the cultural forms they document, collect, and perform as somehow "African" and have long searched for their African origins. Like Raimundo Nina Rodrigues, Puckett was given to citing the Lagosian Renaissance era writings of Colonel A. B. Ellis for information on "original" African forms but is perhaps more oblivious to the fact that those writings are already profoundly shaped by a 19th-century dialogue among West Africans, African Americans, the British, and the French (Puckett 1969[1926]: esp. 2–57; Howe 1930:110; Gordon 1931:219–20; Heyward 1931: 171–79, 181; also Vlach 1990[1978]). Yet Puckett surmised that much of Afro-American cultural distinctiveness might just consist of forgotten and old-fashioned Anglo-Americanisms (1969[1926]:2). Having witnessed the Harlem Renaissance and the "New Negro" movement at first hand, Heyward recognized the potential for Black assimilation. He just doubted that it would make Blacks happy (1931:186–87). And the Society for the Preservation of Spirituals would have had no publicly acknowledgeable function had "education," in the words of its spokesman Smythe, not led Blacks themselves to leave these song forms behind (Society for the Preservation of Spirituals 1955 [recording of Smythe's introduction]). The changes in Black North America wrought by emancipation and education, as well as the *changeability* of race relations by means of protest and legislation, were quite evident to these writers and orators. As partisans in a debate over the appropriateness of those changes, they were not oblivious to the contrary arguments and results.

Like African-Brazilians, Afro-Carolinians have long cultivated the co-operation of missionaries, ethnographers, and rich whites in canonizing and dignifying their cultural forms. For example, Afro-Carolinian low-country basket makers—along with merchants, tourist agencies, journalists, museums, and state agencies—have made the Africanness of their craft into a proud selling point.[4] And, with the benefit of travel to Africa, intermarriage with Africans, and the study of available publications on African arts, Afro-Carolinian craftswomen themselves have begun to "re-Africanize" local forms (Rosengarten 1987[1986]: esp. 16–17, 27, 35–36, 42–43, 46n5; 1997). Yet in basket forms, as in language, these African Americans also duplicate the Afro-Brazilian talent for code-switching. They can speak Gullah and "standard" English; they can make rice fanners and cocktail trays, stacked baskets of Kongo inspiration and "missionary bags."

In the North American ethnography, the Gullah case is typically treated as the exemplary case of Black North American resistance through the "memory" of African culture. That is not to say that such authors are oblivious to the "syncretization" or "creolization" of these African "memories." What I would add, then, to conventional representations of Gullah culture is a sketch of the interregional, interracial, and transnational debate that has shaped Afro-Carolinian self-representation and production since the Civil War, of its parallels with transformative debates elsewhere in the Americas, and of the unexpected intersection of those debates in the Yoruba cultural renaissance of colonial Lagos. The Africanness of the Gullahs, or Geechees, is not simply a hypothesis to be proven or disproven through an archaeology of verbal or material forms. Like the history of the entire black Atlantic world—and of human culture generally—it is preeminently a history of rival political interests, strategic cultural citations, and contestations of collective identity in which translocal actors, including ethnographers themselves, are patently and appropriately involved.

The Origins of the Term "Jeje"

THE TERM "JEJE" (ZHAY-zhee) has appeared in Brazilian documents since 1739, though it is absent from the primer written in Brazil by Peixoto (1943–44[1741]), which evidently documents the same language.[1] Peixoto's primer is based on the speech of so-called Mina slaves in Brazil, a language very similar to the contemporary language called "Fòn" in West Africa (see Rassinoux 1987). The "Mina" slaves' speech and their worship of the *vodun* gods leave little doubt that they had been kidnapped from the region between Mina Castle in Ghana and present-day Yorubaland—for this is the West African distribution of *vodun*-worship.

However, for Brazilians and Brazilianists, the origin of the term "Jeje" remains unclear. Inspired by Ellis's title, *The Ewe-Speaking Peoples of the Slave Coast of West Africa* (1970[1890]), Rodrigues established the Brazilian etymological tradition of identifying the term "Ewè"—the name of the dialect now spoken in southwestern Togo and southeastern Ghana—as the origin of the term "Jeje."[2] Rodrigues also contemplated the possibility that the word "Jeje" came from the "Geng," or "Gen," which today describes the dialect of the Mina people of Togo and southwestern Benin (R. N. Rodrigues 1988[1905]:105–6, 232; Cacciatore 1977:153; Capo 1984:168). Evidently, Rodrigues did not know that the initial *g* in "Geng" and "Gen" is hard, unlike the fricative *j* in the word "Jeje." Nor did he realize that the Jeje language documented in Brazil and in Africa is much more similar to the Fòn dialect than to the Gèn or the Ewè dialect.

Segurola, a lexicographer of the Fòn dialect, denies that the term "Jeje" originated in any way from the Fòn language (Segurola 1968[1963], 1:264). Elbein dos Santos (1993[1975]:31n11) says that the origin of the term "Jeje" is simply unknown, but she credits the colonial French administrators with its initial application to the ethnic groups living around Porto-Novo. Pierre Verger (1970[1957]:19), for his part, declares that "Jeje" comes from the term "Ajá," which designates the people of southeastern Bénin and of eastern Togo who gave birth to the dynasties that ruled Allada, Porto Novo, and Abomey.[3] Costa Lima offers the alternative interpretation that "Jeje" comes from the Yorùbá word for "foreigner" (*àjèjì*), in reference to their linguistically dissimilar Gbè-speaking neighbors (V. Lima 1977:14–15; also Abraham 1962[1946]:38), while Suzanne Blier believes that "Jeje" comes from the name of the town of Adjadji, near Allada. Some identify that town as the origin of the Ajá-Tado dynasty, which reigned in the cities of Allada, Porto Novo, and Abomey (personal communication, 21 October 1997; Blier 1995b:405, 408).

In any case, the use of the term "Jeje" to designate all the speakers of Ewè, Gèn, Ajá, and Fòn originated in Brazil, where it appears as early as 1739, i.e.,120 years before its first appearance anywhere in the extensive historiography of the Gulf of Guinea.

Notes

RDD	*Revue du Dahomey et Dépendances: Organe du Comité du Dahomey* (Paris)
SAL	*Salvador* (Salvador da Bahia)
SMA	*Société des Missions Africaines*
SS	*Suprême Sagesse* (Porto-Novo)
TB	*A Tribuna da Bahia* (Salvador)
TSD	*La Tribune Sociale du Dahomey* (Ouidah)
UH	*Ùltima Hora* (Brasília)
VB	*ViverBahia* (Salvador da Bahia)
VD	*La Voix du Dahomey* (Porto Novo)
VIS	*Visão* (São Paulo)
VSP	*Veja* (São Paulo)
REAL	*Realidade* (São Paulo)

Introduction

1. Though in the past "nation" referred to a group of slaves and freed people sharing approximately the same African origin, the meaning of the term has diversified across the 20th-century contexts of its use. Nowadays, the term means slightly different things in Brazil, Cuba, and Haiti. In Brazil, every temple and its members belong to a particular nation. In Haiti, each god belongs to one or another nation. Finally, in Cuba, each nation is a separate denomination, and each nation may include a different avatar of the same god. In all three places, most nations are assumed to have their headquarters and origins in some eponymous African place.

2. The Brazilian term "Nagô" is cognate with the name of a western Yorùbá subgroup called the Ànàgó, whose name the Fòn speakers, and sometimes the Yorùbá speakers as well, use to describe all Yorùbá speakers. "Kétu" is the name of another western Yorùbá kingdom; hence the Brazilian term "Quêto." "Ìjèṣà" is the name of an eastern Yorùbá ethnic group; hence the Brazilian term "Quêto." The Cuban term "Lucumí" appears to derive from the Yorùbá phrase "my friend" in *Olùkù mi*.

3. In the past, there have been nations called Congo and Cabinda. Elsewhere within Bahia's sphere of influence, several other nations are recognized, such as Efã, a word that appears to derive from the name of the Èkìtì city of Efòn-Aláàyè, near Adó-Èkìtì, in Nigeria. However, it is sometimes identified with the similarly named "Fòn" ethnic group and, therefore, with the Jeje nation. Nonetheless, like the Ijexá nation, it appears to be merging into Quêto.

4. C.I.A. World Fact Book—Brazil (http://www.odci.gov/cia/publications/factbook/print/br.html).

5. For example, Arthur Ramos, Édison Carneiro, Ruth Landes, and Roger Bastide on Afro-Brazilians; Lydia Cabrera on Afro-Cubans; Herskovits himself on Haitians and Surinam Maroons; Gonzalo Aguirre Beltrán on Afro-Mexicans; W.E.B. Du Bois, Joseph Holloway, Sterling Stuckey, Albert Raboteau, and Margaret Creel on Blacks in the United States.

6. On the other hand, 20th-century Òyó mythology and rituals tend to represent the forest and nonurban areas (the closest counterpart to "nature" in the

Ọ̀yọ́ worldview) as dangerous, filthy places, to which the people who eventually became the gods fled and turned into geographical features and atmospheric forces when social contracts went awry. Ọ̀yọ́ religion, then, tends to ritualize the resocialization of these features and forces, through the investment of their ritually prepared icons in human-made vessels.

CHAPTER ONE
THE ENGLISH PROFESSORS OF BRAZIL

1. From Pierson 1942[1939]:293.
2. For example, R. N. Rodrigues 1935[1900/1896]; Rio 1976[1906]:19–20; Pierson 1942:237–251, 272; L. D. Turner 1942; Olinto 1980[1964]; Karasch 1987:194, 322; J. H. Rodrigues 1965; Bastide 1971[1967]; 1978[1960]; Verger 1976[1968]; J. M. Turner 1975[1974]; Manuela Carneiro da Cunha 1985. See Sarracino (1988) and Cabrera (1974:27; 1983[1954]:405–6) on the lesser-known exchanges between Cuba and West Africa. See Métraux (1972[1959]:360) on Haitian king Christophe's little-known recruitment of four thousand free Dahomeans into the police force of postrevolutionary Haiti, a population movement that may have brought cultural influences to Haiti as well. One is tempted to seek in this unusual immigration part of the reason for the prominence of Dahomean culture in Haiti despite the fact that Dahomeans had been vastly outnumbered by other Africans in the 18th-century slave population. See Carr (1989[1955]: esp. 2–3) and Mintz and Price (1992[1976]:56) on the settlement of more than fifteen thousand free Africans into the British West Indies and the extraordinary religious legacy of one *vodun* priest among them. See Bolster (1997), Linebaugh and Rediker (2000), and Gilroy (1993:12ff.) on the influential character of black sailors' mobility generally.
3. In a rare exception, Ramos wrote that missionary-taught literacy among the West African Yorùbá after the mid–19th century might account partly for the rise of Nagô as a lingua franca in Bahia (1942:20). Kubik (1979) and Côrtes de Oliveira (1992) also discuss the role of the slave trade itself in generating the ethnic groupings known as "nations" in black Brazil.
4. See also Blier (1988) and Yai (1992), who offer glimpses of how the Atlantic dialogue has shaped aspects of Dahomean culture that Herskovits mistook for primordial.
5. It is not only in the Americas that black Atlantic ethnic groups have been described as nations. The Reverend Samuel Johnson, writing in the late 19th century, was among those to refer to the West African Yorùbá as a "nation" (1921[1897]), as do contemporary Yorùbá nationalists resisting Hausa-Fulani hegemony in successive Nigerian military governments. The usual term for "nation" in Yorùbá is *orílẹ̀èdè*, which defines peoplehood in terms of shared land and language.
6. Landes 1947:17; "MEC quer preservar a cultura negra na Bahia," *AT*, 9 Sept. 1982 (Pasta [Folder] 324, *AT* archives); "Preservação do Candomblé é defendida em encontro," *AT*, 3 Aug. 1987, p. 3, attachment to the report on the "Encontro brasileiro da Tradição dos Orixá e Cultura: Preparatória da 'IV Conferência

Mundial da Tradição dos Orixá e Cultura'—IV COMTOC-'88," Salvador, Bahia, 31 July–2 Aug. 1987, Local—Ilê Axé Opô Afonjá-Cabula (*TB* archives).

7. Kasinitz coined the term "ethnicity entrepreneurs" to describe those who construct and mobilize ethnic identities in the strategic pursuit of collective or individual interests (1992:195ff.). Palmié tells the story of two particular Cuban Lucumí priests who co-opted the initially anti-African Fernando Ortiz and enlisted him in the promotion of their specific ethnic group (2002:248–59).

8. Secondarily, argued Rodrigues, the widespread usage Nagô language had enhanced the spread of Nagô religion.

9. See João do Rio (1976[1906]) for early references and Prandi 1991:108–9 on the most recent trends in the Brazilian Center-South.

10. On the one hand, Herskovits assumes that the predominance of any given African culture in the Americas has resulted from the numerical predominance of its bearers among the captives brought to its American host region (1958[1941]:52). On the other hand, Bastide writes point-blank: "In any given area there is a dominant African culture; *but the predominance of that culture (whichever it may be) bears no direct relation to the preponderance of such-and-such an ethnic group in the slave-shipments to the area concerned*" (1971[1967]:11; emphasis in original).

11. In what they describe as the exceptional cases of Trinidadian and Cuban religion, Mintz and Price (1992[1976]:60) attribute the strength of "continuities" from Africa to the combination of the recentness of the immigration and the immigrants' immediate or quickly ensuing freedom.

12. See, for example, Creel 1987; Levine 1977; Bastide 1971[1967]:213–22; Georgia Writers' Project 1940; Ramos 1940[1934], 1942; and the numerous studies of Maroon societies. In regard to the association of African "survivals" with rural isolation, Bastide (1983:242–43) describes Bahia as exceptional, being a large city that retained its African religion with relative purity; Herskovits (1958[1941]:115–16, 120, 124) recognizes Paramaribo, Port-au-Prince, and Bahia as similarly exceptional; and Puckett (1969[1926]:10–11) represents New Orleans as exceptional. With so many exceptions, these urban cases would better be regarded as *disproof* of any simple *rule*. See Matory 2000.

13. Rodrigues's opinion also seems to have influenced the famous Cuban folklorist Fernando Ortiz (e.g., Ortiz 1973[1906]:28).

14. Consider also the life of Isadora Maria Hamus (apparently the similarly Anglicized form of the Portuguese name Ramos), who was born in Cachoeira, Bahia, in 1888. At six years of age, she went to Lagos with a relative and remained there eight years, learning Yorùbá and English. In the early 1940s, she was a businesswoman and Candomblé priestess in the town of São Felix, Bahia (L. D. Turner 1942:64).

15. Tata Àngel, personal communication, Chelsea, Massachusetts—28 May 1996; Cosentino 1995:33, 44–55; Simpson 1966:540, 543, 544, 547; Thompson 1983:112; Akintola 1992: esp. 7; Anyebe 1989: esp. 53, 57, 79, 102, 106; Fadipe 1970[1939]:248; Pierson 1942[1939]:259; Comhaire 1949:43; Paul Lola Bamgboșe-Martins, personal communication, 26 Nov. 1995. Mr. Bamgboșe-Martins normally resides in Lagos, but I met him first on one of his many extended sojourns in Rio.

16. Other important priestly travelers between West Africa and Brazil have been Otampê Ojarô, founder of the Candomblé of Alaketu (V. Lima 1977:26–28); Marcos Pimentel, a 19th-century chief priest of the Terreiro de Mocambo on the island of Itaparica (Elbein dos Santos and Santos 1981[1969]:159–60); and, most important, Ìyá Nassô, founder of Casa Branca, whose mother secured her own manumission in Bahia and returned to Africa to enter the priesthood. The daughter was born in Africa but voluntarily moved to Bahia to found this first of the three most famous Candomblé temples in Brazil. Her successor Marcelina is said to have gone from Africa to Bahia voluntarily, returned to Africa for an extended sojourn before returning finally to Bahia to assume the leadership of the Casa Branca Temple (Bastide 1978[1960]:165). Verger (1980) reports that it was Marcelina who first brought to Bahia the famous Bamgboṣe—priest, grandfather of Felisberto Sowzer, and great-great-grandfather of Aizinho of Oxaguiã, chief of the Pilão de Prata Temple in Bahia.

17. During the Nigerian struggle for independence in the 1940s and 1950s, Adeyẹmọ Alakija played a leading role in the Yorùbá nationalist organization founded by Ọbafẹmi Awolọwọ—Egbẹ́ Ọmọ Odùduà.

19. See Côrtes de Oliveira (1992) and Kubik (1979).

20. Some of the travelers from Cuba to Lagos passed first through Rio and/or Bahia, which may have helped to spread iconographic and ritual innovations from one area of the òrìṣà-worshiping diaspora to another. For example, the fact that elaborate multistrand beadwork (e.g., *collares de mazo* in Cuba and *dilogum* and *oxumbeta* in Brazil) and soup tureen altars are common in these priesthoods and rare in the òrìṣà possession priesthoods of the Ọ̀yọ́ region suggests that these innovations spread from an American point of origin and spread across the priesthoods that remained most united by the Atlantic maritime traffic. Among the Afro-Cuban travelers who sojourned in Brazil before "returning" to the Coast was "Vicente Barragão, born in Havana, Nagô, married, profession—commerce, freed, aged 42 years . . . brought in his company his wife by the name of Delfina, manumitted black woman, born on the Mina Coast, aged 48 years" (Arquivo Pùblico do Estado da Bahia, Seção Colonial e Provincial, Polícia, *Registro de entrada de estrangeiros*, 1855–56, book 5667, p. 8, 23 Jan. 1855).

21. Also after the mid–19th century, it became increasingly common in Nigeria and the Bénin Republic to extend the terms "Ànàgó" and "Nàgó" to the same broad group that is now called "Yorùbá." Historian Robin Law is convinced that this parlance "represents a feedback of Brazilian usage" (personal communication, 1998; also Law 1997:11).

22. This passionate defense of Yorùbá language was ironically inspired by the Education Ordinance of 1882, which required students to read and write exclusively in English and forbade instruction in African languages (Omu 1978:107–8).

23. However, what they considered African names, for example, were actually a Euro-African hybrid, in the sense that an etymologically African given name was paired with an etymologically African surname. Like most precolonial peoples, the ancestors of the Yorùbá had no surnames.

24. Concerning these New World blacks' appearance in the Lagosian press, see, for example, Blyden's lecture "Christianity and the Negro Race" (25 Aug.

1883, vol. 1, no. 6, pp. 3–4; 29 Sept. 1883, vol. 1, no.7, pp. 1–3; also in no.9, pp. 1–3) and his article "The Education of the Negro," *LWR*, 2 May 1896, p. 5. See the citation of Blyden's ideas, for example, in *LS*, 1 Apr. 1896, p. 3. See a reprint of Booker T. Washington's article "The Blackman's View," *LS*, 22 Apr. 1896, p. 3. See W. E. B. Du Bois's "The Conservation of the Races," reprinted in *LWR*, 31 July 1897, pp. 2, 5–6.

25. For example, "Dr. Crummell Is a Pure Negro," *LS*, 2 Sept. 1896, p. 3; "Paul Lawrence Dunbar the Young Rising Poet of America Is of Pure African Blood," *LS*, 11 Nov. 1896, p. 3. See also *LS* of 6 Apr. 1898 on "the famous Askia the great"—"a full-blood negro"; and *LWR* of 29 Aug. 1891, p. 2, on the extraordinary success of "pure-blooded Negro youths" in European educational institutions as disproof of theories of Negro inferiority. An article reprinted in the *LWR*, 23 May 1896, p. 5, from a Black North American publication called the *Reformer* advocates pride in the race, racial loyalty, and racial purity as a necessary condition of racial improvement. An introduction to the reprint of Du Bois's "The Conservation of Races" describes Du Bois as "a pure Negro" (*Lagos Weekly Record*, 31 July 1897, p. 2). See also critical remarks by the *Lagos Weekly Record* editor, John P. Jackson, about the ideas and mixed racial background of T. Thomas Fortune, editor of the *New York Age*. In the *Nigerian Chronicle*, 19 Nov. 1909, p. 6, the writer expresses pride in the full-blooded black doctor Africanus Horton and deplores his mistreatment.

26. Among these SMA affiliates are Fathers Noël Baudin, Pierre-Bertrand Bouche, and E. Courdioux.

27. *Iwe Kika Ẹkẹrin* (1944[pre-1896]) is a primer by Yorùbá Protestant missionary Andrew L. Hethersett. Rodrigues reproduces Martiniano do Bonfim's critical reading of this text.

28. See also Angarica (1955:5) on a usage of the Yorùbá Bible in the propagation of "Santería," or Lucumí religion, in Cuba.

29. See *West Africa*, 22 Aug. 1903, vol.6, pp. 211–14.

30. Martiniano do Bonfim also critically appropriates the work of Rev. Andrew L. Hethersett in his project of religious "purification" in Bahia (see Rodrigues 1935[1895]:42–43).

31. Arquivo Público do Estado da Bahia (hereafter APEB), *Registro de Entrada de Estrangeiros*, 1855–56, Seção Colonial e Provincial, livro 5667 (9 Jan. 1855; 21 Mar. 1855; 20–21 Nov. 1855; 11 Jan. 1856; 28 Jan. 1856); APEB, Presidência da Província, Polícia do Porto, Mapas de saída e entrada de embarcações, 1886–93, Seção Colonial e Provincial, Maço 3194-5, File for 1889 [*sic*], 12 June 1869 [n.b., this 1869 item is out of place but was indeed found in the 1889 file]; APEB, Presidência da Província, Polícia do Porto, Mapas de saída e entrada de embarcação, 1873–78, Maço 3194-3, see 31 Jan. 1876. See also APEB, Presidência da Província, Polícia do Porto, Mapas de entradas e saídas de embarcações, 1878–85, Maço 3194-49, file for 1879, 15 July 1879; APEB, Presidência da Província, Polícia do Porto, Mapas de entradas e saída de embarcações, 1878–85, Seção Colonial e Provincial, Maço 3194-4, 20 Dec. 1878; APEB, Presidência da Província, Polícia do Porto, Mapas de saída e entrada de embarcações, 1886–93, Seção Colonial e Provincial, Maço 3194-5, file for 1889, 4 June 1889; APEB, Polícia do Porto, *Registro de Entrada de Embarcações*, 1886–90 (Maço 5975, e.g.,

27 May 1885 [*sic*]; 22 Oct. 1885 [*sic*]), and 1889–92 (Maço 5976, e.g., p. 7), Seçao Colonial e Provincial; APEB, *Livros da Inspetoria da Polícia do Porto—Entradas de Passageiros*, vol. 6, Anos: 4 Dec. 1991 to 21 Mar. 1895, 27 Aug. 1892, p. 80; APEB, *Livros da Inspetoria da Polícia do Porto—Entradas de Passageiros*, vol. 8, Anos: 2 Jan. 1896 to 31 Dec. 1898, 7 Mar. 1898; APEB, Polícia, *Registros de Passaportes*, 1881–85 (Book 5909) and 1885–90 (Book 5910), Seção Colonial e Provincial; Moloney 1889:255–76; L. D. Turner 1942:65; Olinto 1980[1964]: 266–67; Manuela Carnero da Cunha 1985:123, 125; Pierson 1942:238–39; Verger 1976[1968]:464; Lindsay 1994:43–44; 1995 interviews in Lagos, Rio, and Bahia with Paul Lọla Bamgboşe-Martins of Lagos and Rio, Regina Souza of Salvador, Beatriz da Rocha of Salvador, Aïr José Sowzer de Jesus of Salvador, Albérico Paiva Ferreira of Salvador, George Alakija of Salvador, and Yinka Alli-Balogun of Lagos—all descendants of the turn-of-the-century African-Brazilian travelers and, many of them, travelers in their own right.

32. See the journals of Rev. Thomas Benjamin Wright (CA2 O 97), 14 Apr. 1867, CMS Archives, University of Birmingham. I thank J.D.Y. Peel for alerting me to this reference.

CHAPTER TWO
THE TRANS-ATLANTIC NATION

1. On the contrary, the hierarchical logic of race, the phenomenon of political dynasties (such as the Kennedys), and other hereditary modes of authority (such as the British House of Lords and monarchy) have endured at the heart of the territorial nation-state, as have religious definitions of citizenship. It is clear that the territorial nation-state was a new idea in the late 18th century, but we must not exaggerate the purity of the religious forms of community that Anderson says preceded it or of the nationalist forms that followed its invention. Pace Anderson, it is difficult for me to imagine a time or place in prenationalist Europe where Christendom's contrast to either Judaism or Islam was not central to its own conscious imagination of community, and the nationalist period is notoriously fraught with programs that defined citizenship by explicit contrast to the Jewish Other in the midst of the nation-state. Officially Anglican, Catholic, Islamic, and Jewish nation-states have been more the norm than the exception.

2. For example, in the battle for the 1980 Republican presidential nomination, George H. W. Bush branded Ronald Reagan's proposed fiscal policies "voodoo economics." His meaning was clear to hundreds of millions of North Americans who knew absolutely nothing factual about Voodoo, or Vodou. Even in 2003, when tens of thousands of Haitians, many of whom practice Vodou, lived in the New York area, the otherwise highly intelligent editorialists of the *New York Times* chose the same trope to criticize the fiscal policies of the current president, George W. Bush ("Voodoo vs. 'Rubinomics,'" NYT, 16 Feb. 2003, p. 10). Perhaps if Bush and the *New York Times* staff recognized Haitians as their cultural and historical contemporaries (and the practitioners of other Afro-Atlantic religions as a significant part of their potential electorate and subscription-paying readership), they would be no more inclined to repeat the

term "voodoo economics" than, say, "papist economics," "Hebe economics," or "towel-head economics." It is difficult to say whether they are more guilty of racism or of the stagist historical imaginary.

3. Even language categories of this order are variable in their age and in the circumstances of their usage. Terms such as "Ewè-Gèn-Ajá-Fòn" and "Gbè" are the coinages of linguists rather than of native speakers, and they are anachronistic as references to emically recognized sociolinguistic categories during the period under consideration here (Capo 1991).

4. The Haitian Vodou religion is the subject of a particularly rich ethnographic tradition, which includes Métraux (1972[1959]) and K. M. Brown (1991). Nowhere else in Brazil has the Jeje nation been as effective at institutionalizing itself as in the state of Maranhão (Ferretti 1986:13–14). However, my own study concerns Bahian manifestations of this religion, which were the most actively connected to the larger black Atlantic world.

5. The following are among the Africanist historiographical and ethnographic sources I have consulted in search of variants of the "Djedji" ethnic label. The items in which the term appears are marked with asterisks: Ogilby 1670; Barbot 1992[1678–1712, esp. 1688]; des Marchais 1703–6; Bosman 1967[1704]; Labat 1731; Snelgrave 1734; Atkins 1970[1735]; W. Smith 1967[1744]; Hardwicke and Tweeddale 1745, 1746; G. Smith 1751; Norris 1789; Dalzel 1967[1793]; Pires 1957[1800]; McLeod 1820; Duncan 1968[1847]; Forbes 1851; Burton 1966[1861–64]; Borghero 1864:423;* Bouche 1868;* Bourquet 1872:2, 4;* Bourquet 1873;* Desribes 1877:309, 318, 322;* J. B. Wood 1881; Baudin 1885; Bouche 1885:20, 77, 107;* d'Albéca 1889;* Ellis 1970[1890]; Akinṣọwọn 1930[1914]; Labouret and Rivet 1929, including a copy of the *Doctrina Christiana, Y explicación de sus Misterios, en nuestro idioma Español, y en lengua Arda* (1658); Herskovits and Herskovits 1976[1933]; Maupoil 1988[1934–36]; Quénum 1938[1931]; Akindélé and Aguessy 1953; Mercier 1954; Verger 1966; Polanyi 1966; Akinjogbin 1967; and Manning 1982. See also Moreau de Saint-Méry (1958[1797]), the observer of late 18th-century Saint Domingue whose catalog of African ethnic groups on the island includes no mention of "Djedji" or its variants. The reader will note below a number of early-20th-century Africanist sources that do invoke the term "Djedji," but in a sense far narrower than those above.

6. However, the term "Jeje" was not *universally* known in 18th-century Brazil, even among those who studied the language of Afro-Brazilians who had been *born* in the Ewè-Gèn-Ajá-Fòn region of West Africa. Peixoto's extensive vocabulary and phrase book of the "Lingua Geral de Mina" (1943–44[1741]) was based on the speech of "Mina" slaves in Minas Gerais, Brazil, during the mid–18th century. It represents a language remarkably similar to the modern "Fòn" language (cf., e.g., Rassinoux 1987). Yet Peixoto's book makes no mention of the term "Jeje."

7. Consider the similar transplantation of the Brazilian "Nagô" ethnonym, in reference to all the proto-Yorùbá peoples, to what is now the People's Republic of Bénin (Law 1997:11).

8. "Propagande Coloniale: Conférence de M. George François," in *RDD*, 3rd year, nos. 5–6, May–June 1902, pp. 1–2.

9. In 1902, its population was at least 150,000 ("La Région de Porto-Novo," *RDD*, 3rd year, nos. 5–6, May–June 1902, pp. 1–2. Also "Organisation and fonctionnement de l'enseignement au Dahomey." Letter from the Lt. Gov. of Dahomey to the Gov. Gen. of French West Africa, Gorée, 20 August 1904," CARAN, Archives Nationales de France, Paris (hereafter ANFP), 200MI/1151, J26, Letter 696. See also Letters 907 and 192 from the Lt. Gov. to the Inspecteur du Service de l'Enseignement, Office of Gov. Gen. of AOF, Gorée, dated 19 May 1905; Letter 689 from the Lt. Gov. of Dahomey and Dépendances to the Gov. Gen. of AOF, Dakar, concerning 1906; Letter 909 from the Lt. Gov. of Dahomey and Dépendances to the Gov. General's Office of Inspection de l'Enseignement, concerning 1907; and so forth on 200MI/1151.

10. *Rapport Monographique sur le cercle de Porto Novo*, 1921, CARAN, ANFP, 1G353, bobine 200MI/696.

11. The early Dahomean newspapers I was able to identify at the National Library of France, Versailles, are *La Voix du Dahomey* of Porto Novo (1926–36); *Suprême Sagesse* of Porto-Novo (1934?–1938); *La Presse Porto-Novienne* of Porto Novo (1933–37); *Le Phare* of Cotonou (1929–45); *La Tribune Sociale du Dahomey* of Ouidah (1934–?); and *Le Reveil*, a Catholic church publication in Abomey-Calavi (1934?–1938?).

12. *Rapport Monographique sur le cercle de Porto Novo*, 1921, CARAN, ANFP, 1G353, bobine 200MI/696; "La Région de Porto-Novo," *RDD*, 3rd year, nos. 5–6, May–June 1902, pp. 1–2 (Bibliothèque Nationale de France, Paris [hereafter BNFP]); "Un enlèvement sensationnel," *SS*, 5th year, no. 7, 1 July 1938, pp. 1–3, 7 (Bibliothèque Nationale de France, Versailles [hereafter BNFV], Gr. fol-Jo-393, I, no. 8V, no. 9); "Coutumes Nagô et Djèdj (Cercle de Porto-Novo), 1933," in *Coutumiers Juridiques de l'Afrique Occidentale Française* Paris: Librairie Larose, 1939, tome III, série A, no. 10, pp. 475–530. Similarly, the Brazilian historian Braz do Amaral uses the term "Jeje" to describe only the peoples of the littoral, contrasting them with, for example, the core population of the inland Dahomean kingdom (Amaral n.d.:673, 674, cited in Ramos 1946[1937]:301n). See also personal communication, Elysée Soumonni, 6 Mar. 1996; Lima 1977:14–15; R. N. Rodrigues 1945[1905]:176; Verneau 1890–91:251. From around 1914, these non-"Yorùbá" natives of Porto-Novo came more and more often to be called "Ègùn" (in Yorùbá) or "Goun"/"Gùn" (e.g., Akinṣọwọn 1930[1914]; Kiti 1929).

13. On French commercial and linguistic protectionism against a major British and Lagosian threat, see "Protectorat Français à Porto-Novo," in *Chronique de la Mission du Golfe du Bénin, 1861–1867*, no. 2, entry no. 20.497, rubrics 12/8-200, pp. 9–10, Archives of the Society of African Missions, Rome; "La Région de Porto-Novo," in *Revue du Dahomey et Dépendances*, 3rd year, nos. 5–6, May–June 1902, pp. 1–2, BNFP; *Rapport Monographique sur le cercle de Porto-Novo*, 1G353, bobine 200MI/696, pp. 13–19, for 1920 and 1921, CARAN, ANFP; *Chronique Locale: Porto-Novo*, in VD, 2nd year, no. 14, 1 Mar. 1928, p. 2. Between 4 and 5 February of that year, "un indigène revenant de Lagos (Nigeria) et chargé de marchandises aurait été tué par un coup de fusil tiré par les agents de Douane qui gardent la frontière"; "Rapport Monographique sur le Cercle de Allada," dated 18 Apr. 1921, and ". . . sur

Abomey," dated 17 Apr. 1921, pp. 17–19—both from CARAN, ANFP, 1G353, bobine 200MI/696; "La Région de Porto-Novo," in *RDD*, 3rd year, nos. 5–6, May–June 1902, pp. 1–2. They would exempt the currencies of the "union latine" from these prohibitive tariffs. See also *Le Phare du Dahomey*, June 1933, article by F. Jourdier, BNFV.

14. "Rapport d'Ensemble sur le Service de l'Enseignement pendant l'année 1913," 200MI/1151, J26, CARAN, ANFP.

15. Letter from Courdioux in Porto Novo to the Superior General, 29 Apr. 1868 (20.374, 12/80200); letter from Bouche in Porto Novo to Planque, 1 June 1868 (20.381, 12/80200); letter from Chausse in Porto Novo to Superior, 23 Oct. 1872; letter from Bourguet in Lagos to Planque, 26 June 1872 (17047, 12/80200); letter from Cloud to Superior, 11 Oct. 1872 (17057, 14/802.00, 1872); all in Archives of the Society of African Missions, Rome.

16. "La Région de Porto-Novo," *RDD*, 3rd year, nos. 5–6, May–June 1902, pp. 1–2, BNFP; *Rapport Monographique sur Cercle de Porto Novo*. 1G353, bobine 200MI/696, CARAN, ANFP.

17. Langley 1973:286–325; Ballard 1965; Kaké 1982:204–5. For an instance of the new use of the term "Fòn" in the 1930s, see Quénum 1938[1931].

18. The Brésiliens were particularly aggrieved by being subjected to the jurisdiction of courts of "customary law." This objection might be interpreted as an assertion of special privilege, like the similar assertions of Haitian mulattoes before the Haitian Revolution and of Louisiana Creoles of color. On the other hand, the demand to be treated according to the ethnically neutral laws of the liberal state was precisely the demand of the white creole revolutionaries who founded the New World nation-states. Despite their pretensions of lofty, liberal principle, the white creoles who founded the United States and the postimperial Brazilian nation (not to mention the rulers of Old World settler colonies like South Africa and Israel) never intended to grant equal rights to their compatriots of other races.

19. See, for example, *O Alabama*, 22 Sept. 1868, p. 2; 16 Feb. 1869, p. 2. I thank Dale Graden for alerting me to these references.

20. *O Alabama*, 2 May 1867, p. 3; 16 Feb. 1869, p. 2.

21. Ludovina Pessoa is said to have left the city of Bogum in Africa to found the original temple in the Bahian town of Cachoeira de São Felix. She was still ruling in 1869. She was succeeded by Mãe Valentina, who ruled sometime between Ludovina's death and Valentina's own death around 1936. Succeeding Valentina around 1937 was Mãe Emiliana (Emilia Maria da Piedade), who was born and reared on the Salvador temple grounds. Emiliana ruled for nineteen years and died there in 1956, at the age of 131. Her successor was Mãe Ruinhó (Valentina Maria dos Anjos), who ruled for approximately eighteen years, until her death in 1975, and was succeeded by her daughter Nicinha (Evangelista dos Anjos Costa). Mãe Nicinha ruled for approximately twenty years, until her death in 1995. Upon my last visit to Bahia, in 1995, Nicinha had died. Mãe Iracema, or "Ìndia," was appointed in 2003 as her successor. Valeria Auada, "Projeto Terreiros: A rica história dos terreiros de Candomblé da Bahia que o tempo ameaça destruir," *TB, Variedades* section, 28 Mar. 1987 (*TB* archives); "Jornalista estuda Bogum há 33 anos," *AT*, 14 Jan. 1986 (Pasta 324, *AT* archives); Biaggio Talento, "Àrvore ritual do candomblé divide Salvador," *ESP*,

12 July 1990, p. 17 (*ESP* archives); "Bogum quer tombamento para preservar o seu bissecular Terreiro," *AT*, Segundo Caderno' section, 24 July 1986 (*AT* archives?); Jehová de Carvalho, "Mundo Jeje comemora cinquentenário de sua mãe-de-santo," *AT, Lazer e Informação* (*AT* archives?); "Sucessão," *AT*, 30 Dec. 1975, p. 1 (*AT* archives); "'Cirrum' começou no Bogum e 'Gamo' é a nova yalorixá," *AT*, 30 Dec. 1975, p. 3; *O Alabama*, 16 Feb. 1869, p. 2; Jeje Pai-de-santo Vicente Paulo dos Santos, interview, 3 Jan. 1996.

If there were no abnormally long gaps between Ludovina's death and Emiliana's succession, the end of Ludovina's rule (from 1869 onward) and the entirety of Valentina's would together have taken up around sixty six years. Assuming that Ludovina was halfway through an average term in office in 1869, it seems likely that Bogum and its leadership structures ceased to function for a period of more than thirty seven years, sometime between 1879 and 1917.

22. That is, Rodrigues noted but one possible "vestige" of Dã worship, in the form of a snakelike metal sculpture found in the temple of Mae Livaldina, but the priestess denied that it represented Dã or any snake god. For her, it represented the god of iron, Ogum (R. N. Rodrigues 1945[1905]:368). In support of her claim, see also Peel 1995.

23. One exceptional case concerns a prosperous Afro-Brazilian woman who, upon her death in 1842, bequeathed the following to a Mr. João Simoens Coimbra: "my large water snake (*gibóia*), that is eight and a half palms in length [and] a crucifix also of gold, featuring [an image of the] Holy Lord" (Côrtes de Oliveira 1988:48–49, 50–51; translation mine).

A 1982 oral history recounts the presence of an unnamed serpent god in the town of São Félix, in the Bahian Recôncavo, during the mid–19th century (see Wimberly 1998:83).

Finally, anthropologist Luis Nicolau reports archival information which indicates that, in 1870, two people were initiated in the worship of the snake god Besseim (whose name neither Nicolau nor I have encountered in the Bénin Republic [the former Dahomey]) in Cachoeira, a city in the Bahian Recôncavo region. However, in his extensive review of 19th-century archives in the Recôncavo, he has never found any mention of the EGAF snake god Dã (personal communication, 18 Sept. 2000), diminishing the likelihood that migrations from or contact with the Recôncavo was the source of the Jeje renaissance in Salvador during the 1930s and 1940s.

Among the sources on Brazilian Jeje religiosity from which mention of the snake gods is surprisingly absent, consider Reis 1983:73; 1986:112; Peixoto 1943–44[1741]; and "Desacato à realeza—Tiramos da 'Gazeta' de 28," *PSP*, 30 Sept. 1879 (Pasta 16.017, *ESP* Archives).

24. Objects collected by Arthur Ramos in 1927 included an initiate's bracelet in the form of a snake biting its tail, as well as a sword with a snake head at the end and designed in "typically Dahomean" style. Ramos declared that some of these objects had clear counterparts among the designs depicted in Herskovits's 1938 *Dahomey* (Ramos 1940[1934]:55–56; 1946[1937]:303). In the late 1930s, a series of songs mentioning the term *vodum* was recorded (Edmundo Correia Lopes 1939, cited in Ferraz 1941:272; Ramos in Pereira 1979[1947]:13); one *babalao* was observed keeping a pet snake; and instances of snake veneration

were observed in one Pernambucan house, as well as two Bahian houses—those of Pai Joaozinho da Goméia and Pai Ciríaco (Ferraz 1941:273; Ramos in Pereira 1979[1947]:12, 13; Carneiro 1948:51).

25. Raimundo Nina Rodrigues indentifies the "Maís" as a small minority of the Jejes he counted in Bahia at the turn of the century (1988[1905]:106).

26. With the term "diacritica," I refer to Barth's (1969) sense that ethnic groups regularly employ relatively small cultural markers to distinguish them-selves from neighboring ethnic groups that are otherwise culturally similar. He argues that neighboring ethnic groups are distinguished more substantially by their divergent occupations and economic niches than by their overall cultures.

27. Archives of the Society of African Missions, Rome, Rev. Père I. Pélofy, "Pétronille," manuscript in Pélofy file, 3D34; Pélofy, "La femme chez les Minas (2)," n.d., p. 13.

28. Pélofy, manuscript of "Notice sur Agoué," 1937, pp. 8–9, Archives of the Society of African Missions, Rome, Pélofy file, 3D34. Also published in *EMAL*, Jan., Feb., and March 1938, as "Agoué au Dahomey." See also Verger 1976[1968]:475–77 (esp. 467), 533; J. M. Turner 1975[1974]:75, 102–3, 133.

29. D'Almeida may or may not have traveled personally to Cuba, and it is un-clear whether these slaves had belonged to him for a lengthy period or were sim-ply there awaiting commissioned sale on his behalf. What is minimally certain, however, is that he was in a central position among Brazilian, Cuban, and Coastal traders establishing the common "trademarks" according to which African captives were being sold and permanently labeled in both Brazil and Cuba (see also Eltis 1987:157–58).

30. Archives reveal that some Afro-Cuban travelers resided briefly in Bahia and Rio on their way to West Africa or repeatedly visited Bahia. For example, a Ha-vana-born "Nagô" businessman and his "Mina" wife passed through Rio on 21 January 1855 and arrived in Bahia on 23 January. These "freed blacks" declared their intention to reside temporarily in Bahia and then proceed to the "Coast of Africa" ("Vicente Barragão, born in Havana, Nagô, married, profession—business, freeman, 48 years of age. . . . brought with him his wife Delfina, manu-mitted black, born on the Mina Coast, 48 years of age . . . "; Arquivo Pùblico do Estado da Bahia, Seção Colonial e Provincial, Polícia, *Registro de entrada de es-trangeiros*, 1855–56, book 5667, p. 8, 23 Jan. 1855). The other police registers of foreigners' entry, which might reveal additional cases of this sort, have not been preserved at the Bahian State Archives. Extant police passport records in Bahia, however, reveal the arrival on 3 April 1886 of a Juan Antones on his way to the "Coast of Africa." This African-born man had apparently acquired his Spanish given name in Cuba and had gone to live in Lagos long enough to receive an English passport. Just having arrived for a stopover in Bahia, he informed the police that he intended ultimately to return to the Coast of Africa (*Polícia—Passaportes, 1885–1890* reads, "Juan Antones, black, African with English pass-port—Visa good to proceed to the Coast of Africa" (Arquivo Público do Estado da Bahia, Seção Colonial e Provincial, book 5910, p. 66, 3 Apr. 1886; translation mine).

31. The *-no* suffix in "Magino" and "Caviano" corresponds to the Fòngbè suffix *-nu*, which means "inhabitant of" (Rassinoux 1987:173).

32. So important is this principal item in the cuisine and rituals of Candomblé that one way of describing a person as a member of this religion is to say that he or she is "in the oil" (*no azeite*).

33. This "Asante" is not to be mistaken for the Akan kingdom headquartered in Kumasi (see, e.g., Bergé 1928:719).

34. Consider also the case of João de Oliveira, who was born in the "interior of the Slave Coast," taken to Bahia in 1733, and eventually returned, still a slave, to settle on the Coast near Ouidah. He then sent money to Bahia to purchase his freedom. He became a great protector of and middleman on behalf of the Portuguese on the Coast and played a critical role in establishing the slave trade in Porto-Novo in the 1750s and in Lagos in the 1760s. He maintained a close relationship with his former owner's family. While living on the Coast, he heard that his former owner had died, leaving his widow destitute, and he is said to have "helped her for as long as she was alive" (Verger 1976[1968]: 477–78n18; Law and Mann 1999:317–18, 318n23).

35. Maurille de Souza, first secretary for cultural affairs, Embassy of Bénin, Washington, D.C., personal communication, 18 Feb. 2003.

36. See, also the unusually well-documented case of João José de Medeiros in *Le Phare du Dahomey*, 15 Mar. 1931, p. 9, BNFV. Born in Ouidah, he studied at the "Lycée de Bahia" before returning to visit the colony of Dahomcy. Only the death of his father prevented him from continuing his studies in Bahia.

37. While Landes understands the designation of certain people as "Africans" to mean simply that they are blacks without wealth or education, the people chosen by her Bahian guide to represent this ethnic and class category were the retail merchants of Salvador's "Lower City," whose ancestors and characteristic wares were "from the west coast of Africa" (Landes 1947:138–41).

38. That the Bahian empire of Dã is a novel formulation is demonstrated not only by the long delay in the appearance of Dã worship in Bahia but also by the contrasting case of Maranhão. The city of São Luís in the Brazilian state of Maranhão is home to a religion called Tambor de Mina, with extensive similarities to the Candomblé of Bahia and to the Xangôs of other Northeastern states. Yet the most exemplary nation in Tambor de Mina religion, as well as in its offshoots in the states of Rio and Pará, is not Nagô but Mina-Jeje, or Jeje (Ferretti 1986:14). Consistent with Maranhão's lack of 20th-century contact with the snake-worshiping Djedji Coast, Maranhão's Mina-Jeje nation is distinguished by the virtual absence of the snake god Dã from its pantheon and national emblems

The names "Dambala Uedo," "Dambellah," and "Dágêbe" appear in some few Tambor de Mina stories and songs, but not with any prominence or with intentional reference to any snake gods (Ramos in Pereira 1979[1947]:16; Pereira 1979[1947]:206, 219, 231, 233; Ferretti 1986; also Leacock and Leacock 1975). Only Eduardo reports, without further details, "By the Dahomeans, Dã is believed to be a serpent [citing Herskovits 1938, vol. 2, chap. 30, esp. p. 248], and this belief has been maintained in Maranhão" (Eduardo 1981[1948]:78).

39. Hence it is surprising that Ong (1999) regards "graduated sovereignty" (whereby nation-states concede various elements of their sovereignty over various parts of their territory to transnational actors) as novel or contrary to the antecedent reality of most nation-states.

40. This case also embodies the pan-American and indeed global tension within nationalism. (Cf. Wafer and Santana's [1990] contrast between the "centrifugal" and "centripetal" movements of the Euro-Brazilian intelligentsia, which are embodied in the contrast between the Jeje-Nagô gods and the *caboclo* spirits.)

41. For example, in addition to predicting a transnational end to the nation-state, Appadurai writes of the United States, "Although its forms of public violence are many and worrisome, its state apparatus is not generally dependent on forms of torture, imprisonment, and violent repression" (1996:169). Such an observation is clearly a matter of the observer's perspective and is likely to encounter disagreement among large segments of the U.S. population. The United States has one of the highest rates of imprisonment in the world, and African Americans are particularly likely to perceive the U.S. state as a violently repressive actor in their communities. Moreover, the experiences of Rodney King and Abner Louima amply illustrated for African Americans in the 1990s that agents of the U.S. state frequently employ torture in the exercise of their state-appointed duties.

42. For example, England in the 19th century was at once a kingdom, a territorial nation-state, an empire, and a transnational community.

CHAPTER THREE
PURITY AND TRANSNATIONALISM

1. Pierson 1942[1939]:293.

2. Landes 1947:28.

3. I regard the terms "primitive" and "modern" as faulty metaphors insinuating that peoples who are our cultural and historical contemporaries somehow belong to another time period and have a past that is somehow shorter than our own. Like other evolutionist historical paradigms, including the representation of transnationalism as a recent "rupture," it conceals the long-developing and ongoing processes that have been part of every people's history.

4. Anthropologists like Abner Cohen (1969) distinguish ethnic "groups" (which are largely self-defined and corporate in their interests or conduct) from ethnic "categories" (which are lumped together and named chiefly by others).

5. The title of Òyó's head priest of Ṣàngó is, in the current Yorùbá orthography, "Ìyá Nàsó."

6. Martiniano published in a Bahian newspaper his own account of the new offices that he had introduced into the bureaucracy of the Ilê Axé Opô Afonjá temple under the administration of Mãe Aninha. He claims to have established the role of the "Twelve Ministers of Xangô" based on its precedent in the Òyó kingdom. Since he had never traveled to the inland kingdom of Òyó, his innovation, or "revival," must have been based on what he had read, heard, or inferred along the Bahian-Lagosian migratory axis. See "Os Ministros de Xangô," *Estado da Bahia*, 5 May 1937. Martiniano's account of the history and role of the *mamgbas*, or *mógbà* (in modern Yorùbá orthography), is idiosyncratic, undocumented elsewhere in the ethnographic and historical literatures on the Òyó palace and priesthoods. The temple's ultimate use of the term *obá* (cognate with the Yorùbá term *ọba*, for "monarch") is strikingly at odds with any conceivable precedent in Òyó.

On the other hand, Martiniano's emphasis on the number 12 and the division of the *mógbà* nonpossession priests into those of the left and of the right do mark his debt to another idiosyncratic account that we know at least one of Afro-Brazilian voyager had taken to Bahia and shared with Martiniano's student Raimundo Nina Rodrigues. That account is Ellis's *Yoruba-Speaking Peoples of the Slave Coast of West Africa*, which reports: "The Magba, or chief priest of Shango, has twelve assistants, who are termed, in order of authority, right-hand (*Oton*), left-hand (*Osin*), third, fourth, fifth, and so on" (1964[1894]:97).

7. "To this day," says Roberto Motta, some ritual songs (*toadas*) in the Xangô religion of Pernambuco "are still attributed to Bonfim" (personal communication, 8 Nov. 1996; also Pierson 1942[1939]:295).

8. For example, Azevedo Santos 1993:108. *Santera* Marta Vega, director of the Caribbean Cultural Center in New York City, and *babalao* Antonio Castañeda Márquez, president of the Yorùbá Cultural Association of Cuba, have questioned the itinerant Nigerians' claims to superior religious knowledge and Abimbọla's efforts to establish Ifẹ's centrality in the international organization of the *òrìṣà* priesthood.

9. In the United States, both Adefunmi I of Oyotunji, S.C., and significant numbers of Black and white Americans have followed variants of this pattern. New World priests whose power had not been so well established in the Brazilian, Cuban, Puerto Rican, and other U.S. *òrìṣà*-worshiping communities have been willing to place "purity" squarely above American genealogical seniority. Like Mãe Aninha in the 1930s, they have tried to jump to the head of the queue of American priestly leadership by going to Africa for initiation or by seeking instruction directly from itinerant Nigerian priests. Among contemporary Brazilian priests, the most famous exemplar of this phenomenon is Juana Elbein dos Santos, who is said to have been initiated in the worship of the god Odùdùa (nowadays the foremost emblem of bourgeois nationalism among West African Yorùbá/people) in West Africa. She is the author *Os Nàgô e a Morte* (1976), an "African purist" interpretation Quêto/ Nagô religion that is found on the bookshelves of many Candomblé priests.

10. In the *limpezas de egum* I have observed, one sacrificial item is not food: twine is used to wrap up the patient and is then cut off to remove the encumbrances placed in the patient's life by the *egum*. The same procedure takes place in many *limpezas de Exú* (Exú cleansings) as well.

11. In Nigerian Yorùbá, *àmàlà* is the term for rehydrated yam or plantain flour, which is typically eaten with pureed okra and a meat sauce.

12. The *resguardo* usually lasts for the number of days corresponding to the emblematic number of the god being honored.

13. Some observers note that these prohibitions have declined in effectiveness over time and, understandably, that they apply more often to heterosexual than to homosexual relations. After all, being nonprocreative, homosexual relations usurp the symbolic grounds of Candomblé's *initiatic* reproduction to a lesser degree.

14. Similarly, a Lebanese-American initiate of "Santería" chided me in 1984 for touching his head, the portal of the *oricha*'s entry. In explanation of his objection, he asked rhetorically, "You wouldn't touch a woman's private parts, would you?"

15. While I know of one "great house"—Gantois—where succession to the highest office has, to date, remained within the same biological family, the most highly purist temple, Opô Afonjá, appears to forbid hereditary succession. In all houses, the legitimation of successors is based more on divination by a neutral party and thus selection by the gods than on heredity.

16. Matory 1994b:70–73. Depending on which political faction one studies from the mid–19th century onward, the royalist, Ọ̀yọ́-biased Ṣàngó might seem the most important or culturally emblematic deity of the Yorùbá people. On the other hand, the metropolitan Yorùbá bourgeoisie appears to favor a range of Ifẹ̀-biased gods: the Ifẹ̀-centered Ọ̀runmìlà, god of the highly literary Ifá divination priesthood and favorite of the Lagosian Cultural Renaissance; Odùdùa, mythical founder of Ilé-Ifẹ̀ and favorite of Awolọwọ and his political followers; or Ọbàtálá, a quasi-Christ-like god headquartered at Ifẹ̀ and an apparent favorite of apologists for Yorùbá religion in the 20th-century mission churches, such as Idowu and Awolalu. Local bourgeoisies resisting the dominance of the Ìjẹ̀bú in commerce and party politics, as well as local royal dynasties resisting the dominance of Ọ̀yọ́ in chieftaincy politics, tend to build up the importance of local gods, and especially local variants of the "white gods" among them. Hardly any major leadership faction in Yorùbá cultural politics is without a clear and nearly predictable consensus on the matter of which gods matter most.

17. Ocha, or Lucumí religion, is the site of a similar struggle in the United States (Matory, forthcoming a).

18. I do not know when Opô Afonjá began to perform this ceremony or when it was named the "Water [or Waters] of Oxalá" in contrast to the name given to it in Opô Afonjá's elder sister temple Gantois—that is, the "New Yam." Reliable sources suggest, however, that Opô Afonjá has been performing this rite at least since 1934 (Santos 1962:96).

19. For glimpses of the role of popular literacy in Candomblé members' and "African" ethnics' cultural self-articulation in Bahia from the 1890s, see R. N. Rodrigues 1935[1900/1896]:35, 42–43; Pierson 1942[1939]:242–43, 259, 272, 293; Landes 1947:29–30, 112–15, 174; and Martiniano's "Os Ministros de Xangô," *Estado da Bahia*, 5 May 1937. Consider also the publications of *egum* priest Deoscoredes ("Didi") dos Santos (1962), Maria Stella de Azevedo Santos ("Mãe Stella" [1993]), and Juana Elbein dos Santos (1976), the last of which several Candomblé priests have recommended to me as an excellent philosophical analysis of their religion. It and Verger's volume (1981) are found on the shelves and in the drawers of many Candomblé priests. Since the 1980s, many priests have taken book-based courses in Yorùbá language at the Center of Afro-Oriental Studies of the Federal University of Bahia.

CHAPTER FOUR
CANDOMBLÉ's NEWEST NATION: BRAZIL

1. Landes 1947:17; "MEC quer preservar a cultura negra na Bahia," *AT*, 9 Sept. 1982 (Pasta 324, *AT* archives); "Preservação do candomblé é defendida em encontro," *AT*, 3 Aug. 1987, p. 3, attachment to the report "Encontro brasileiro

da Tradição dos Orixá e Cultura: Preparatória da 'IV Conferência Mundial da Tradição dos Orixá e Cultura'—IV COMTOC-'88," Salvador, Bahia, 31 July–2 Aug. 1987, Local—Ilê Axé Opô Afonjá-Cabula (*TB* archives).

2. Peter Fry sagaciously contrasts the appropriation of Afro-Brazilian cultural products (such as the pork-and-bean-based *feijoada*, samba, and Candomblé) as emblems of the Brazilian *national* "mass culture," with the use of Black-invented dishes similar to *feijoada* in the United States as emblems of specifically *Black* cultural difference (1982:52).

3. Though Freyre's intention to contrast Brazilian race relations with the North American ones he had observed firsthand is well known, he entered into the dialogue about *regionalism* in the United States as well. Sometime after the 1926 Congress in Recife, he attended a "Regionalist Conference" organized in Charlottesville, Virginia, with the support of President Franklin D. Roosevelt (Mendonça Teles 1987:344).

4. These introspective, nationalist monographs include Sérgio Buarque de Holanda's *Raízes do Brasil* (1936), Vianna Moog's *Bandeirantes e Pioneiros* (1954), and Caio Prado Jr.'s *Formação do Brasil contemporâneo* (1942).

5. Among the other Regionalists were José Lins do Rego, Graciliano Ramos, Rachel de Queiros, and Jorge Amado, who wrote Realist fiction set on the sugar plantations, in the arid backlands (*sertōcs*), and in the dark urban neighborhoods of the Northeast. Édison Carneiro, Aydano de Couto Ferraz, and other participants in the Afro-Brazilian Congresses of the 1930s articulated the allied vision of a *negro* and *mulato* folk at the core of Northeastern and, ultimately, of Brazilian distinctiveness.

6. The passage continues:
There have been others who have hinted at the possibility that the inclination to colored women to be observed in the son of the family in slave-holding countries is a development out of the intimate relations of the white child with its *Negro wet-nurse*. The psychic importance of the act of suckling and its effects upon the child is viewed as enormous by modern psychologists, and it may be that Calhoun is right in supposing that these effects are of great significance in the case of white children brought up by Negro women.

7. This is the Northeastern term for a sexy body odor, associated especially with *mulatas*. It is also called *budum* (see Freyre 1986[1933]:279n5).

8. On the term *ciclo do negro* (black cycle), see Jehová de Carvalho, "Mundo Jeje comemora cinquentenário de sua mãe-de-santo," *AT*, 24 July 1988, *Lazer e Informação* (*AT* archives). Carvalho is an *ogã*, or non-possession priest and sponsor, of the "great" Bogum temple of the Jeje nation.

9. Amid Europeans' 19th-century racialization of the emergent hierarchies in the global political economy, Brazilians were not the only nonwhite nation to jockey for status by arguing their racial superiority to other nonwhite nations. Dikötter (1997) details a range of similar East Asian cases.

10. In this case, it was the Casa Branca do Engenho Velho temple, or Ilê Iyá Nassô, in Salvador.

11. Landes 1947:34, 61; also Duarte 1974:12; Ângela Peroba, "Candomblé Recupera o Seu Passado," *JAB*, n.d. (*TB* archives); Valéria Auada, "'Projeto Terreiro': Com ajuda dos orixá e dos homens é preciso salvar os terreiros de

candomblé," *TB*, 27 Mar. 1987, Variedades, p. 1 (*TB* archives); Antônio Jorge Moura, "Encontro cria Conselho do Candomblé," *JOB*, 3 Aug. 1987 (Pasta 22.844, *ESP* archives); Vito Hugo Soares, "Candomblé recupera os objetos sacros," *JOB*, 16 Mar. 1988 (Pasta 22.844, *ESP* archives).

12. Replacing the union, the Federation of the Afro-Brazilian Cult was founded in 1946 and became a public utility in 1960. In 1967, the state governor declared the Candomblé a form of folklore rather than a religion, thus exempting it from certain constitutional protections. With that justification, he placed the religion under police supervision, but priests organized no official protest until 1973, when the Casa Branca temple founded the Confederation of Candomblé for that purpose. This move brought the federation back to life and persuaded Governor Santos to liberate Candomblé from police regulation in 1976 (Barbosa 1984:70–71).

13. "Mãe Menininha," *VB*, pp. 3–8, Pasta C-37 [no.1], Bahiatursa office, Salvador.

14. In contrast to recent writings, which tend to represent Carneiro and other "intellectuals" as crusaders coming to the rescue of a persecuted and passive Candomblé (e.g., Dantas 1988:192ff.), Ruth Landes, a contemporary of the events and personalities, writes of priests who battled fiercely to protect their houses long before Carneiro convened the Bahian Congresso and who *protected* the intellectuals more than the intellectuals protected them (1947: e.g., 24, 72, 81). For reports that Mãe Aninha commanded the resistance to the persecution of the Candomblé during this period and sheltered on her temple grounds numerous persecuted politicians, see "Candomblé ganha o 1° museu," *JOB*, 20 Feb. 1989 (Pasta 22.844, *ESP* archives).

15. Roberto Fernandes, "Votos e orixás levam Waldir e Josaphat ao candomblé," JOB, 21 Sept. 1986, Primeiro Caderno, p. 12ff.

16. For example, Mãe Stella of Opô Afonjá recalls that one of her predecessors, Mãe Aninha, lived in Rio at one point in the 1930s. At the time, cabinet minister Oswaldo Aranha was also the highest of the sponsors and nonpossession personnel (*chefe da Casa Civil*). Through him and *ogã* Jorge Manuel da Rocha, she secured an interview with President Vargas that resulted in Decree No. 1202, which guaranteed "liberty for the practice of the orixá religion" (Azevedo 1993:13; Barbosa 1984:70; "Terreiro Faz 300 Anos Hoje," *FSP*, 6 June 1985 (*ESP* archives); Valéria Auada, "'Projeto Terreiros': A rica história dos terreiros de Candomblé da Bahia que o tempo ameaça destruir," *TB*, 28 Mar. 1987, Variedades p. 1 (*TB* archives); Hamilton Vieira, "A história do Axé Opô Afonjá na homenagem a Mãe Stella," *AT*, 14 Sept. 1989, Caderno 2, p. 3 (Pasta 324, *AT* archives); Ângela Peroba, "Federação teme os aproveitores," *JAB*, n.d., p. 9 (*TB* archives). In the neighboring state of Maranhão, the state governor also put an end to the persecution of Afro-Brazilian religions in the 1930s. "Encontro brasileiro da Tradição dos Orixá e Cultura: Preparatória da 'IV Conferência Mundial da Tradição dos Orixá e Cultura'—IV COMTOC-'88," Salvador, Bahia, 31 July–2 Aug. 1987, Local—Ilê Axé Opô Afonjá-Cabula, p. 46 (*TB* archives).

17. Samba schools had to register, have officers, and design their floats and songs on approved themes of Brazilian history. They were and still are judged in

official competitions by upper-class judges, who thereby determine their eligibility for funding during the next Carnaval season.

18. In addition to the earlier-cited reports of his intervention to protect the Candomblé, there are reports of his intervention to protect Umbanda as well. "A umbanda cruza fronteiras," VIS, 3 Oct. 1983 (Pasta 862, AT archives).

19. On Freyre's arrest for his association with the Democratic Left and with the First Afro-Brazilian Congress, see Dantas 1988:163n, citing Mor 1977.

20. "Candomblé: Progresso ou Sacrilégio?" [TB?], n.d., (TB archives).

21. "Projeto Jardins das Folhas Sagradas," project description, Secretaria Municipal do Meio Ambiente, Prefeitura Municipal do Salvador, collected December 1995 from the Secretaria, Salvador, Bahia.

22. See "O governo vai à festa de Oxossi," ESP, 27 Mar. 1973 (Pasta 16.017, ESP archives); Roberto da Matta, "Uma Religião Democrática," MRJ, 28 Sept. 1983 (Pasta 324, AT archives); Elias Fajardo da Fonseca, "Uma festa da umbanda: Os mediuns desfilam, a Riotur patrocina," OG, 15 May 1979 (Pasta 862, AT archives). The archives of Bahiatursa (the Bahian state tourist board) in the Upper City of Salvador contain printed materials going back to at least 1975, identifying Candomblé temples, diviners, and festivals for interested tourists.

23. "Olga de Alaketu, rainha por herança de sangue," FSP, 22 July 1974 (Pasta 324, AT archives); "Sete décadas da estrela do Afonjá," TB, 27 Apr. 1995, Recorte de Jornais, Bahiatursa office, Salvador.

24. Roberto da Matta, "Uma Religião Democrática," MRJ, 28 Sept. 1983 (Pasta 324, AT archives); Leo Gilson Ribeiro, "O Outro Brasil," JT, 12 Mar. 1988, Caderno de Sábado, pp. 1–2 (ESP archives); "Umbanda, Protesto contra as Tradições," UH, 8 July 1973 (ESP archives); Reynivaldo Brito, "As sacerdotisas do Candomblé," AT, 15 Oct. 1983 (Pasta 324, AT archives); "Umbanda: Em vez de política, 'alívio' ao povo," ESP, 20 Nov. 1983, p. 1 (ESP archives); Valéria Auada, "Projeto Terreiro: Com ajuda dos orixá e dos homens é preciso salvar os terreiros de Candomblé," TB, 27 Mar. 1987, Variedades, p. 1 (TB archives).

25. See chapter 6; Dantas 1988:174–81; Rodrigué 1991; "A festa dos orixás: Venerada na Bahia, Mãe Menininha chega aos 90 anos como a primeira dama de uma religião que emergiu das sombras," VSP, 8 Feb. 1984 (Pasta 22.844, ESP archives).

26. Luiz Antônio Novaes, "Na terra do misticismo, quem reina é o presidente Sarney," FSP, 9 Apr. 1989 (TB archives); Luiz Antônio Novaes, " Brasília mistura seitas e política," FSP, 9 Apr. 1989 (TB archives); "Líder da Umbanda vai à Àfrica manter contatos," UH, 22 July 1970 (ESP archives); "Chefe de umbanda, só com carteirinha," JT, 11 Dec. 1986 (Pasta 16.017, ESP archives); "O peso do terreiro: Teses estudam os umbandistas e a política," IE, 1 June 1983, pp. 40–42 (Pasta 862, AT archives); also D. D. Brown and Bick 1987:83. The early-20th-century rumors are documented, for example, in Duarte 1974:12 and Bastide 1978[1960]:164 (concerning the violent reaction to rumors that the governor of the state of Alagoas and his powerful friends "protected, frequented or even took part in sessions of the Xangô religion") and Landes 1947:192 (which reports rumors that powerful Bahian politicians consulted with Candomblé priests). See Reis 1986 on an early-19th-century instance.

27. See "Líder da Umbanda vai à Àfrica manter contatos," *UH*, 22 July 1970 (*ESP* archives); "O boom umbandista," *VSP*, 7 Jan. 1981, pp. 40–41 (*AT* archives); Alexandre Kadunc, "Jamil Rachid: Este homem é o pai de 350 mil crentes," [*VSP* or *MRJ*?], 11 Dec. 1976, pp. 122–24 (Pasta 862, *AT* archives); Joana Angélica, "Umbanda, não. Umbandas: elas são muitas. Quatro, pelo menos," *OG*, 9 Jan. 1978 (*AT* archives). Also Prandi 1991.

28. The exceptional Pai Aizinho de Oxaguiã of the Pilão de Prata temple, however, still appears to conduct a brisk wholesale trade between Lagos and Bahia. He tells me that he personally frequents Lagos and makes his purchases with the assistance of relatives there. Pai Aizinho is the grandson of the famous *babaláwo*, Felisberto Sowzer, who conducted a similar trade during the 1930s.

29. On Cleusa, see Symona Gropper, "Menininha de Gantois: A Festa dos 80 Anos," *JOB*, 13 Feb. 1974 (Pasta 22844, *ESP* archives).

30. "Memorial Mãe Menininha do Gantois," posthumous commemorative booklet, n.d. [published after Jan. 1992; collected 22 Dec. 1995], Pasta sobre "Candomblé," Bahiatursa office, Salvador; Symona Gropper, "Menininha de Gantois: A Festa dos 80 Anos," *JOB*, 13 Feb. 1974 (Pasta 22844, *ESP* archives); "O Gantois e sua origem escrava," *ESP*, 15 Aug. 1986 (Pasta 22.844, *ESP* archives); "Há guerra pelo poder na macumba do Recife," 24 Oct. 1971 (Pasta 16.017, *ESP* archives); "E Salvador pára no enterro de Menininha," *ESP*, 15 Aug. 1986, p. 14 (*ESP* archives). Also "Espíritas-católicos, 'à brasileira,'" [*FSP*?], n.d. [12 Feb. 1977?] (*TB* archives); "Festa na Casa Branca homenageia Mãe Teté," *AT*, 1 Oct. 1985 (Pasta 324, *AT* archives).

31. "Líder da Umbanda vai à Àfrica manter contatos," *UH*, 22 July 1970 (*ESP* archives); "O peso do terreiro: Teses estudam os umbandistas e a política," *IE*, 1 June 1983, pp. 40–42 (Pasta 862, *AT* archives); D. D. Brown and Bick 1987:83.

32. See Roberto Fernandes, "Votos e orixás levam Waldir e Josaphat ao candomblé," *JOB*, 21 Sept. 1986, Primeiro Caderno, p. 12ff. Antônio Carlos Magalhães is identified as the *Chefe político incontestе* in "A festa dos orixás: Venerada na Bahia, Mãe Menininha chega aos 90 anos como a primeira dama de uma religião que emergiu das sombras," *VSP*, 8 Feb. 1984 (Pasta 22.844, *ESP* archives). On the front cover of the São Paulo–based national news magazine *Isto É* (12 Aug. 1992), ACM is called both "The Chief of Brazil" (*O Cacique do Brasil*) and the "owner of the government" (*dono do governo*).

33. Valéria Auada, "'Projeto Terreiros': A rica história dos terreiros de candomblé da Bahia que o tempo ameaça destruir," *TB*, 28 Mar. 1987, Variedades, p. 1 (*TB* archives); "Espíritas-católicos, 'à brasileira,'" [*FSP*?], n.d. [12 Feb. 1977?] (*TB* archives); Luiz Antônio Novaes, "Políticos pedem auxílio a Pai Paiva," *FSP*, 1 Apr. 1989 (*TB* archives); "A festa dos orixás: Venerada na Bahia, Mãe Menininha chega aos 90 anos como a primeira dama de uma religião que emergiu das sombras," *VSP*, 8 Feb. 1984 (Pasta 22.844, *ESP* archives); "Terreiro Casa Branca, uma decisão histórica da Sphan," *ESP*, 1 June 1984 (Pasta 22.844, *ESP* archives); "O Gantois e sua origem escrava," *ESP*, 15 Aug. 1986 (Pasta 22.844, *ESP* archives); "Uma religião cortejada pela política," *VSP*, 7 Jan. 1981, p. 41 (*AT* archives); "Memorial Mãe Menininha do Gantois," posthumous commemorative booklet, n.d. [published after Jan. 1992; collected 22 Dec. 1995], Pasta sobre "Candomblé," Bahiatursa office, Salvador.

34. History of "Axé Opô Afonjá," n.d. [post-1976, before change from five-digit phone numbers, and before Bahiatursa moved from Rua da Gamboa de Cima], Pasta A-25(2), Bahiatursa office, Salvador.

35. Risério and Gil 1988; Dzidzienyo 1985; Olinto 1980:259–65, 295–313; Santos 1967; Beatriz da Rocha, interview in Salvador, 1 Jan. 1996; "Encontro brasileiro da Tradição dos Orixá e Cultura: Preparatória da 'IV Conferência Mundial da Tradição dos Orixá e Cultura'—IV COMTOC-'88," Salvador, Bahia, 31 July–2 Aug. 1987, Local—Ilê Axé Opô Afonjá-Cabula (TB archives), p. 55; "Mãe Estela exige louvor aos Orixás," CB, 8 July 1987, Cidade, p. 5, attachment to the report "Encontro brasileiro da Tradição dos Orixá e Cultura: Preparatória da 'IV Conferência Mundial da Tradição dos Orixá e Cultura'—IV COMTOC-'88," Salvador, Bahia, 31 July–2 Aug. 1987, Local—Ilê Axé Opô Afonjá-Cabula (TB archives); Regina Penteado, "Olga, uma mãe de Santo na Ponte Aérea," FSP, 12 Feb. 1978, Foletim (ESP archives); Anísio Felix and Lana Zanata, "Faraimará Mãe Stella, rainha do Axé Opô Afonjá," BH, 27 Apr. 1995, p. 5, Recorte de Jornais, Bahiatursa office, Salvador.

36. However, it should be noted that Mãe Stella's campaign against syncretism is, at this point, an avant-garde movement. Several famous Nagô priests disagree with its objectives (e.g., Francisco Ribeiro Neto, "O monge e a mãe-de-santo: 'Somos Filhos Do Mesmo Deus," MRJ, n.d. [1973–74] [Pasta 324, AT archives]; Reynivaldo Brito, "Candomblé é uma opção de cura," AT, 10 Aug. 1982 [Pasta 324, AT archives]). And, like all such movements, this one is more purist than literally pure. Her own temple features a gigantic outdoor crucifix and lithographs of Roman Catholic saints hanging in the Oxalá house.

37. See Azevedo 1986; Brown and Bick 1987:87; "Na Bahia, bispo e mães-de-santo juntos contra o sincretismo," ESP, 30 July 1983, p. 9 [ESP archives]; José Maria Mayrink, "Cardeal defende diálogo com candomblé," ESP, 15 Jan. 1989, p. 17 [ESP archives].

38. Howard French, "At African Heart of Voodoo, Pride over Heritage," NYT, 10 Mar. 1996, International, p. 3.

39. Anísio Felix and Lana Zanata, "Faraimará Mãe Stella, rainha do Axé Opô Afonjá," BH, 27 Apr. 1995, p. 5, Recorte de Jornais, Bahiatursa office, Salvador.

40. "Negros apelam para a Justiça," TB, 28 Aug. 1989, Cidade, p. 4 (TB archives); "Vaticano autoriza rito afro-brasileiro," FT, 23 Feb. 1990 (ESP archives); Carter Anderson, "Negros não querem cultura afro entre católicos," FSP, 9 Apr. 1990 (ESP archives); "Igreja terá missa com ritos do candomblé," JT, 24 Feb. 1990, p. 11 (ESP archives).

41. Alberto Sobral, "Os dogmas de Roma contra os mistérios do candomblé da Bahia," CB, 19 Sept. 1980, p. 6 (TB archives); Marilene Felinto, "Discriminação também atinge branco na BA," FSP, 13 May 1991 (TB archives); "[Report and Program by the] Comissão de Candomblé do Estado de São Paulo" (cataloged 23 July 1983 in Pasta 324, AT archives); Luis Antônio Giron, "Bloco Olodum prega pós-utopia da libertação," FSP, 28 Dec. 1990, p. E4 (ESP archives); Hamilton Vieira, "O suingue didático dos blocos afros," AT, 23 Feb. 1990, Caderno 2, p. 1 (Pasta 4995, ESP archives); "Encontro brasileiro da Tradição dos Orixá e Cultura: Preparatória da 'IV Conferência Mundial da Tradição dos Orixá e Cultura'—IV COMTOC-'88," Salvador, Bahia, 31 July–2 Aug. 1987,

Local—Ilê Axé Opô Afonjá-Cabula, pp. 43, 59 (*TB* archives); "Palestras no Encontro da Tradição dos Orixá," *AT*, 2 Aug. 1987, p. 3, attachment to the report "Encontro brasileiro da Tradição dos Orixá e Cultura: Preparatória da 'IV Conferência Mundial da Tradição dos Orixá e Cultura'—IV COMTOC-'88," Salvador, Bahia, 31 July–2 Aug. 1987, Local—Ilê Axé Opô Afonjá-Cabula (*TB* archives); Anísio Felix and Lana Zanata, "Faraimará Mãe Stella, rainha do Axé Opô Afonjá," *BH*, 27 Apr. 1995, p. 5, Recorte de Jornais, Bahiatursa office, Salvador; Henfrey 1981:89; Antônio Carlos dos Santos [Vovô], director-president of the Carnaval club Ilê Aiyê, interview in Salvador, 13 July 1992; notice to Bahian media about the *decá* ceremony of São Paulo priest Alzira Rufino at a *caboclo*/Angola terreiro, April 1990 (Pasta 5887, *AT* archives).

42. The priests of the purist Candomblé do not automatically approve of all the ways of Black nationalists. For example, I have heard Mãe Stella playfully threaten to cut off people's dreadlocks. However, I am one of many African-American researchers I know around the Candomblé who have been given special access because we are the "Black brothers and sisters" (*irmãos negros*) of the priest or priestess.

43. "Mãe-de-Santo recebe comenda do Itamaraty," *JT*, 21 Apr. 1977 (Pasta 16.017, *ESP* archives); Reynivaldo Brito, "As sacerdotisas do candomblé," *AT*, 15 Oct. 1983 (Pasta 324, *AT* archives); "Olga de Alaketu, rainha por herança de sangue," *FSP*, 22 July 1974 (Pasta 324, *AT* archives).

44. Regina Penteado, "Olga, uma mãe de Santo na Ponte Aérea," *FSP*, 12 Feb. 1978, Foletim, p. 5 (*ESP* archives).

45. "O boom umbandista," *VSP*, 7 Jan. 1981, p. 41 (*AT* archives).

46. "Mãe Stella de Oxóssi," *TB*, 27 Apr. 1995, p. 1, Recorte de Jornais, Bahiatursa office, Salvador; "Candomblé em Festa," *TB*, 27 Apr. 1995, Segundo Caderno, p. 1, Recorte de Jornais, Bahiatursa office, Salvador.

47. Julivaldo Freitas, "A Última Geração do Candomblé," *TB*, 15 June 1976, Segundo Caderno (*TB* archives); "Candomblé: Progresso ou Sacrilégio?" [*TB*?], n.d., (*TB* archives); "O Candomblé não Será Mais Controlado pela Polícia," *TB*, 17 Oct. 1975 (*TB* archives); "Candomblé na Bahia é agora religião como as outras e dispensa licença policial," *JOB*, 16 Jan. 1976 (Pasta 324, *AT* archives); "Terreiros vão bater em homenagem ao Governador do Estado," *AT*, n.d. [1976?] (Pasta 324, *AT* archives); "Polícia fora dos terreiros," [*AT*, 1976?] (Pasta 324, *AT* archives); "Candomblé vai à Igreja com Governador," *AT*, 4 Sept. 1976, p. 1 (*AT* archives); Albénsio Fonseca, "Candomblé, uma bricolage cultural," *AT*, 30 July 1987, attachment to the report "Encontro brasileiro da Tradição dos Orixá e Cultura: Preparatória da 'IV Conferência Mundial da Tradição dos Orixá e Cultura'—IV COMTOC-'88," Salvador, Bahia, 31 July–2 Aug. 1987, Local—Ilê Axé Opô Afonjá-Cabula (*TB* archives).

48. "INPS cuidará também de chefes de umbanda," *ESP*, 7 Oct. 1972 (Pasta 16.017, *ESP* archives).

49. Risério and Gil 1988:233–47; "Em fase final restauração do Gantois," *TB*, 9 Dec. 1988, Cidade, p. 2 (*TB* archives); Valéria Auada, "'Projeto Terreiro': Com ajuda dos orixá e dos homens é preciso salvar os terreiros de candomblé," *TB*, 27 Mar. 1987, Variedades, p. 1 (*TB* archives); Valéria Auada, "'Projeto Terreiros': A rica história dos terreiros de candomblé da Bahia que o tempo ameaça

destruir," *TB*, 28 Mar. 1987, Variedades, p. 1 (*TB* archives); "Convênio garante verbas para obras no Gantois," *TB*, 7 Aug. 1989 (*TB* archives); Anísio Felix and Lana Zanata, "Faraimará Mãe Stella, rainha do Axé Opô Afonjá," *BH*, 27 Apr. 1995, p. 5, Recorte de Jornais, Bahiatursa office, Salvador.

50. Juca Ferreira, municipal secretary of the environment, Salvador, interview, 15 Dec. 1995.

51. See also "Projeto Jardins das Folhas Sagradas," project description, Secretaria Municipal do Meio Ambiente, Prefeitura Municipal do Salvador, collected December 1995 from the Secretaria, Salvador, Bahia.

52. A term used, for example, in "Revitalização de terreiros recebe verba de R$85 mil," *AT*, 22 Dec. 1994, p. 3 (in booklet of news clippings on "Projeto Jardins das Folhas Sagradas," Secretaria Municipal do Meio Ambiente, Prefeitura de Salvador).

53. "Terreiros de candomblé serão recuperados," *TB*, 22 Dec. 1994; Juca Ferreira, municipal secretary of the environment, Salvador, interview, 15 Dec. 1995. Part of the money has been used already to help Mãe Stella of Opô Afonjá make a movie, apparently based on her book *Meu Tempo É Agora* (1993). "Cadastro dos Candomblés da Cidade do Salvador," computer printout on houses receiving support from the Projeto Jardins das Folhas Sagradas and the support they received, provided by Juca Ferreira, Secretário, Secretaria do Meio Ambiente, Prefeitura Municipal do Salvador, collected Dec. 1995.

54. Beneficiaries included the usual Casa Branca, Gantois, Opô Afonjá, Bogum, and Ilê Axé Ibá Ogum. Several lesser-known Nagô houses entered this select list—the Terreiro do Cobre, Ilê Axé Obá Tony, and Ilê Iyá Odé, as well as two Angola houses—the famous Bate-Folha and the Manso Dandalunga Cocuazenza.

55. "Terreiro fez festa para comemorar a utildade pública," *AT*, 10 Oct. 1977 (*AT* archives).

56. Valéria Auada, "'Projeto Terreiro': Com ajuda dos orixá e dos homens é preciso salvar os terreiros de candomblé," *TB*, 27 Mar. 1987, Variedades, p. 1 (*TB* archives); Valéria Auada, "'Projeto Terreiros': A rica história dos terreiros de candomblé da Bahia que o tempo ameaça destruir," *TB*, 28 Mar. 1987, Variedades, p. 1 (*TB* archives); "Vaticano autoriza rito afro-brasileiro," *FT*, 23 Feb. 1990 (*ESP* archives); "Terreiro de candomblé é tombado," *ESP*, 5 Aug. 1982 (Pasta 22.844, *ESP* archives); "Terreiro Casa Branca, uma decisão histórica da Sphan," *ESP*, 1 June 1984 (Pasta 22.844, *ESP* archives).

57. Aureliano Biancarelli, "Estado tomba terreiro de candomblé de SP," *FSP*, 3 May 1990 (*ESP* archives); Aureliano Biancarelli, "Estado tomba terreiro de candomblé de SP," *FSP*, 3 May 1990 (Pasta 22.844, *ESP* archives).

58. "Terreiro vibra com decreto do prefeito," *AT*, 24 July 1985 (Pasta 324, *AT* archives); "Disputa no terreiro," *ESP*, 16 July 1983, p. 11 (*ESP* archives).

59. Rose Gonçalves, "Final feliz para Oxumaré," *TB*, 12 Apr. 1988, Cultura, p. 1 (*TB* archives).

60. The first temple of my friend Pai Francisco was expropriated by the state, but the precise reasons are as yet unclear to me.

61. Risério and Gil 1988:246; "Prefeito desapropria Terreiro Casa Branca," *AT*, 6 June 1985 (Pasta 324, *AT* archives); "Na Bahia, morre por abandono uma àrvore sagrada," *OG*, 9 Mar. 1983 (Pasta 324, *AT* archives); Emília Magalhães,

"Folhas Sagradas," *BH*, 11 Aug. 1993 (from the files of the Secretaria Municipal do Meio Ambiente); "Encontro do Parque São Bartolomeu," letter endorsing the preservation of the park and signed by Maria Stella de Azevedo Santos, Àngelo Agnelo Pereira, Helena Santana da Silva (BF), Cleusa Millet, and Juca Ferreira, apparently on 5 June 1993 (from the files of the Secretaria Municipal do Meio Ambiente).

62. Risério and Gil 1988:42; Valéria Auada, "'Projeto Terreiros': A rica história dos terreiros de candomblé da Bahia que o tempo ameaça destruir," *TB*, 28 Mar. 1987, Variedades, p. 1 (*TB* archives); "Terreiro Casa Branca, uma decisão histórica da Sphan," *ESP*, 1 June 1984 (Pasta 22.844, *ESP* archives); "O tombamento do terreiro Casa Branca," *ESP*, 31 May 1984, p. 18 (*ESP* archives).

63. The priestesses normally claimed that their houses were two hundred to four hundred years old (Régis 1984:28; "Convênio garante verbas para obras no Gantois," *TB*, 7 Aug. 1989 (*TB* archives); "Terreiro Faz 300 Anos Hoje," *FSP*, 6 June 1985 (*ESP* archives); "Candomblé: o rei chegou," *FSP*, 7 Apr. 1977 (*ESP* archives); Ney Bianchi and Gervásio Baptista, "O Candomblé da Bahia: O futuro do Brasil," *MRJ*, n.d. [1982–83], last page (*AT* archives); "Orixás em risco: Candomblé pode perder sua meca nacional," 4 Aug. 1982 (*AT* archives); "Olga de Alaketu, rainha por herança de sangue," *FSP*, 22 July 1974 (Pasta 324, *AT* archives); "Terreiro Casa Branca espera desapropriação," *AT*, 4 June 1985 (Pasta 324, *AT* archives); Gioconda Mentoni, "O Tombamento do Mais Antigo Terreiro do Brasil," *JOB*, 2 Aug. 1982, Caderno B (Pasta 324, *AT* Archives); "Zumbis e Êguns, o Culto aos Mortos no Haití e Bahia," *JAB*, 8 July 1977, JBa2 (Pasta 324, *AT* archives); "Prefeito desapropria area da 'Casa Branca,'" *AT*, 7 June 1985 (Pasta 324, *AT* archives); "Bogum quer tombamento para preservar o seu bissecular Terreiro," *AT*, 24 July 1986, Segundo Caderno (*AT* archives); Valéria Auada, "Projeto Terreiros: A rica história dos terreiros de candomblé da Bahia que o tempo ameaça destruir," *TB*, 28 Mar. 1987, Variedades (*TB* archives). However, none of these temples appears to have been founded before the middle of the 19th century. "Em fase final restauração do Gantois," *TB*, 9 Dec. 1988, Cidade, p. 2 (*TB* archives); "Salvador ja fez preservação," *FSP*, 3 May 1990 (*ESP* archives); Aureliano Biancarelli, "Estado tomba terreiro de candomblé de SP," *FSP*, 3 May 1990 (Pasta 22.844, *ESP* archives); Jolivaldo Freitas, "Êguns fazem mistério e adiam 'apocalipse,'" *AT*, 26 June 1980 (*AT* archives).

64. "Salvador ja fez preservação," *FSP*, 3 May 1990 (*ESP* archives); "Terreiro ameaçado na Bahia recebe defensores," *ESP*, 22 July 1982 (Pasta 22.844, *ESP* archives); "Terreiro de candomblé é tombado," *ESP*, 5 Aug. 1982 (Pasta 22.844, *ESP* archives); "Terreiro Casa Branca, uma decisão histórica da Sphan," *ESP*, 1 June 1984 (Pasta 22.844, *ESP* archives); Aureliano Biancarelli, "Estado tomba terreiro de candomblé de SP," *FSP*, 3 May 1990 (Pasta 22.844, *ESP* archives); "Terreiro vibra com decreto do prefeito," *AT*, 24 July 1985 (Pasta 324, *AT* archives).

65. Alternately, they were made into *utilidades pùblicas*. "Terreiro fez festa para comemorar a utildade pública," *AT*, 10 Oct. 1977 *AT* archives. There was one case that the Candomblé priests lost. In the mid-1980s, a private neurological clinic complained that a *gameleira branca*, or *Lôco*, a tree especially sacred

to the Jeje nation, might soon fall and damage the clinic. The tree had apparently been damaged by fire from votive candles set in its trunk, though at least one Jeje official publicly blamed unscrupulous Protestants. Despite years of complaints by the clinic owners and staff, pressure from the Jeje Bogum temple *delayed* any move to cut the tree down. When the tree was cut down, temple officials got the minor satisfaction of divining before giving their consent and demanding that the city tree cutters abstain from sex and alcohol before performing the operation. See Biaggio Talento, "Àrvore ritual do candomblé divide Salvador," *ESP*, 12 July 1990, p. 17 (*ESP* archives). Bogum ultimately failed in another of its major efforts as well. In 1986, despite her claims that Jeje is the most "matriarchal" of nations, Bogum's efforts to become a historical landmark apparently met with no official response in 1985 and were repeated, apparently without success in 1986 "Terreiros querem proteção para manter culto a orixás," *AT*, 10 Dec. 1985, Segundo Caderno. (Pasta 324, *AT* archives); "Bogum quer tombamento para preservar o seu bissecular Terreiro," *AT*, 24 July 1986, Segundo Caderno (*AT* archives). Hence, in the 1980s, even the best-known Jeje temple clearly lacked the clout of the "great" Nagô houses.

66. Gioconda Mentoni, "O Tombamento do Mais Antigo Terreiro do Brasil," *JOB*, 2 Aug. 1982, Caderno B (Pasta 324, *AT* Archives); "O tombamento do terreiro Casa Branca," *ESP*, 31 May 1984, p. 18 (*ESP* archives).

67. "'Casa Branca' recebe apoio de blocos afro," *AT*, 27 July 1982 (Pasta 324, *AT* archives).

68. Casa Branca's advocates included novelist Jorge Amado, plastic artist Carybé, composer Dorival Caymmi, singer Maria Bethânia, actress Zezé Motta, Catholic priest Timóteo Amoroso Anastácio, former mayor Edvaldo Brito, anthropologist Olympio Serra, anthropologist Pedro Agostinho da Silva, several *blocos Afro* (Afrocentric Bahian Carnaval clubs), Mayor Renan Baleeiro, and his successor, Mayor Manoel Castro, as well as Mário Cravo and José Carlos Capinã.

69. "Terreiro ameaçado na Bahia recebe defensores," *ESP*, 22 July 1982 (Pasta 22.844, *ESP* archives); "Terreiro de candomblé é tombado," *ESP*, 5 Aug. 1982 (Pasta 22.844, *ESP* archives); "Terreiro Casa Branca, uma decisão histórica da Sphan," *ESP*, 1 June 1984 (Pasta 22.844, *ESP* archives); "Terreiro ameaçado na Bahia recebe defensores," *ESP*, 22 July 1982 (Pasta 324, *AT* archives); "MEC quer preservar a cultura negra na Bahia," *AT*, 9 Sept. 1982 (Pasta 324, *AT* archives); "Prefeito desapropria area da 'Casa Branca,'" *AT*, 7 June 1985 (Pasta 324, *AT* archives); "Festa na Casa Branca homenageia Mãe Teté," *AT*, 1 Oct. 1985 (Pasta 324, *AT* archives); "Terreiro Casa Branca espera desapropriação," *AT*, 4 June 1985 (Pasta 324, *AT* archives); Gioconda Mentoni, "O Tombamento do Mais Antigo Terreiro do Brasil," *JOB*, 2 Aug. 1982, Caderno B (Pasta 324, *AT* archives); "'Casa Branca' recebe apoio de blocos afro," *AT*, 27 July 1982 (Pasta 324, *AT* archives); "O tombamento do terreiro Casa Branca," *ESP*, 31 May 1984, p. 18 (*ESP* archives).

70. Aureliano Biancarelli, "Estado tomba terreiro de candomblé de SP," *FSP*, 3 May 1990 (Pasta 22.844, *ESP* archives). I attended the conferral of CONDEPHAAT's official recognition in 1992.

71. Marcia Moreira, "A indústria do bozó," *TB*, 15 Oct. 1989 (*TB* archives); Symona Gropper, "Menininha de Gantois: A Festa dos 80 Anos," *JOB*, 13 Feb.

1974 (Pasta 22844, *ESP* archives); "A festa dos orixás: Venerada na Bahia, Mãe Menininha chega aos 90 anos como a primeira dama de uma religião que emergiu das sombras," *VSP*, 8 Feb. 1984 (Pasta 22.844, *ESP* archives); Orixás em risco: Candomblé pode perder sua meca nacional," 4 Aug. 1982 (*AT* archives); "Olga de Alaketu, rainha por herança de sangue," *FSP*, 22 July 1974 (Pasta 324, *AT* archives); Regina Penteado, "Olga, uma mãe de Santo na Ponte Aérea," *FSP*, 12 Feb. 1978, Foletim (*ESP* archives); Ayrton Centeno, "Imigrantes se rendem ao culto afro," *ESP*, 10 June 1990, p. 23 (*ESP* archives); Anísio Felix and Lana Zanata, "Faraimará Mãe Stella, rainha do Axé Opô Afonjá," *BH*, 27 Apr. 1995, p. 5, Recorte de Jornais, Bahiatursa office, Salvador; "Mãe Stella: 'Se nós não preservamos a natureza viva, termina tudo,'" *AT*, 30 April 1995, Segundo Caderno, p. 1, Recorte de Jornais, Bahiatursa office, Salvador; "Sete décadas da estrela do Afonjá," *TB*, 27 Apr. 1995, Recorte de Jornais, Bahiatursa office, Salvador.

72. Marcia Moreira, "A indústria do bozó," *TB*, 15 Oct. 1989 (*TB* archives); Aureliano Biancarelli, "Mãe Sylvia atende até em Nova York," *FSP*, 3 May 1990 (*ESP* archives); "Milhares de discos vendidos, brigas pela autoria dos 'pontos': A Umbanda na Parada de Sucessos," *OG*, 13 Mar. 1977 (Pasta 16.017, *ESP* archives); Nélson Blecher, "No Vale dos Orixás, surge o promissor marketing da fé," *FSP*, 27 Jan. 1985 (Pasta 16.017, *ESP* archives); Luiz Carlos Correa, "Indústria da religião volta a crescer," *ESP*, 6 May 1990, p. 18 (*ESP* archives); Rodney Mello, "São Paulo: Na Linha de Umbanda," *OC*, 16 May 1973, pp. 66–68 (*AT* archives); Biaggio Talento, "Na Bahia, vendas ainda caem," *ESP*, 6 May 1990, p. 18 (*ESP* archives); Carlos Navarro, "Crise ou não, ninguém vai dispensar o ori," *ESP*, 6 May 1990, p. 18 (*ESP* archives).

73. For example, "Falta fé nas missas, mas há moda em dia," *ESP*, 20 Nov. 1983, p. 23 (*ESP* archives).

74. "Candomblé ganha o 1° museu," *JOB*, 20 Feb. 1989 (Pasta 22.844, *ESP* archives); "Tradição dos Orixá abre seu encontro na Bahia," *TB*, 1 Aug. 1987, Cidade, p. 3[?], attachment to the report "Encontro brasileiro da Tradição dos Orixá e Cultura: Preparatória da 'IV Conferência Mundial da Tradição dos Orixá e Cultura'—IV COMTOC-'88," Salvador, Bahia, 31 July–2 Aug. 1987, Local—Ilê Axé Opô Afonjá-Cabula (TB archives); "O boom umbandista," *VSP*, 7 Jan. 1981, pp. 40–41 (*AT* archives).

75. "A festa dos orixás: Venerada na Bahia, Mãe Menininha chega aos 90 anos como a primeira dama de uma religião que emergiu das sombras," *VSP*, 8 Feb. 1984, p. 67 (Pasta 22.844, *ESP* archives); Reynivaldo Brito, "Candomblé é uma opção de cura," *AT*, 10 Aug. 1982 (Pasta 324, *AT* archives); "Na Bahia, bispo e mães-de-santo juntos contra o sincretismo," *ESP*, 30 July 1983, p. 9 (*ESP* archives).

76. "Ilê Axé Apô Afonjá [*sic*]: O trabalho social dos terreiros," *SAL*, 2(18) July–Sept. 1982 (Pasta 324, *AT* archives); "Ministro da Cultura garante apoio a projeto baiano: Foram liberados US$100 mil para concluir restauração dos terreiros da cidade," *CB*, 22 Dec. 1994, p. 11[?] (in booklet of news clippings on "Projeto Jardins das Folhas Sagradas," Secretaria Municipal do Meio Ambiente, Prefeitura de Salvador, collected Dec. 1995).

77. See Duarte 1974:12; Bastide 1978[1960]:164, the Alagoas governor whose apparent support of the Afro-Brazilian Xangô religion in his state inspired, in

1912, a rebellion among his white opponents. During the 1970s, the São Paulo newspapers ranged from sarcastic to vituperative in their comments on the politicians' public attentions to Candomblé and Umbanda. See "O governo vai à festa de Oxossi," *ESP*, 27 Mar. 1973 (Pasta 16.017, *ESP* archives); "Governo estadual oficializa outra festa de macumba," *ESP*, 29 Apr. 1973 (Pasta 16.017, *ESP* archives); Regina Penteado, "Olga, uma mãe de Santo na Ponte Aérea," *FSP*, 12 Feb. 1978, Foletim (*ESP* archives).

78. "Templos não pagam imposto," *AT*, 9 Sept. 1982 (Pasta 324, *AT* archives).

79. "Terreiros querem proteção para manter culto a orixás," *AT*, 10 Dec. 1985, Segundo Caderno (Pasta 324, *AT* archives).

80. State Constitution, chap. XV, Article 275, published in *Diário do Legislativo*, Caderno 3, p. 37; translation mine.

81. Remembered to a lesser degree are police officer Adolfinho Pontes and federally appointed mayor Landulfo Alves de Almeida.

82. "'Casa Branca' recebe apoio de blocos afro," *AT*, 27 July 1982 (Pasta 324, *AT* archives); "Mãe Menininha," *VB*, pp. 3–8, Pasta C-37 [no.1], Bahiatursa office, Salvador.

83. Ângela Peroba, "Federação teme os aproveitores," *JAB*, n.d., p. 9 (*TB* archives).

84. "Candomblé ganha o 1° museu," *JOB*, 20 Feb. 1989 (Pasta 22.844, *ESP* archives); Ângela Peroba, "Candomblé Recupera o Seu Passado," *JAB*, n.d. [*TB* archives]; Valéria Auada, "'Projeto Terreiros': A rica história dos terreiros de candomblé da Bahia que o tempo ameaça destruir," *TB*, 28 Mar. 1987, Variedades, p. 1 (*TB* archives); Valéria Auada, "Projeto Terreiro: Com ajuda dos orixá e dos homens é preciso salvar os terreiros de candomblé," *TB*, 27 Mar. 1987, Variedades, p. 1 (*TB* archives). See also Stella de Azevedo Santos 1993:13.

85. Ferreira 1984:60, 66; Carvalho 1991[1989]:33–34; Jehová de Carvalho, "Mundo Jeje comemora cinquentenário de sua mãe-de-santo," *AT*, 24 July 1988, Lazer e Informação (*AT* archives). Another report, current in São Paulo, suggests that it was Angola priest Joãozinho da Goméia who intervened with Vargas and ended police persecution. See "[Report and Program by the] Comissão de Candomblé do Estado de São Paulo" (cataloged 23 July 1983 in Pasta 324, *AT* archives).

Chapter Five
Para Inglês Ver

1. This Brazilian term, written in Portuguese orthography, is pronounced not "ah-day," like the similar-looking Yorùbá word for "crown," but with a short "e" sound, as one would pronounce *adẹ* in Yorùbá orthography.

2. Chief priests (*pais*- and *mães-de-santo*) are almost always possession priests who, through the excellence of their ritual and political skills, have founded or assumed control over an independent temple.

3. Likewise, he probably would have known that, in every community of Egúngún masquerade priests, who communicate with the world of the dead, there is a female office-holder. Yet Martiniano is quoted as condemning women's

communication with the dead in the famous Casa Branca do Engenho Velho temple, the so-called womb of the Nagô nation in Brazil (Landes 1947:31).

4. In such scenarios, including the southern U.S. case, the sexual exploiters have often been quick to attribute sexual indiscipline to the exploited rather than to themselves.

5. Unlike Landes's other informants, Martiniano objects to pressing the hair not because of the "heat" but because it masks one's African features (Landes 1947:23).

6. Indeed, Landes's 1940 article not only asserts the numerical importance of "passive homosexuals" in the Candomblé but also identifies a dozen such men by name and describes them in the most demeaning terms possible. However, it was not her violation of these men's privacy and good name that upset Ramos.

7. Carneiro also thought highly enough of their leadership to invite Pais Bernardino and Manuel Falefá, among others, to the second Afro-Brazilian Congress, which Carneiro organized in 1937, before meeting Landes (Corrêa 2002:12).

8. One step in Brazil's progressive abolition of slavery, this law freed all the children born of slave mothers after its promulgation.

9. Elias Fajardo da Fonseca, "Uma festa da umbanda: Os mediuns desfilam, a Riotur patrocina," OG, 15 May 1979 (Pasta 862, AT archives).

10. For the "Black Mother"–like images projected upon Mãe Menininha, see "Três anos sem Menininha do Gantois," TB, 11 Aug. 1989, Cidade, p. 2 (TB archives); Symona Gropper, "Menininha de Gantois: A Festa dos 80 anos," JOB, 13 Feb. 1974 (Pasta 22844, ESP archives); "A festa dos orixás: Venerada na Bahia, Mãe Menininha chega aos 90 anos como a primeira dama de uma religião que emergiu das sombras," VSP, 8 Feb. 1984 (Pasta 22.844, ESP archives); "O Gantois e sua origem escrava," ESP, 15 Aug. 1986 (Pasta 22.844, ESP archives); "Mãe Menininha," VB, pp. 3–8, Pasta C-37 [no. 1], Bahiatursa office, Salvador; "Mocidade Independente de Padre Miguel apresenta Mãe Menininha do Gantois," Carnaval parade program, 1976, Pasta C-37(12), Bahiatursa office, Salvador; "Memorial Mãe Menininha do Gantois," posthumous commemorative booklet, n.d. [published after Jan. 1992; collected 22 Dec. 1995], Pasta sobre "Candomblé," Bahiatursa office, Salvador.

11. Barboza and Alencar 1985:24, 26, 148–49, 213; Symona Gropper, "Menininha de Gantois: A Festa dos 80 Anos," JOB, 13 Feb. 1974 (Pasta 22844, ESP archives); "O Gantois e sua origem escrava," ESP, 15 Aug. 1986 (Pasta 22.844, ESP archives).

12. "Memorial Mãe Menininha do Gantois," posthumous commemorative booklet, n.d., p. 3 [published after Jan. 1992; collected 22 Dec. 1995], Pasta sobre "Candomblé," Bahiatursa office, Salvador.

13. "Mãe Menininha," VB, pp. 3–8, Pasta C-37 [no. 1], Bahiatursa office, Salvador; "Mocidade Independente de Padre Miguel apresenta Mãe Menininha do Gantois," Carnaval parade program, 1976, Pasta C-37(12), Bahiatursa office, Salvador; "Memorial Mãe Menininha do Gantois," posthumous commemorative booklet, n.d. [published after Jan. 1992; collected 22 Dec. 1995], Pasta sobre "Candomblé," Bahiatursa office, Salvador. Also Marlise Simons, "In Brazil, High and Low Mourn a Cult Priestess," NYT, 13 Oct. 1986.

14. For this and other such images of Mãe Stella, see Hamilton Vieira, "A história do Axé Opô Afonjá na homenagem a Mãe Stella," AT, 14 Sept. 1989, Segundo

Caderno, p. 3 (Pasta 324, *AT* archives); Gideon Rosa, "Faraimará: União de fé," *AT*, 28 Apr. 1995, Segundo Caderno, p. 6, Recorte de Jornais, Bahiatursa office, Salvador; Anisio Felix and Lana Zanata, "Faraimará Mãe Stella, rainha do Axé Opô Afonjá," *BH*, 27 Apr. 1995, p. 5, Recorte de Jornais, Bahiatursa office, Salvador.

15. On their description as "one of the last," see, for example, "Anúncio com Mãe Menininha provoca polêmica," *UH*, 12 May 1978 (Pasta 22.844, *ESP* archives); Landes 1947:28.

16. On their description as "one of the last," see, for example, "Anúncio com Mãe Menininha provoca polêmica," *UH*, 12 May 1978 (Pasta 22.844, *ESP* archives); Landes 1947:28. On the disciplinarian theme, see, for example, "Memorial Mãe Menininha do Gantois," posthumous commemorative booklet, n.d. [published after Jan. 1992; collected 22 Dec. 1995], Pasta sobre "Candomblé," Bahiatursa office, Salvador; Symona Gropper, "Menininha de Gantois: A Festa dos 80 anos," *JOB*, 13 Feb. 1974 (Pasta 22844, *ESP* archives).

17. Reynivaldo Brito, "As sacerdotisas do Candomblé," *AT*, 15 Oct. 1983 (Pasta 324, *AT* archives).

18. For example, see "Olga de Alaketu, rainha por herança de sangue," *FSP*, 22 July 1974 (Pasta 324, *AT* archives); "Memorial Mãe Menininha do Gantois," posthumous commemorative booklet, n.d. [published after Jan. 1992; collected 22 Dec. 1995], Pasta sobre "Candomblé," Bahiatursa office, Salvador.

19. "Faraimará," tribute and ad for the "I Feira de Cultura Africana Afonjá'," the "Seminário Livre de Cultura Negra," and the "Show da Família Caymmí," *AT*, 28 Sept. 1995, p. 7, Recorte de Jornais, Bahiatursa office, Salvador; Anísio Felix and Lana Zanata, "Faraimará Mãe Stella, rainha do Axé Opô Afonjá," *BH*, 27 Apr. 1995, p. 5, Recorte de Jornais, Bahiatursa office, Salvador; "Sete décadas da estrela do Afonjá," *TB*, 27 Apr. 1995, Recorte de Jornais, Bahiatursa office, Salvador.

20. For example, see "Terreiro Casa Branca, uma decisão histórica da Sphan," *ESP*, 1 June 1984 (Pasta 22.844, *ESP* archives); Aureliano Biancarelli, "Estado tomba terreiro de candomblé de SP," *FSP*, 3 May 1990 (Pasta 22.844, *ESP* archives); "Bogum quer tombamento para preservar o seu bissecular Terreiro," *AT*, 24 July 1986, Caderno 2 (*AT* archives).

21. See, for example, "Olga de Alaketu, rainha por herança de sangue," *FSP*, 22 July 1974 (Pasta 324, *AT* archives); Hamilton Vieira, "A história do Axé Opô Afónjá na homenagem a Mãe Stella," *AT*, 14 Sept. 1989, Segundo Caderno, p. 3 (Pasta 324, *AT* archives); "Sete décadas da estrela do Afonjá," *TB*, 27 Apr. 1995, Recorte de Jornais, Bahiatursa office, Salvador.

22. Also "E Salvador pára no enterro de Menininha," *ESP*, 15 Aug. 1986, p. 14 (*ESP* archives). Pai Francisco says that Mãe Vitória's funeral too was enormous and blocked major roads near her home.

23. All the above testimony by Mãe Stella is taken from Hamilton Vieira, "A história do Axé Opô Afonjá na homenagem a Mãe Stella," *AT*, 14 Sept. 1989, Segundo Caderno, p. 3 (Pasta 324, *AT* archives); "Mãe Stella: 'Se nós não preservamos a natureza viva, termina tudo,'" AT, 30 Apr. 1995, Segundo Caderno, p. 1, Recorte de Jornais, Bahiatursa office, Salvador; "Sete décadas da estrela do Afonjá," *TB*, 27 Apr. 1995, Recorte de Jornais, Bahiatursa office, Salvador.

24. Cf. Wikan (1977) on the role of the homosexual *xanath* as a foil to the ideal woman in Oman.

25. Carneiro's *Candomblés da Bahia* was originally published in 1948 by the State Museum of Bahia, a division of the Secretariat of Education and Health of the state of Bahia.

26. With reference to a 1960–69 survey of 992 temples in urban Salvador da Bahia, Lima (1977[1971]:4, 171) reported that twenty eight out of the thirty four male priests surveyed declared themselves or were widely known to be "*homossexuais*" ("homosexuals"—I am quoting Lima's own terminology). On the other hand, in personal communication with me, he has questioned whether the incidence of male homosexuality is any greater in Candomblé than in psychiatry or the Roman Catholic priesthood. The rarity of non-normative studies of homosexuals' eminence in psychiatry and the Catholic priesthood is perhaps a measure of the academy's prurience about black sexuality and its instinctual deference to white institutions. My argument, however, addresses not homosexuals' statistical presence in Candomblé but the cultural history underlying members' and neighbors' *assumption* that any male possession priest is a particular, locally recognized *type* of homosexual, and making that type of person feel welcome in this religion.

27. Of course, this contrast between penetrators and penetrated is not the only idiom of sexual classification available in Brazil, particularly in recent decades (Jackson 2000:950–60; Kulick 1998; Parker 1991, 1998; Green 1999). However, this particular idiom of sexuality and power remains central to most working-class Brazilians' vocabulary of social classification and to Brazilian men's and boys' daily negotiation of respect.

28. The reader should not misinterpret "transvestism," which means "cross-dressing," as a euphemism for or type of homosexuality.

29. Many Cuban and Puerto Rican adherents of similar traditions think so, too. A category of men known as *maricas* or *addodis* has for decades been identified as common in the Yorùbá-affiliated denomination of Afro-Cuban religion called Regla de Ocha or Lucumí. They are said to be protected by the goddesses Yemayá and Ochún, who love them dearly (Cabrera 1983[1954]:56; Lachatañeré 1992[1939]:223–24). The earlier of these written accounts dates from the same period as Landes's observations about homosexuals in the Brazilian Candomblé, the late 1930s.

30. Fry 1986:138. Fry's analysis is far removed from the psychological framework and pathologizing conclusions of Landes, Carneiro, and their successors Bastide (e.g., 1961:309) and Ribeiro (1969:109–20). Fry analyzes local images of "magical power" and the role of *inversion* within them. However, this model prioritizes nationalist logics of respectability and normalcy over the distinctly Afro-Brazilian and Afro-Atlantic forms of symbolism, logic, hierarchy, and planning that shape these religions. Afro-Brazilian culture ends up looking like a form of "letting loose," a sort of compartmentalized abandon, and an upside-down version of Euro-Brazil rather than a right-side-up version of itself (see also V. Turner 1983).

31. As Andrews (1992:252) shows, men of any given social race earn more on average than the women of that social race. It has also been observed that, in the

contexts where light-skinned gay men successfully conceal their sexuality, they possess considerable economic and political advantages over women as a group and blacks as a group (*VSP*, 12 May 1993, pp. 52–59). For an explanation of the term "social race," which I use in the absence of an alternative generic term for the Brazilian color and status categories, see Wagley 1963[1952]:14; Degler 1971:105.

32. Leão Teixeira 1987:43–45, 48; also Azevedo Santos 1993:52–54 on the servile status of women consecrated to female *orixás*.

33. See Hamilton Vieira, "A história do Axé Opô Afonjá na homenagem a Mãe Stella," *AT*, 14 Sept. 1989, Segundo Caderno (Pasta 324, *AT* archives); "Mãe Stella: 'Se nós não preservamos a natureza viva, termina tudo,'" *AT*, 30 Apr. 1995, Segundo Caderno, p. 1, Recorte de Jornais, Bahiatursa office, Salvador.

34. E.g., Herskovits 1958[1941]:214; 1966c[1945]: 57–58.

35. Similarly, among the West African cousins of the *vodum*-worshiping Jeje nation, male and female senior priests are called "*mothers* of the gods" (*vodun nɔ*).

36. Indeed, next to the Aláàfin (or king of Ọ̀yọ́) himself, the highest-ranking priest of Ṣàngó is a woman—the very same Ìyá Nàsó officeholder after whom the founder of the Casa Branca was named or modeled.

37. In Yorùbá, the term *ọkọ*, which Yorùbá speakers of English translate as "husband," refers not just to the male spouse and procreational partner of the wife but also to the husband's female and male consanguineal relatives. Nonetheless, the male spouse is the paradigm of the category. It should be pointed out, though, that in the real world, *ìyàwó* (wives) are not always or in every matter subordinate to their male spouses or in-laws. Some wives are wealthier or better-connected than their spouses or simply possess indomitable personalities. Nonetheless, most Yorùbá people normatively expect wives to be subordinate to their husbands and, at the very least, to feign such subordination on public and ceremonial occasions. This ideal scenario is reinforced by the marked tendency of women (in a society in which seniority confers higher rank) to marry men who are older than they.

38. My friend Pai Francisco, on the other hand, says that he has never heard *iaô* glossed as "son," "daughter," or "slave"—only as "wife" (*esposa*).

39. In much of West Africa, women weavers conventionally weave on broad looms, whereas men conventionally weave on narrow looms. Tailors then stitch together the narrow strips of cloth to form garments. Ìgbòhoans disrupt the usual gendered division of labor in cloth manufacture in West Africa.

40. Oyewumi (1997:117–18) was told by one West African male possession priest that he braids his hair because initiates are forbidden to cut their hair, a testimony that Oyewumi believes sufficient to demonstrate the absence of gender symbolism in the sartorial presentation of possession priests. Whatever the contemporary variation in the interpretation of the hairstyle itself, its meaning for most audiences will be shaped by its contemporary syntagmatic context, which Oyewumi pointedly ignores in her response to my argument—that is, the simultaneous details of priestly jewelry, cosmetics, and clothing that are worn only by women and male possession priests. Male possession priests who dress in this way are also the only men who are described as *ìyàwó*, or "brides." Contrary to

Oyewumi's argument, the overall *ensemble* of sartorial items conveys a power-fully and unmistakably gendered message.

Similarly, but without the benefit of any expert local testimony, Oyewumi (1997:37) asserts that the lateral self-prostration (*ìyíkàá*), which is restricted to women, came about because a pregnant woman cannot prostrate herself frontally, or *dòbálè*. She presents this speculation also as evidence that no con-cept of gender shapes Yorùbá thought or action, though she fails to explain why contemporary women who have never been pregnant and women who are not pregnant also conventionally perform lateral self-prostration and why men, who are fully capable of either form of self-prostration, do not conventionally per-form the lateral form.

41. In a contrasting case equally revealing of the sexual implications of the verb, the term *mágùn* (lit. "don't mount") refers to a "medicine," or magical ap-plication, that kills the paramour of a married woman at the moment he at-tempts to penetrate her.

42. Duly warned by my colleague Wande Abimbola, I acknowledge that the English gloss "to climb" better captures the fact that many *òrìṣà* (though not Ṣàngó) are regarded as *rising* from the ground rather than descending from above (Abimbola 1997a:152–54). But this gloss fails to encode the equestrian and sexual implications that are implicit in the terms *ẹṣin* (horse), *ìyàwó* (bride or wife), and *gùn*. It also unintentionally de-emphasizes the fact that the divine agent is understood to end up *on top*—that is, is a position symbolizing his or her control over the priestly medium. The main virtue of the gloss "to climb" is not its greater semantic precision as a translation but its sublimation of the equestrian and sexual implications of the folk terminology, which might other-wise appear to stigmatize the religion in the eyes of mightier religions and na-tions. Hence, it is not my aim (nor is it within my competency) to contradict Abimbola, who is a widely traveled *babaláwo* diviner, spokesperson of the Ifá priesthood at its Ifè heartland, and university professor. Rather, it is to examine both the historical roots of cultural interpretations like his own and the cos-mopolitan cultural politics that shape them.

43. Carneiro himself writes elsewhere, "The initiates, after final admission into the community, are considered the chief's spiritual children. It is held that everybody, whether within or without the *candomblé*, has a protector, who pre-sumably dwells in him, and 'mounts' him as though he were a 'horse'" (Carneiro 1940:273).

44. Peter Fry (2002:27) poses a considerable challenge to my explanatory model. Citing Landes, he argues that *adé* priests are more common in the "cabo-clo" temples than in the "traditional," African-oriented temples of Candomblé, a view that would appear to undermine the argument that the West African prece-dents substantially account for the prominence of *adé* priests in Bahia.

I do not regard the Quêto/Nagô nation and the *caboclo*-worshipers as contrast-ing nations or as isomorphic with the Yorùbá in contrast to Brazilian Indians or *mestiços*. Every nation of Candomblé is dominated by Ọ̀yọ́-Yorùbá models of spirit possession. Local African-American religious complexes often use nationhood to mark perceived moral differences between culturally indistinct local populations—for example, by Cuban understandings, the lucumís are to piety as congos are to "black magic" (Palmié 2002), though the reality is far more hybrid and complex. Similarly, Landes and Carneiro invented the distinction between the Nagô and *caboclo* nations to insulate the Nagô from the taint of the *adés* notwithstanding their indistinguishable priestly genealogies and the omnipresence of *adé* priests in both "nations."

Few temples actually describe themselves as "*caboclo*" temples. Most temples that outsiders now identify as "*caboclo*" call themselves "Angola" (Santos 1995:80–81), which might, if we assume that this moral binary has African roots, send us searching for non-Yorùbá precedents for the *adé* priest. Sweet (1996) identifies West-Central African transvestites, including homosexual transvestites, as an important precedent of the *adés* in Candomblé. To Sweet, these men and the West African *elégùn* illustrate a pan-African set of "core beliefs." Even if the *adé's* antecedents are not limited to West Africa, it would be difficult to ignore the unique breadth, historical depth, demographic foundations, and articulacy of his parallels with the Ọ̀yọ́ ọkùnrin, or male possession priest. I simply observe that, like the practitioners of Candomble in the 1930s, the West African peoples (including the ancestors of the Yoruba) who preeminently shaped today's Candomblé also clearly connect male transvestism with spirit possession. They also verbally describe spirit possession and its practitioners by analogy to sexual intercourse and marriage. If such analogies once existed among West-Central Africans, they are absent from the 16th- and 17th-century archival record and from the 20th-century ethnographic record. In the present day, the shared Yoruba, Ewè/Gèn/Ajá/Fòn (EGAF), and Brazilian Candomblé imagery of marriage to the divinity, who then episodically displaces its bride's personality and consciousness, is foreign to West-Central African religions (Wyatt MacGaffey, personal communication, 1996).

Hence, the historical, ethnographic, and linguistic evidence leaves little doubt of a possibly pan-African but far more demonstrably West African cultural model historically shaped the *adé's* symbolic role in Candomblé. Landes's and others' relegation of male priests to the "caboclo" nation is a moral judgment rather than a demographic observation, and one that rival priests, in the 1930s and now, have been willing to wield against each other.

45. I owe this insight to journalist and anthropologist Erica Jane de Hohenstein (personal communication, Sept. 1987). I have noticed this phenomenon regularly since she called it to my attention.

46. That is, among the West African Yorùbá, *men* prostrate themselves flat on the ground, whereas the *ìyíkàà* (lying first on one side and then on the other) is the more appropriate gesture for *women* in sacred contexts. In Cuban Ocha and Brazilian Candomblé, by contrast, it is the gender of one's divinity, or "saint," that determines the appropriate style of self-prostration. Whether male or female, a person governed by a male saint salutes elders and altars with the *ìdòbálè*, whereas a male or female person governed by a female saint performs the *ìyíkàà*.

47. At the time of my research in Ìgbòho, I had never heard of a named or symbolically marked category of men who are penetrated *sexually* by other men, but, in sum, I could see that those who are regularly penetrated *spiritually* by the gods have a great deal in common (sartorially, professionally, and symbolically) with the Brazilian *passivo, bicha,* or *adé* category. Imagine my surprise

when I met an art historian from the city of Ọ̀yọ́ whose extended family included many Ṣàngó priests in that West African cultural capital. During his time among *oricha* worshipers in the United States, this scholar too became aware of the importance of men who love men in the New World priesthoods. Without having read my work, he had concluded that male-male sexual conduct among New World priests was a *continuation* rather than a mere reinterpretation of West African religious traditions. He told me that, on two occasions between 1968 and 1973, he witnessed possessed male Ṣàngó priests anally penetrating unpossessed male priests in an Ọ̀yọ́ shrine. He does not know, however, if this practice was widespread or whether it represented a tradition or norm. Nor do I. As yet, I would extend my case no further based on this unique testimony, which the original observer has shared with me privately but has himself hesitated to publish.

48. While African nationalists are often wont to dismiss homosexuality as a "white man's disease," there is evidence of its practice in contemporary and pre-colonial Africa. There is nothing "un-African" about same-sex intercourse. For example, it is widely reported among Hausa elites, and the 1999 trial of Yorùbá cadets in the Nigerian Defence Academy suggests that it occurs among some Yorùbá men as well. Some Ìjẹ̀bú people report the existence of a refined class of male homosexual entertainers known as *wọ̀wọ̀* and a more general class of anally penetrated men called *àdódìí* in mid-20th-century Ìjẹ̀bú society. See Matory 1994b:2 for additional citations on same-sex sexual practices in the recent and distant past of sub-Saharan Africa. See Sweet 1996 on West-Central Africa.

49. In the service of this point, Oyewumi argues that intercultural translation is fraught with difficulties and must be undertaken with care, a point made by dozens of scholars before her, including students of Yorùbá gender and society (e.g., Matory 1991:535–42; 1994b:219–25; Drewal 1992:x; Apter 1992:226). As in her main argument about gender, however, Oyewumi asserts the originality of her point by misrepresenting the arguments of earlier scholars.

50. For a more comprehensive critique of Oyewumi's *Invention of Women*, see Matory (forthcoming b).

51. For an excellent overview of the anthropological literature on gender in sub-Saharan Africa (and one greatly at odds with Oyewumi's caricature of "Western" views of gender in Africa and elsewhere), see Potash 1989; Kaplan 1997. Cross-cultural variation in the construction of gender roles has been a central theme in anthropology since the 1920s (e.g., Mead 1928,1963[1935]; Kessler and McKenna 1978; Collier and Yanagisako 1987; Atkinson and Errington 1990).

52. Danforth documents the importance of similar conferences, linking scholars and nationalist laypeople, in the overseas propagation of Greek nationalism. See Danforth 1995:90, 92, 95.

53. The author of the position that each group should speak for itself was not advocating the truth of Oyewumi's argument but was defending the autonomy of each national tradition within the Yorùbá-Atlantic world. This person was, from the beginning, an important leader of the movement to reunite the international community of òrìṣà worshipers but came to resist the emergent principle of West African Yorùbá supremacy and the apparent male supremacy that she encountered among West African collaborators in the project.

54. Mọlara Ogundipẹ at "Roundtable: The Invention of Woman: Theorizing African Women and Gender Now and into the Future," African Studies Association, Washington, DC, 6 Dec. 2002.

55. For example, Mikelle Smith Omari, reports that a priestess modified her altar after hearing from Omari that it differed from African altars that Omari had seen (1984:54n50; 55n64). In the course of his research among the Saramacca maroons of Dutch Guiana, Herskovits amply shared the products of his own and other ethnographers' research to teach the Surinam maroons about "the people of the West Coast of Africa—Ashanti and the Gold Coast, Dahomey, Nigeria, Benin, and Cameroons, Loango, and the Congo," as well as Haiti (Herskovits 1934:10, 23, 50, 102, 156, 200, 258–62, 265; also 57, 282, 305). He told Anansi stories that he had heard in Africa, played sacred music he had recorded in Haiti, and showed photographs of shrines and ceremonies shot by other researchers. Herskovits shows that Saramacca leaders greatly appreciated his knowledge of what they regarded as their ancestral homeland, that this knowledge facilitated his access to information about the Saramacca, and that he avoided revealing information that he thought might offend his hosts. Herskovits's Saramacca friends, like my Candomblé friends, regarded him not only as a source of information about Africa but as a potential means of their own access to their beloved ancestral homeland (Herskovits 1934:57, 305).

CHAPTER SIX
MAN IN THE "CITY OF WOMEN"

1. Hip-hop and Jamaican Rastafarianism have been more popular, but Yorùbá religion has displayed an unparalleled longevity and purchase among class and educational elites as well.

2. The titles *pai-* and *mãe-de-santo* generically identify a male chief priest and a chief priestess. Each nation, however, uses a more specific set of terms as well. In Jeje, a male chief priest consecrated to a male god is called a *doté*; a chief priestess consecrated to a male god is called a *doné*. Whether male or female, a chief priest consecrated to a goddess is called a *guaiacú*. As the senior and highest-ranking initiate in his temple, Pai Francisco is also called a *rumbondo*. Like him, anyone who has been initiated for more than seven years and has undergone the rite officially recognizing this senior status is called a *deré*.

3. In West Africa, "Nàgó" and "Ànàgó" are virtually interchangeable terms. According to Burton (1966[1861–64]:162–63), "Ànàgó" is the Ouidah-Fòn dialect term for the same people whom the Fòn speakers of Abomey call "Nàgó," that is, the Yorùbá generally.

4. In the Jeje nation, the avatars, or *marcas*, of each god also have nation-specific names. See also Querino (1988[1938]:35); Ramos (1940[1934]:53–54), mentioned previously; and numerous others.

5. See also Aureliano Biancarelli, "Pai-de-santo faz ligação com orixás," *FSP*, 3 May 1990 (*ESP* archives). On the contrast between the "Africanist" and the "white" forms of Umbanda, see Joana Angelica, "Umbanda, não. Umbandas: elas são muitas. Quatro, pelo menos," *OG*, 9 Jan. 1978 (*AT* archives).

6. "Líder da Umbanda vai à Àfrica manter contatos," *UH*, 22 July 1970 (ESP archives); "O boom umbandista," *VSP*, 7 Jan. 1981, pp. 40–41 (*AT* archives); Alexandre Kadunc, "Jamil Rachid: Este homem é o pai de 350 mil crentes," [*VSP* or *MSP*], 11 Dec. 1976, pp. 122–24 (Pasta 862, *AT* archives); Joana Angélica, "Umbanda, não. Umbandas: elas são muitas. Quatro, pelo menos," *OG*, 9 Jan. 1978 (*AT* archives). Also Prandi 1991.

7. "Líder da Umbanda vai à Africa manter contatos" and "Roteiro," *UH*, 22 July 1970 (*ESP* archives).

8. See also "O peso do terreiro: Teses estudam os umbandistas e a política," *IE*, 1 June 1983, pp. 40–42 (Pasta 862, *AT* archives); "O boom umbandista," *VSP*, 7 Jan. 1981, pp. 40–41 (*AT* archives); Alexandre Kadunc, "Jamil Rachid: Este homem é o pai de 350 mil crentes," [*VSP* or *MSP*], 11 Dec. 1976, pp. 122–24 (Pasta 862, *AT* archives).

9. "'Cirrum' começou no Bogum e 'Gamo' é a nova yalorixá," *AT*, 30 Dec. 1975, p. 3 (*AT* archives).

10. "Bogum quer tombamento para preservar o seu bissecular Terreiro," *AT*, 24 July 1986, Segundo Caderno (*AT* archives). "Matriarchy" is not a term that would conventionally describe the possession priesthoods of the 20th-century West African EGAF region. There, many possession priests and chief priests are male, though the ranking priests, male and female, are indeed called "mothers" (*nɔ*) of the gods (see, e.g., Herskovits 1976[1933]; Rosenthal 1998). Nonetheless, in the late 1940s, even students of the Jeje nation in Maranhão had embraced the matriarchal standard of priestly legitimacy. By the time of Pereira's and Ferretti's writings, the praise of "matriarchy," as well as the equal and opposite dispraise of "male homosexuality," had entered the canon of a widely exported Bahian ethnography. Pereira 1979[1947]:22; Ferretti 1986:47, 60.

11. Pai Vicente is of the Jeje Caviano subnation, which he says is identical to the Jeje-Marrim; Pai Francisco is of the closely related Jeje Savalú subnation.

12. See chapter 4. Pai Francisco is among those men who say that Joãozinho was in fact responsible for intervening with Vargas.

13. On the inference that she was Jeje, see R. Pires 1992:65.

14. Note that, in his two autobiographical narrations to me, Pai Francisco identified the gods exclusively by their Quêto/Nagô names. Note also that, in Brazil, being "white" does not necessarily denote the complete absence of African ancestry or features.

15. An indirect way of identifying a Candomblé member is to say that he or she is "in the [red palm] oil [*no azeite*]." Recall that red palm oil was an important item in the trans-Atlantic trade linking Bahia and Nigeria in the 1930s.

16. Thus, seventy years after Carneiro and Landes argued falsely that men's presence in the possession priesthood was new, the view that men's presence is novel is now well established.

17. Pai Oscar himself now commands an enormous and scenic temple in Bahia. He travels extensively and sells the finest Nigerian jewelry, clothing, and ritual materials, shipped by his Lagosian cousin, to select clients in Bahia. He himself regularly wears entire outfits of costly handwoven "cloth from the Coast" (*pano da Costa*).

18. Dr. Alakija appreciates Candomblé in much the same way as other highly educated Bahians—that is, as a beautuiful aspect of the regional and national legacy. He is not a member of any temple, but, as a Brazilian government representative at the 1977 World Black and African Festival of Arts and Culture in Lagos, Dr. Alakija presented his study on a theme that the psychiatrists of Northeastern Brazil, ever since Dr. Raymundo Nina Rodrigues founded Afro-Brazilian studies, have made a major object of investigation—"The Trance State in 'Candomblé.'"

19. As in the 1930s, many Candomblé priestesses are still street vendors.

20. Note that literacy and newspaper reading were, judging by this case, normal features of life for such elderly priestesses in the mid–20th century.

21. Deka is the cognate Fòn expression for "prayer calabash," which, in contemporary Bénin, is used to pour libations to the ancestors and gods.

22. Pai Francisco tells me that his godfather was a light mulatto.

23. The Brazilian currency of that time was the *cruzeiro* (CR$). Regina Pires (1992:66) reports a purchase price of CR$9,000,000, or US$767,000 according to the exchange rates of that time. Pai Francisco told me on 11 April 2003 that the figure was actually a tenth of Pires's figure. Though unable to estimate the dollar equivalence of the figure in those days, he insists that it was an enormous sum.

24. Several of my *santero* friends keep copies of my *Sex and the Empire That Is No More* (1994) or texts of sacred panegyrics photocopied from it on their altars.

25. In Brazil, "Onirê" is the appellation of the specific avatar (*marca*), or version, of Ogum. On the other hand, in Yorùbáland, "Onírè" means "Owner of the [Èkìtì-Yorùbá] town of Ìrè" and simply highlights the Yorùbá god of iron's mythical association with that town.

26. This contempt has old precedents among the transnational Africans of Bahia in the 1930s. Historically, it appears to have been inspired not by the commonplace mutual resentment of rival oppressed groups but by the tendency of many pre-Freyrean white Brazilian nationalists (much like the nationalist whites of the Dominican Republic and the United States) to emphasize the creole nation's Indian roots in order to conceal its African ones. In these nations, where Indians were long ago reduced by disease and extermination to minuscule percentages of the population, some advocates of the descendants of African slaves have perceived the political and literary Indianisms of nationalist whites as complicit in the continued oppression of blacks. Consider, for example, the words of Pierson's informant, quoted in chapter 4 of this book (from Pierson 1942:272). Whether for this very reason we may never know for sure, but the trans-Atlantic *babaláwo* Martiniano do Bonfim also reportedly reviled the *caboclo* nation (Landes 1947:32).

27. See also Marilene Felinto, "Discriminação também atinge branco na BA," "Minoria branca se concentra em poucos bairros," "Harlem da Bahia é a rua do Curuzu," and "Negros ainda não votam em negros," all in *FSP*, 13 May 1991 (*TB* archives); as well as Henfrey 1981:89.

28. See "Sete décadas da estrela do Afonjá," *TB*, 27 Apr. 1995, Recorte de Jornais, Bahiatursa office, Salvador. What Mãe Stella calls "immediatism" is the impious practice of using Candomblé's technologies for the pursuit of immediate material benefits in lieu of long-term piety. On the other hand, fans of the *caboclos* praise them for the immediacy of their work.

29. "Anúncio com Mãe Menininha provoca polémica," *UH*, 12 May 1978 (Pasta 22.844, *ESP* archives); Antonieta Santos, "Anúncio com Mãe Menininha causa novos protestos," *FSP*, 18 May 1978 (Pasta 16.017, *ESP* archives).

30. Antonieta Santos, "Anúncio com Mãe Menininha causa novos protestos," *FSP*, 18 May 1978 (Pasta 16.017, *ESP* archives)

31. See, for example, Marcia Moreira, "A indústria do bozó," *TB*, 15 Oct. 1989 (*TB* archives).

32. Rodney Mello, "São Paulo: Na Linha de Umbanda," *OC*, 16 May 1973, pp. 66–68 (*AT* archives); Luiz Carlos Correa, "Indústria da religião volta a crescer," *ESP*, 6 May 1990, p. 18 (*ESP* archives); Nélson Blecher, "No Vale dos Orixás, surge o promissor marketing da fé," *FSP*, 27 Jan. 1985 (Pasta 16.017, *ESP* archives); Alexandre Kadunc, "Paulo Pergunta, Fim de Semana: 'O Candomblé e a Umbanda representam hoje uma força religiosa, social, política incontestável. Por que?'" *JT*, 1 Dec. 1979 (Pasta 22.844, *ESP* archives); Aureliano Biancarelli, "Estado tomba terreiro de candomblé de SP," *FSP*, 3 May 1990 (*ESP* archives).

33. See "Gardens of Sacred Leave [*sic*]," translation of description and budget of this $500,000 temple renovation project, p. 17, Secretaria Municipal do Meio Ambiente, Prefeitura Municipal do Salvador, collected December 1995 from the Secretaria, Salvador, Bahia; "Revitalização de terreiros recebe verba de R$85 mil," AT, 22 Dec. 1994, p. 3 (in booklet of news clippings on "Projeto Jardins das Folhas Sagradas," Secretaria Municipal do Meio Ambiente, Prefeitura de Salvador); Antônio Risério and Gilberto Gil, "Fundação Gregório de Mattos: Roteiro de uma intervenção político-cultural," *O Poético e o Político e outros escritos*, pp. 242–43, Paz e Terra, Rio, de Janeiro, and São Paulo.

34. Which he says neighbors the Jeje nation.

35. Pierson (1942:296–97) notes also the case of Pai Procópio in this regard.

36. See "Milhares de discos vendidos, brigas pela autoria dos 'pontos': A Umbanda na Parada de Sucessos," *OG*, 13 Mar. 1977 (Pasta 16.017, *ESP* archives); "Babalorixá baiano não quer que turista visite condomble [*sic*]," *AT*, 25 June 1974 (*AT* archives); "Egungun," phonograph record produced by SECNEB, Salvador, with Didí's cooperation in 1982; "Candomblé: Joãozinho da Goméia," phonograph record produced by Fermata, S.P., released some time after his death in 1971; "Candomblé da Bahia: Toques, Cantos e Saudações aos Orixás," phonograph record produced by Gravações Elétricas, R.J., in 1989.

37. Waldemiro de Xangô, personal communication, 14 Feb. 1992, São Paulo; liner notes on "Candomblé: Joãozinho da Goméia," phonograph record produced by Fermata, S.P., released some time after his death in 1971; "Espíritas-católicos, à brasileira,'" [*FSP*?], n.d. [12 Feb. 1977?] (*TB* archives); "O Gantois e sua origem escrava," *ESP*, 15 Aug. 1986 (Pasta 22.844, *ESP* archives).

38. Hélio Carneiro and Reynivaldo Brito, "Candomblé: Deu branco na fé?" *MSP*, 1 Dec. 1984 (Pasta 324, *AT* archives); "Causa sagrada": Pais-de-santo vêem abuso no jogo de bùzios," *VSP*, 9 Jan. 1985 (Pasta 324, *AT* archives); Reynivaldo Brito, "Candomblé é uma opção de cura," *AT*, 10 Aug. 1982 (Pasta 324, *AT* archives).

39. "Candomblé: O rei chegou," *FSP*, 7 Apr. 1977 (*ESP* archives). A Nigerian author and publisher resident in São Paulo, Sikiru Salami, has also played an

important role in guiding former Umbanda priests in São Paulo into a respectable position within the Candomblé.

40. "Candomblé: Progresso ou Sacrilégio?" [*TB*?], n.d., (*TB* archives).

41. "O Candomblé não Será Mais Controlado pela Polícia," *TB*, 17 Oct. 1975 (*TB* archives); "Desrespeito aos princípios do culto afro-brasileiro," *AT*, 25 Aug. 1976 (*TB* archives); "'Terreiros' que não tenham requisitos poderão fechar," *AT*, 12 Nov. 1976 (Pasta 324, *AT* archives); "Espíritas unidos aos candomblés na luta contra falsos terreiros," 19 Sept. 1978 (*AT* archives); "Babalorixá [*sic*] do Gantois contra o charlatanismo no candomblé," *AT*, 12 Dec. 1978, p. 12 (*AT* archives).

42. "Umbanda Proibida de Fazer Despachos," *NP*, 12 Dec. 1986 (*ESP* archives); "Chefe de umbanda, só com carteirinha," *JT*, 11 Dec. 1986 (Pasta 16.017, *ESP* archives).

43. Luiz Sérgio Barbosa, "O Que Precisa uma Baiana para Vender Acarajé," *JJAB*, Ano I, Apr. 1995, p.2; "Muitos farsantes na 'tradição' de Omolú," *AT*, 22 July 1981 (Pasta 324, AT archives); "Desrespeito aos princípios do culto afro-brasileiro," *AT*, 25 Aug. 1976 (*TB* archives).

44. Mãe Stella de Azevedo Santos, interview at Opô Afonjá, 7 Sept. 1992; Pai Ari de Oliveira Mascarenhas (social director and technical coordinator of the federation), interview at headquarters, 3 Jan. 1996.

45. "O Candomblé não Será Mais Controlado pela Polícia," *TB*, 17 Oct. 1975 (*TB* archives); "Desrespeito aos princípios do culto afro-brasileiro," *AT*, 25 Aug. 1976 (*TB* archives); "'Terreiros' que não tenham requisitos poderão fechar," *AT*, 12 Nov. 1976 (Pasta 324, *AT* archives); "Espíritas unidos aos candomblés na luta contra falsos terreiros," 19 Sept. 1978 (*AT* archives); "Babalorixá [*sic*] do Gantois contra o charlatanismo no candomblé," *AT*, 12 Dec. 1978, p. 12 (*AT* archives).

46. Pai Ari de Oliveira Mascarenhas (social director and technical coordinator of the federation), interview at headquarters, 3 Jan. 1996. See also *JJAB*, Ano I, Apr. 1995.

47. Pai Ari de Oliveira Mascarenhas (social director and technical coordinator of the Federação), interview at headquarters, 3 Jan. 1996.

48. Pai Ari de Oliveira Mascarenhas (social director and technical coordinator of the Federação), interview at headquarters, 3 Jan. 1996.

49. Along with her poet and journalist *ogã* Jehová de Carvalho, the deceased *doné*, or chief priestess, Nicinha, kept the dignity of tree worship in the public eye. Her own head was consecrated to the tree god Lôco (Pai Francisco prefers the Anglicized spelling "Loko"), which is said to be of the "family" of the Jeje thunder god Caviano. The tree god O Zanador, which corresponds to (or is syncretized with) Melchior, the black king among the Three Wise Men, is often mentioned in the frequent newspaper coverage of this house, which has been occasioned by various priestly successions, a visit to the temple by Mayor Edvaldo Britto, state funding of its renovation, and temple officials' fight to cultivate and protect the sacred trees of the city. Contrary to newspaper reports, Pai Francisco vows that Mãe Nicinha had not died at the time of the planting but was simply sick.

50. "Replantio de àrvore marca homenagens para 'Doné' Nicinha," *CB*, 11 Oct. 1994, p. 10 (sheaf of clippings on the replanting of the *gameleira*, 10 Oct.

1994, Estrada de São Lazaro, assembled by the Projeto Jardins das Folhas Sagradas, Secretaria do Meio Ambiente, Prefeitura do Salvador).

51. "Cerimônia especial para o replantio de gameleira," AT, 11 Oct. 1994, p. 3 (stapled sheaf of clippings on the replanting of the *gameleira*, 10 Oct. 1994, Estrada de São Lazaro, assembled by the Projeto Jardins das Folhas Sagradas, Secretaria do Meio Ambiente, Prefeitura do Salvador [collected Dec. 1995]).

52. *Garantindo espaço para o arreio de ebós*—Juca Ferreira, quoted in "Adeptos reinauguram espaco sagrado," BH, 10 Oct. 1994, Cidade, p. 3 (sheaf of clippings on the replanting of the *gameleira*, 10 Oct. 1994, Estrada de São Lazaro, assembled by the Projeto Jardins das Folhas Sagradas, Secretaria do Meio Ambiente, Prefeitura do Salvador).

53. See contents of the stapled sheaf of clippings on the inauguration of the *praça* at the Bate-Folha temple, Secretaria Municipal do Meio Ambiente (collected Dec. 1995), cited earlier, esp. "Prefeita inaugura praça em terreiro de candomblé," AT, 8 May 1995. Also "Revitalização de terreiros recebe verba de R$85 mil," AT, 22 Dec. 1994, p. 3 (in booklet of news clippings on "Projeto Jardins das Folhas Sagradas," Secretaria Municipal do Meio Ambiente, Prefeitura de Salvador); "Cadastro dos Candomblés da Cidade do Salvador," computer printout on houses receiving support from the Projeto Jardins das Folhas Sagradas and the support they received; provided by Juca Ferreira, Secretário, Secretaria do Meio Ambiente, Prefeitura Municipal do Salvador.

CHAPTER SEVEN
CONCLUSION

1. See *Oceans Connect: Mapping a New Global Scholarship*, proceedings of an international workshop, Duke University, Durham, NC, 29–31 Oct. 1998.

2. See, for example, Otero 2000.

3. See, for example, Sally Cole (1994) on Ruth Landes's reflexive ethnography. Like Bourdieu (1985[1972]) and Connerton (1995[1989]), Herskovits and his followers have sought in learned motor habits a central mode of sociocultural reproduction. See Matory (1999c, 1999d) for an examination of the 19th-century transnationalism documented early on by the likes of Verger (1976[1968]).

4. Consider also Fernando Ortiz's tandem notion of "transculturation" (Ortiz 1995[1947]) and Bastide's "interpenetration of civilizations" (1978[1960]).

5. Of course, the literature addressing sociocultural reproduction in terms of such tropes as "cultural memory," "social memory," and "collective memory" *beyond* the black Atlantic is also broad and deep, addressing agency and strategy to variable degrees. That literature addresses, among other themes, the political and technical conditions under which history is narrated and debated (e.g., Appadurai 1981; Malkki 1995); the rituals, bodily practices, and modes of dress that retain traces of past events (e.g., Stoller 1995; Comaroff and Comaroff 1987; Connerton 1995[1989]); the marks that past crises have left upon our current vocabulary and legal procedures (Schudson 1997); the use of monuments in the selective commemoration of past heroes and the concomitantly selective

authorization of present-day leaders (e.g., Werbner 1998); and the role of literal dialogue—that is, conversation among particular people—in shaping those people's recollection and articulation of past events (Middleton and Edwards 1990).

6. See Trouillot 1995:14–16 for another critique of the "memory" metaphor.

7. See my 1999a article for a more exhaustive account of the trans-atlantic phenomena that I call "dialogue."

8. Among the telltale signs of widespread Anglophone influence in the Afro-Atlantic religions of Cuba and Brazil is the frequent appearance of *w*'s and *k*'s in Brazilian Nagô and Jeje orthographies, and *w*'s, *k*'s, and double *g*'s in Cuban Lucumí orthography.

9. Roach (1996) employs "dialogue" as an analytic metaphor in a sense quite different from my own. He briefly analogizes culture to a *scripted* dialogue, which is reproduced or "remembered" as each new actor replaces a deceased one in the same dramatic role. Though the performance is necessarily somewhat modified, the endurance of the role or space left by the dead or by the apparently forgotten constitutes a collective cultural memory. On the other hand, the feature of "dialogue" that I wish to highlight in my analytic metaphor is the fact of the interlocutors' coevalness and diversity of interests in the interpretation and reproduction of the past.

APPENDIX A
GEECHEES AND GULLAHS

1. Timothy Carrier, *Family across the Sea* [film], 1991 (South Carolina ETV Network, Columbia, SC: South Carolina E.T.V., 1990, distributed by San Francisco Newsreel).

2. Born to the southern aristocracy, DuBose Heyward became a notable participant in the Harlem Renaissance. He wrote the novel *Porgy* (1925) and helped to adapt it for the stage in 1935, when it became the nationally famous *Porgy and Bess*. With degrees from Mississippi College and Yale, Puckett published numerous books on southern Black folklore, including *Folk Beliefs of the Southern Negro* (1969[1926]). Unlike the others, Harris had actually grown up poor, the illegitimate son of an Irish day laborer. But, like many Irish immigrants and their children, he had a great investment in America's guarantee of social superiority for those who could call themselves white. Thus, he became a dramatic spokesman for the southern gentry, the goodness of its ways, the wisdom and loving submission of the old black slave. Harris is best known as the creator of Uncle Remus, through whose fictional mouth he narrated the folktales he had heard from African Americans. See, for example, Smythe et al. 1931; *Who Was Who in America* 1968: vol.4, p. 768; May 1983: vol. 108, p. 218; Trosky and Olendorf 1992: vol. 137, pp. 191–37. See also Ignatiev 1995 on the violent efforts of Irish immigrants to prove their whiteness in the 19th- and 20th-century United States.

3. While the society's transcriptions of Gullah, like Puckett's and Harris's transcriptions of Black English generally, seem intent on conveying the ignorance of the original speakers—such as "wuz" for "was" and "frum" for "from"—the introductory presentation of Augustine T. Smythe and the performance of this

white choir, as recorded in 1955, struck me as both sensitive and warm. My further sources of information on the Society for the Preservation of Spirituals include letters from Dale Rosengarten to the author (dated 5 Sept. 1996 and 12 Oct. 1996) and a letter from Mary Julia Royall to Dale Rosengarten, n.d., kindly copied to me by Ms. Rosengarten on 5 Sept. 1996.

4. For example, one tourist brochure entitled "Sweetgrass Baskets of Mt. Pleasant, S.C.: A Proud Tradition, a Valuable Investment . . ." identifies the baskets as "one of the oldest art forms of African origin in the United States." According to the brochure, funding for its production was provided by the Mount Pleasant Town Council, the South Carolina State Development Board, and the South Carolina Grant Consortium. Also see Rosengarten 1987[1986]:61–63.

Appendix B
The Origins of the Term "Jeje"

1. For mentions of the term "Jeje" in Brazil during the 18th century, see Verger 1976[1968]:6, 7, 17, 381, 450, 462, 593ff.; R. N. Rodrigues 1945[1905]:176; Reis 1983.

2. It is very unlikely that "Ewè" is the etymological origin of "Jeje." The only similarity between their pronunciations is the first vowel sound. However, Ewè does belong to the same Gbè dialect group as the West African language called "Jeje" in the 19th century and the beginning of the 20th.

3. See also *Rapport Monographique sur le Cercle de Porto-Novo*, by the Administrateur en chef, 17 Apr. 1921, 1G353, bobine 200MI/696, CARAN, ANFP.

Bibliography

Abimbọla, Wande. 1997a. *Ifa Will Mend Our Broken World: Thoughts on Yoruba Religion and Culture in Africa and the Diaspora*. Interviews with an introduction by Ivor Miller. Roxbury, MA: Aim Books.

———. 1997b. "Images of Women in the Ifa Literary Corpus." In *Queens, Queen Mothers, Priestesses, and Power*, ed. Flora E. S. Kaplan, pp. 401–14. New York: New York Academy of Sciences.

———. 2000. "Wanted: An Afro-Atlantic Identity." Thirtieth Anniversary Celebration, Department of Afro-American Studies, Harvard University, Cambridge, MA, 8 April.

Abiọdun, Rowland. 1989. "Women in Yorùbá Religious Images." *African Languages and Cultures* 2 (1): 1–18.

Abraham, R. C. 1962[1946]. *Dictionary of Modern Yoruba*. London: Hodder and Stoughton.

Abu-Lughod, Janet. 1997. "Going beyond Global Babble." In *Culture, Globalization and the World-System*, ed. Anthony D. King, pp. 131–37. Minneapolis: University of Minnesota Press.

Adas, Michael. 1989. *Machines as the Measure of Man: Science, Technology, and Ideologies of Western Dominance*. Ithaca, NY: Cornell University Press.

Adefunmi, Ọba Ọfuntọla Oseijeman Adelabu, I. 1982. *Olorisha: A Guidebook into Yoruba Religion*. Sheldon, SC: Orisha Academy of the Yoruba Village of Oyotunji.

Aderibigbe, A. B. 1975. "Early History of Lagos to about 1850." In *Lagos: The Development of an African City*, ed. A. B. Aderibigbe and J. F. Ade Ajayi, pp.1–26. London: Longmans Nigeria.

Adetugbo, A. 1967. "The Yoruba Language in Western Nigeria: Its Major Dialect Areas." Ph.D. diss. Columbia University.

Aguirre Beltrán, Gonzalo. 1989a. *Cuijla: Esbozo etnográfico de un pueblo negro*, 2nd ed. corrected and augmented. Mexico, City: Universidad Veracruzana; Instituto Nacional Indigenista; Gobierno del Estado de Veracruz; Fondo de Cultura Económica. First edition published in 1958.

———. 1989b. *La población negra de México: Estudio etnohistórico*, augmented and corrected edition. Mexico City: Universidad Veracruzana; Instituto Nacional Indigenista; Govierno del Estado de Veracruz; Fondo de Cultura Económica. First edition published in 1946.

Ajayi, J.F.A. 1960. "How Yoruba Was Reduced to Writing." *Odu* 8:49–58.

———. 1961. "Nineteenth-Century Origins of Nigerian Nationalism." *Journal of the Historical Society of Nigeria* 2 (2):196–210.

Akindélé, A., and C. Aguessy. 1953. "Contribution à l'étude de l'histoire de l'ancien royaume de Porto-Novo." *Mémoires de l'Institut Français d'Afrique Noire*, 25. Dakar: IFAN.

Akinjogbin, I. A. 1967. *Dahomey and Its Neighbours 1708–1818*. Cambridge: Cambridge University Press.

Akinṣọwọn, Aki[n]dele (aka Benjamin Coker). 1930[1914]. *Iwe itan ajaṣẹ*. 2nd ed. Lagos: Ifẹ-Olu Printing Works.

Akintọla, Akinbọwale. 1992. *The Reformed Ogboni Fraternity (R.O.F.): Its Origins and Interpretation of Its Doctrines and Symbolism*. Ìbàdàn, Nigeria: Valour Ventures.

Alakija, George. 1977. "The Trance State in Candomblé/L'état de transe dans le Candomblé." Colloquium on Black Civilisation and Science and Technology, Second World Black and African Festival of Arts and Culture, Lagos and Kaduna, Nigeria, 15 Jan. to 12 Feb.

Alba, Richard D. 1990. *Ethnic Identity: The Transformation of White America*. New Haven, CT: Yale University Press.

d'Albéca, Alexandre L. 1889. *Les établissements français du Golfe de Bénin: Géographie, commerce, langues*. Paris: Librairie Militaire de l. Baudoin et Cie.

Amadiume, Ifi. 1987. *Male Daughters, Female Husbands: Gender and Sex in an African Society*. London: Zed Books.

Amado, Jorge. 1940. "Elogio de um Chefe de seita." In *O Negro no Brasil*, Trabalhos apresentados ão 2° Congresso Afro-Brasileiro, 1937, Bahia, pp. 325–28. Rio: Civilizaçao Brasileira.

Anderson, Benedict. 1991[1983]. *Imagined Communities: Reflections on the Origin and Spread of Nationalism*. Rev. ed. London: Verso.

———. 1998. *The Spectre of Comparisons: Nationalism, Southeast Asia, and the World*. London: Verso.

Andrade Lima, Lamartine de. 1984. "Roteiro de Nina Rodrigues." Ensaios/Pesquisas 2 do Centro de Estudos Afro-Orientais, Universidade Federal da Bahia, Abril, segunda tiragem.

Andrews. George Reid. 1991. *Blacks and Whites in São Paulo, Brazil, 1888–1988*. Madison: University of Wisconsin Press.

———. 1992. "Racial Inequality in Brazil and the United States: A Statistical Comparison." *Journal of Social History* 26:229–63.

Angarica, Nicolás Valentín. 1955a. *Manual de Orihaté: Religión Lucumí*. N.P. [Venezuela].

1955b. *Manual de Orihaté: Religión Lucumí*. Havana: n.p.

Anyebe, A. P. 1989. *Ogboni: The Birth and Growth of the Refomed Ogboni Fraternity*. Ikeja, Lagos State, Nigeria: Sam Lao Publishers.

Appadurai, Arjun. 1981. "The Past as a Scarce Resource." *Man*, n.s., 16:201–19.

———. 1990. "Disjuncture and Difference in the Global Cultural Economy." *Public Culture* 2 (2): 1–24.

———. 1996. *Modernity at Large: Cultural Dimensions of Globalization*. Minneapolis: University of Minnesota Press.

Apter, Andrew. 1992. *Black Critics and Kings*. Chicago: University of Chicago Press.

Asad, Talal. 1986. "The Concept of Cultural Translation in British Social Anthropology." In *Writing Culture*, ed. James Clifford and George E. Marcus, pp. 141–64. Berkeley and Los Angeles: University of California Press.

Atanda, J. A. 1970. "The Changing Status of the Alafin of Oyo under Colonial Rule and Independence." In *West African Chiefs*, ed. Michael Crowder and Obaro Ikime, pp. 212–30. Ilé-Ifẹ̀, Nigeria: University of Ife Press.

Atkins, John. 1970[1735]. *A Voyage to Guinea, Brazil, and the West Indies*. London: Frank Cass.

Atkinson, Jane, and Shelly Errington, eds. 1990. *Power and Difference: Gender in Island Southeast Asia*. Stanford, CA: Stanford University Press.

Awde, Nicholas. 1996. *Hausa: Hausa-English, English-Hausa Dictionary*. New York: Hippocrene.

Awẹ, Bọlanlẹ 1977. "The Iyalode in the Traditional Yoruba Political System." In *Sexual Stratification*, ed. Alice Schlegel, pp. 144–60. New York: Columbia University Press.

Awolalu, J. Ọmọṣade. 1979. *Yoruba Beliefs and Sacrificial Rites*. London: Longman.

Ayandele, E. A. 1966. *The Missionary Impact on Modern Nigeria 1842–1914*. London: Longmans, Green.

Azevedo, Thales de. 1975. *Democracia racial*. Petrópolis, Rio de Janeiro: Vozes.

Azevedo Santos, Maria Stella de. 1986. "Sincretismo e Branqueamento." Lecture presented at the Third International Congress of Orişa Tradition and Culture, New York City, Oct. 6–10.

———. 1993. *Meu Tempo È Agora*. São Paulo: Editora Oduduwa.

Bailyn, Bernard. 1979. *The New England Merchants in the Seventeenth Century*. Cambridge, MA: Harvard University Press.

———. 1996. "The Idea of Atlantic History." Working Paper No. 96-01, International Seminar on the History of the Atlantic World, 1500–1800, Harvard University.

Baker, Lee D. 1998. *From Savage to Negro: Anthropology and the Construction of Race, 1896–1954*. Berkeley and Los Angeles: University of California Press.

Bakhtin, M. M. 1981. *The Dialogic Imagination*. Ed. Michael Holquist. Austin: University of Texas Press.

Ballard, J. A. 1965. Les Incidents de 1923 à Porto-Novo: La Politique à l'Èpoque Coloniale. *Études Dahoméenes*, n.s., 5 (Oct.): 69–87.

Barber, Karin. 1990. "Discursive Strategies in the Texts of Ifá and in the 'Holy Book of Odù of the African Church of Ọrunmìlà." In *Self-Assertion and Brokerage*, ed. P. F. de Moraes Farias and Karen Barber, pp. 196–224. Birmingham University African Studies Series 2. Birmingham: Centre of West African Studies, University of Birmingham.

Barbosa, Luiz Sérgio. 1984. "A Federação Baiana do Culto Afro-Brasileiro." In Vivaldo da Costa Lima et al, *Encontro de nações-de-candomblé*, pp. 69–72. Salvador: Ianamá, Centro de Estudos Afro-Orientais of the Federal University of Bahia, and Centro Editorial e Didático.

Barbot, Jean. 1992. *Barbot on Guinea: The Writings of Jean Barbot on West Africa 1678–1712*, 2 vols. Ed. P.E.H. Hair, Adam Jones, and Robin Law. London: Hakluyt Society.

Barboza, Marília T., and Vera de Alencar. 1985. *Caymmi: Som Imagem Magia: Biografia de Dorival Caymmi*. Rio de Janeiro: Sargaço Produções Artísticas.

Barickman, B. J. 1994. "A Bit of Land, Which They Call Roça: Slave Provision Grounds in the Bahian Recôneavo, 1780–1860." *Hispanic American Historical Review* 74 (4):649–87.

Barnet, Miguel, ed. 1994 [1966]. *Biography of a Runaway Slave.* Translated from the Spanish by W. Nick Hill. Willimantic, CT: Curbstone Press.

Barth, Fredrik. 1969. "Introduction." In *Ethnic Groups and Boundaries,* ed. Fredrik Barth, pp. 9–38. Boston: Little, Brown.

Bascom. William R. 1969. *Ifa Divination.* Bloomington: Indiana University Press.

———. 1972. *Shango in the New World.* Austin: African and Afro-American Research Institute, University of Texas at Austin.

Bastide, Roger. 1961. *O Candomblé da Bahia.* São Paulo: Editora Nacional.

———. 1967. *Les Amériques Noires.* Paris: Payot.

———. 1971[1967]. *African Civilizations in the New World.* Trans. Peter Green. New York: Harper and Row.

———. 1978[1960]. *The African Religions of Brazil.* Trans. Helen Sebba. Baltimore: Johns Hopkins University Press.

———. 1983. *Estudos Afro-Brasileiros.* São Paulo: Editora Perspectiva.

Baudin, Noel. 1885. *Fetichism and Fetich Worshipers.* New York: Benziger Bros.

Beauvoir, Simone de. 1968[1963]. *Force of Circumstance (La Force des choses).* Trans. Richard Howard. Harmondsworth, UK: Penguin.

Bergé, J.A.M.A.R. 1928. "Étude sur le Pays Mahi (1926–1928)." *Bulletin du Comité d' Études Historiques et Scientifiques de l'Afrique Occidentale Française* 11 (4):708–55.

Bernal, Martin. 1996. "The Afrocentric Interpretation of History: Bernal Replies Lefkowitz." *Journal of Blacks in Higher Education* 11 (Spring): 86–94.

———. 1997. *Black Athena: The Afroasiatic Roots of Classical Civilization.* Vol. I. New Brunswick, NJ: Rutgers University Press.

Bharati, Agehananda. 1972. *The Asians in East Africa.* Chicago: Professional-Technical Series, Nelson-Hall.

Birman, Patricia. 1985. "Identidade social e homossexualismo no Candomblé." *Religião e Sociedade* 12 (1): 2–21.

Blier, Suzanne Preston. 1988. "Melville J. Herskovits and the Arts of Ancient Dahomey." *Res* 16 (Autumn):124–142.

———. 1995a. *African Vodun.* Chicago: University of Chicago Press.

———. 1995b. "The Path of the Leopard: Motherhood and Majesty in Early Danhomè." *Journal of African History* 36:391–417.

———. 1995c. "Vodun: West African Roots of Vodou." *In Sacred Arts of Haitian Vodou,* Donald J. Cosentino, pp. 61–87. Los Angeles: UCLA Fowler Museum of Cultural History.

Boddy, Janice. 1989. *Wombs and Alien Spirits: Women, Men, and the Zar Cult in Northern Sudan.* Madison: University of Wisconsin Press.

Bolster, W. Jeffrey. 1997. *Black Jacks: African American Seamen in the Age of Sail.* Cambridge, MA: Harvard University Press.

Borghero, Francesco. 1864. "Missions du Dahomey . . . Adressée par M. l'abbé Borghero, à M. l'abbé Augustin Planque, Supérieur des Missions africaines à Lyon." *Annales de la Propagation de la Foi* 36:419–44.

Bosman, William. 1967[1704]. A *New and Accurate Description of the Coast of Guinea*. London: Frank Cass.

Bouche, Pierre. 1868. Letter to Supérieur des Missions africaines at Lyon, written from Porto-Novo. Entry 20.358, Rubric 12/80200. Société des Missions Africaines Archives, Rome.

———. 1885. *Sept Ans en Afrique Occidentale: La Côte des Esclaves et le Dahomey*. Paris: Librairie Plon.

Bourdieu, Pierre. 1984[1979]. *Distinction*. Trans. Richard Nice. Cambridge, MA: Harvard University Press.

———. 1985[1972]. *Outline of a Theory of Practice*. Trans. Richard Nice. Cambridge: Cambridge University Press.

Bourgeois, Jean-Louis, Carollee Pelos, and Basil Davidson. 1996[1983]. *Spectacular Vernacular: The Adobe Tradition*. 2nd ed. New York: Aperture.

Bourguet, R. Père. 1872. Bourguet to Planque, 26 June 1872, entry no. 17.047, rubrics no. 12/80200, Lagos, pp. 2, 4, Société des Missions Africaines Archives, Rome.

———. 1873. Retyped copy of a letter from Bourguet in Lagos to Planque, 26 June 1873, entry no. 17047, rubrics no. 12/80200, Société des Missions Africaines Archives, Rome.

Bovill, Edward William. 1999[1958]. *The Golden Trade of the Moors*. Princeton, NJ: Marcus Wiener.

Bowen, T. J. 1968[1857]. *Adventures and Missionary Labors: In several countries in the Interior of Africa* 2nd ed. London: Frank Cass.

Braga, Julio. 1995. *Na Gamela do Feitiço: Repressão e Resistência nos Candomblés da Bahia*. Salvador: Editôra da Universidade Federal da Bahia.

Braudel, Fernand. 1992[1949]. *The Mediterranean and the Mediterranean World in the Age of Philip II*. Translated by Sian Reynolds and abridged by Richard Ollard. London: HarperCollins.

Brookshaw, David. 1986. *Race and Color in Brazilian Literature*. Metuchen, NJ: Scarecrow.

Brown, David Hilary. 1989. "Garden in the Machine: Afro-Cuban Sacred Art and Performance in Urban New Jersey and New York." 2 vols. Ph.D. diss., Yale University

Brown, Diana DeG., and Mario Bick. 1987. "Religion, Class, and Context: Continuities and Discontinuities in Brazilian Umbanda." *American Ethnologist* 14 (1):73–93.

Brown, Karen McCarthy. 1987. "Alourdes: A Case Study of Moral Leadership in Haitian Vodun." In *Saints and Virtues*, ed. John S. Hawley, pp. 144–67. Berkeley and Los Angeles: University of California Press.

———. 1989. "Systematic Remembering, Systematic Forgetting." In *Africa's Ogun*, ed. Sandra Barnes, pp. 65–89. Bloomington: Indiana University Press.

———. 1991. *Mama Lola*. Berkeley and Los Angeles: University of California Press.

Buarque de Holanda, Sérgio. 1995[1936]. *Raízes do Brasil*, 26th ed. São Paulo: Companhia das Letras.

Buck-Morss, Susan. 2000. "Hegel and Haiti." *Critical Inquiry* 26 (Summer): 821–65.

Buckley, Anthony D. 1997[1985]. *Yoruba Medicine*. New York: Athelia Henrietta Press.

Burton, Richard. 1966[1861–64]. *A Mission to Gelele, King of Dahome*. London: Routledge and Kegan Paul.

Butler, Kim D. 1998. *Freedoms Given, Freedoms Won: Afro-Brazilians in Post-Abolition São Paulo and Salvador*. New Brunswick, NJ: Rutgers University Press.

Cabrera, Lydia. 1974. *Yemayá y Ochún*. Madrid: Colección del Chicherekú en el exilio.

———. 1980. *Koeko Iyawó: Aprende Novicia*. Miami: Colección del Chicherekú en el exilio.

———. 1983[1954]. *El Monte*. Miami: Colección del Chicherekú.

———. 1986[1957]. *Anagó: Vocabulario Lucumí*. Miami: Ediciones Universal/Colección del Chicherekú.

Cacciatore, Olga Gudolle. 1977. *Dicionário de Cultos Afro-Brasileiros*. 2nd ed. Rio de Janeiro: Forense Universitário.

Capo, Hounkpati. 1984. "Elements of Ewe-Gen-Aja-Fon dialectology." In *Peuples du Golfe du Bénin (Aja-Ewé)*, ed. François de Medeiros, pp. 167–78. Paris: Karthala and Centre de Recherches Africaines.

———. 1991. *A Comparative Phonology of Gbe*. Berlin and New York: Foris and Garome; Bénin: Labo Gbe (Int.).

Carelli, Mario. 1993. *Cultures croisées*. Paris: Nathan.

Carneiro, Édison. 1940. "The Structure of African Cults in Bahia." *Journal of American Folk-Lore* 53:271–78.

———. 1948. *Candomblés da Bahia*. Salvador da Bahia: Publicações do Museu do Estado, No. 8.

———. 1967. Candomblé da Bahia. In *Antologia do Negro*, ed. Édison Carneiro, pp. 263–72. Rio de Janeiro: Edições de Ouro.

———. 1986[1948]. *Candomblés da Bahia*. 7th ed. Rio de Janeiro: Civilização Brasileira.

———. 1991. *Religiões Negras e Negros Bantos*. 3rd ed. Rio: Civilização Brasileira. Previously published as *Religiões Negras* (Rio: Civilização Brasileira, 1936) and *Negros Bantos* (Rio: Civilização Brasileira, 1937).

Carney, Judith Ann. 2001. *Black Rice: The African Origins of Rice Cultivation in the Americas*. Cambridge, MA: Harvard University Press.

Carr, Andrew. 1989[1955]. *A Rada Community in Trinidad*. Port-of-Spain, Trinidad: Paria Publishing Company.

Carvalho, Jehová de. 1984. "Nação-Jeje." In Vivaldo da Costa Lima et al *Encontro de nações-de-candomblé*, pp. 49–58. Salvador: Ianamá: Centro de Estudos Afro-Orientais of the Federal University of Bahia, and Centro Editorial e Didático.

———. 1991[1989]. *Reinvenção do Reino dos Voduns: Poemas em torno do Candomblé do Bogum votivos aos Voduns*. Salvador, Bahia: Littera.

Caulfield, Mina. 1969. "Culture and Imperialism: Proposing a New Dialectic." In *Reinventing Anthropology*, ed. Dell Hymes, pp. 182–212. New York: Pantheon/Random.

Chakrabarty, Dipesh. 1997. "Postcoloniality and the Artifice of History: Who Speaks for 'Indian' Pasts?" In *A Subaltern Studies Reader, 1986–1995*, ed. Ranajit Guha, pp. 263–93. Minneapolis: University of Minnesota Press.

Chatterjee, Partha. 1992. *The Nation and Its Fragments*. Princeton, NJ: Princeton University Press.

Chaudhuri, K. N. 1990. *Asia before Europe: Economy and Civilization of the Indian Ocean from the Rise of Islam to 1750*. Cambridge: Cambridge University Press.

Clapperton, Hugh. 1829. *Journal of a Second Expedition in the Interior of Africa, from the Bight of Benin to Soccatoo*. London: John Murray.

Clarke, William H. 1972. *Travels and Explorations in Yorubaland 1854–1858*. Edited with an introduction by J. A. Atanda. Ìbàdàn, Nigeria: University of Ibadan Press.

Clifford, James. 1988. *The Predicament of Culture*. Cambridge, MA: Harvard University Press.

———. 1997. *Routes: Travel and Translation in the Late Twentieth Century*. Cambridge, MA: Harvard University Press.

Cohen, Abner. 1969. *Custom and Politics in Africa*. Berkeley and Los Angeles: University of California Press.

Cole, P. D. 1975. "Lagos Society in the 19th Century." In *Lagos: The Development of an African City*, ed. A. B. Aderibigbe and J. F. Ade Ajayi, pp. 27–58. London: Longmans Nigeria.

Cole, Sally. 1994. "Introduction—Ruth Landes in Brazil: Writing, Race, and Gender in 1930s American Anthropology." In *The City of Women*, by Ruth Landes. Albuquerque: University of New Mexico Press.

Collier, Jane Fishburne, and Sylvia Junko Yanagisako, eds. 1987. *Gender and Kinship*. Stanford, CA: Stanford University Press.

Comaroff, John L., and Jean Comaroff. 1987. "The Madman and the Migrant: Work and Labor in the Historical Consciousness of a South African People." *American Ethnologist* 14 (2): 191–209.

———. 1991. *Of Revelation and Revolution: Christianity, Colonialism, and Consciousness in South Africa*. Vol. 1. Chicago: University of Chicago Press.

———. 1997. *Of Revelation and Revolution: The Dialectics of Modernity on a South African Frontier*. Vol. 2. Chicago: University of Chicago Press.

Comhaire, Jean. 1949. "À Propos des 'Brésiliens' de Lagos." *Grands Lacs* 64 (7), n.s. no. 119:41–43.

Connerton, Paul. 1995[1989]. *How Societies Remember*. Cambridge: Cambridge University Press.

Conrad, Robert Edgar. 1986. *World of Sorrow: The African Slave Trade to Brazil*. Baton Rouge: Louisiana State University Press.

Cooper, Frederick. 2001. "What Is the Concept of Globalization Good For? An African Historian's Perspective." *African Affairs* 100: 169–213.

Corrêa, Mariza. 2000. "O Mistério dos Orixás e das Bonecas: Raça e Gênero na Antropologia Brasileira." *Etnográfica* 4 (2): 233–65.

Correia Lopes, Edmundo. 1943. "Exéquias no 'Bôgúm' do Salvador." *O Mundo Português* 109:559–65.

Côrtes de Oliveira, Maria Inês. 1988. *O Liberto: O seu mundo e os outros*. São Paulo: Corrupio; Brasília: CNPq.

———. 1992. "Retrouver une identité: Jeux sociaux des Africains de Bahia (vers 1750–vers 1890)." Ph.D. diss., University of Paris, Sorbonne (Paris IV).

Cosentino, Donald J. 1995. "Imagine Heaven." In *Sacred Arts of Haitian Vodou*, ed. Donald J. Cosentino, pp. 25–55. Los Angeles: UCLA Fowler Museum of Cultural History.

Cossard-Binon, Gisèle. 1970. "Contribution à l'étude des Candomblés au *Brésil: Le Candomblé Angola.*" Ph.D. diss., University of Paris.

———. 1981. "A Filha de Santo." In *Olóòrìṣà*, ed. Carlos Eugênio Marcondes de Moura, pp. 127–51 São Paulo: Àgora.

Creel, Margaret. 1987. *"A Peculiar People": Slave Religion and Community-Culture among the Gullahs.* New York: New York University Press.

Cunha, Euclides da. 1944[1902]. *Rebellion in the Backlands.* Trans. Samuel Putnam. Chicago: University of Chicago Press.

Cunha, Manuela Carneiro da. 1985. *Negros, estrangeiros.* São Paulo: Brasiliense.

Cunha, Marianno Carneiro da. 1985. *Da Senzala ao Sobrado: Arquitetura na Nigéria e na República Popular do Benim.* São Paulo: Nobel.

Curtin, Philip D. 1969. *The Atlantic Slave Trade: A Census.* Madison: University of Wisconsin Press.

Curtin, Philip D., Steven Feierman, Leonard Thompson, and Jan Vansina. 1978. *African History.* Boston: Little, Brown.

Dalzel, Archibald. 1967[1793]. *The History of Dahomy: An Inland Kingdom of Africa.* London: Frank Cass.

Da Matta, Roberto. 1991[1979]. *Carnivals, Rogues, and Heroes: An Interpretation of the Brazilian Dilemma.* Trans. John Drury. Notre Dame, In: University of Notre Dame Press.

Danforth, Loring M. 1995. *The Macedonian Conflict: Ethnic Nationalism in a Transnational World.* Princeton, NJ: Princeton University Press.

Dantas, Beatriz Góis. 1982. "Repensando a pureza Nagô." *Religião e Sociedade* 8:15–20.

———. 1988. *Vovó Nagô e Papai Branco: Usos e Abusos da Àfrica no Brasil.* Rio de Janeiro: Graal.

Daramọla, Olu., and Adebayọ Jeje. 1975 [1967]. *Àwọn Àṣà àti Òrìṣà Ilẹ̀ Yorùbá.* Ìbàdàn, Nigeria: Onibon-Oje Press & Book Industries.

Dayan, Joan. 1995. *Haiti, History, and the Gods.* Berkeley and Los Angeles: University of California Press.

Degler, Carl N. 1971. *Neither Black Nor White: Slavery and Race Relations in Brazil and the United States.* Madison: University of Wisconsin Press.

Deleuze, Gilles, and Félix Guattari. 1987[1980]. *A Thousand Plateaus: Capitalism and Schizophrenia.* Trans. Brian Massumi. Minneapolis: University of Minnesota Press.

DesMarchais, Chevalier Étienne Renaud. 1703–6. Iovrnal de Navigation de la Coste de Gvinnée Isles de LAmerique et Indes DEspagne švr Le Vaisseav dv roy le Favcon François armé par lordre de Sa Maisté povr la Royalle Compagnie de LAssiente. N.P.: 1900–91[?]. Microfilm, London: British Museum, 1973.

Deren, Maya. 1983[1970]. *Divine Horsemen: The Voodoo Gods of Haiti.* New Paltz, NY: Documentext, McPherson.

DaMatta, Roberto. 1991[1979]. *Carnivals, Rogues, and Heroes: An Interpretation of the Brazilian Dilemma.* Trans. by John Drury. Notre Dame, IN: University of Notre Dame Press.

Desribes, E. 1877. *L'Évangile au Dahomey et à la Côte des Esclaves, ou, Histoire des Missions Africaines de Lyon.* Clermont-Ferrand: Imprimerie Centrale, Meneboode.

Dietler, Michael. 1994. "'Our Ancestors the Gauls': Archaeology, Ethnic Nationalism, and the Manipulation of Celtic Identity in Modern Europe." *American Anthropologist* 96:584–605.

Dikötter, Frank, ed. 1997. *The Construction of Racial Identities in China and Japan.* Honolulu: University of Hawai'i Press.

Doortmont, Michel. 1990. "The Invention of the Yorubas: Regional and Pan-African Nationalism versus Ethnic Provincialism." In *Self-Assertion and Brokerage: Early Cultural Nationalism in West Africa,* ed. P. F. de Moraes Farias and Karin Barber, pp. 101–08. Birmingham University African Studies Series 2, Birmingham, UK: Centre of West African Studies, University of Birmingham.

Dorsainvil, J. C. 1931. *Vodou et Névrose.* Port-au-Prince: Imprimerie "La Presse."

Douglas, Mary. 1984[1966]. *Purity and Danger.* London: Routledge and Kegan Paul.

Drewal, Margaret Thompson. 1992. *Yoruba Ritual: Performers, Play, Agency.* Bloomington: Indiana University Press.

Du Bois, W. E. Burghardt. 1939. *Black Folk: Then and Now. An Essay in the History and Sociology of the Negro Race.* New York: Henry Holt.

Duara, Prasenjit. 1997. "Nationalists among Transnationals: Overseas Chinese and the Idea of China, 1900–1911." In *Ungrounded Empires: The Cultural Politics of Modern Chinese Transnationalism,* ed. Aihwa Ong and Donald M. Nonini, pp. 39–60. New York: Routledge.

Duarte, Abelardo. 1974. *Catálogo Ilustrado da Coleção Perseverança.* Museum of the Instituto Histórico e Geográfico de Alagoas. Maceió: Departamento de Assuntos Culturais—SENEC.

Duncan, John. 1968[1847]. *Travels in Western Africa in 1845 and 1846.* London: Frank Cass.

Dzidzienyo, Anani. 1985. "The African Connection and the Afro-Brazilian Condition." In *Race, Class and Power in Brazil,* ed. Pierre-Michel Fontaine, pp. 135–53. Los Angeles: University of California, Center for Afro-American Studies.

Ebron, Paulla A. 2002. *Performing Africa.* Princeton, NJ: Princeton University Press.

Eco, Humberto. 1986. *Travels in Hyperreality,* trans. by William Weaver. San Diego, New York, London, Harcourt Brace Jovanovich.

Eduardo, Octavio da Costa. 1981[1948]. *The Negro in Northern Brazil.* London: African Publication Society.

Elbein dos Santos, Juana. 1993[1975]. *Os Nàgô e a Morte.* Petrópolis, Brazil: Vozes.

Elbein dos Santos, Juana, and Deoscoredes M. dos Santos. 1981[1969]. "O Culto dos Ancestrais na Bahia: O Culto dos Égun." In *Olóòrìsà: Escritos sobre a Religiao dos Orixás.* São Paulo: Agora.

Ellis, A. B. 1964[1894]. *The Yorùbá-Speaking Peoples of the Slave Coast of West Africa.* Chicago: Benin Press.

———. 1970[1890]. *The Ewe-Speaking Peoples of the Slave Coast of West Africa.* Oosterhout, N.B., Netherlands: Anthropological Publications.

Eltis, David. 1987. *Economic Growth and the Ending of the Transatlantic Slave Trade*. New York: Oxford University Press.

Epega, D. Onadele. 1931. *The Mystery of the Yoruba Gods*. Lagos: Hope Rising Press.

———. 1971[1932]. *The Basis of Yoruba Religion*. Ed. D. Olarimwa Epega. Lagos: Ijamido Printers.

Fabian, Johannes. 1983. *Time and the Other: How Anthropoloy Makes Its Object*. New York: Columbia University Press.

Fabunmi, M. A. 1969. *Ifẹ Shrines*. Ifẹ̀, Nigeria: University of Ifẹ̀ Press.

Fadipẹ, N. A. 1970[1939]. *The Sociology of the Yorùbá*. Ìbàdàn, Nigeria: Ibàdàn University Press.

Fernandez, James W. 1986. *Persuasions and Performances*. Bloomington: Indiana University Press.

Ferraz, Aydano do Couto Ferraz. 1941. "Vestígios de um culto Dahomeano no Brasil." *Revista do Arquivo Municipal (São Paulo), ano VII, vol LXXVI, pp. 271–74.*

Ferretti, Sergio. 1986. *Querebentan de Zomadonu: Etnologia da Casa das Minas*. São Luis, Brazil: Universidade Federal do Maranhao.

Fitz, Earl E., and Irwin Stern. 1988. "Regionalism." In *Dictionary of Brazilian Literature*, ed. Irwin Stern, pp. 275–80. New York: Greenwood.

Foner, Nancy. 1997. "What's New about Transnationalism? New York Immigrants Today and at the Turn of the Century." *Diaspora* 6 (3):355–75.

Forbes, Frederick E. 1851. *Dahomey and Dahomans, Being the Journals of Two Missions to the King of Dahomey, and Residence at His Capital, in the Years 1849 and 1850*. London: Longman, Brown, Green, and Longmans.

Frazier, E. Franklin. 1942. "The Negro Family in Bahia, Brazil." *American Sociological Review* 7 (4): 465–78.

———. 1943. "Rejoinder [to Herskovits] by E. Franklin Frazier." *American Sociological Review* 8(4):394–402.

———. 1974 (1964). *The Negro Church in America*. (Published in tandem with C. Eric Lincoln, *The Black Church since Frazier*.) New York: Schocken Books.

Freeman, Derek. 1983. *Margaret Mead and Samoa: The Making and Unmaking of an Anthropological Myth*. Cambridge, MA: Harvard University Press.

Freyre, Gilberto. 1955[1926]. *Manifesto Regionalista de 1926*. Rio de Janeiro: Ministério da Educação e Cultura/Serviço de Documentação.

———. 1959[1945]. *New World in the Tropics*. New York: Knopf.

———. 1963. *The Mansions and the Shanties: The Making of Modern Brazil*. New York: Knopf.

———. 1986[1933]. *The Masters and the Slaves*. Trans. Samuel Putnam. Berkeley and Los Angeles: University of California Press.

———. ed. 1988[1937]. *Novos Estudos Afro-Brasileiros*. Recife: Fundação Joaquim Nabuco/Editora Massangana.

Frigerio, Alejandro. 1983. "*The Search for Africa*." M.A. thesis, University of California, Los Angeles.

Frobenius, Leo. 1968[1913]. *The Voice of Africa*. Vol. 1. New York: Benjamin Blom.

Fry, Peter. 1982. *Para Inglês Ver*. Rio de Janeiro: Zahar.
————. 1986. "Male Homosexuality and Spirit Possession in Brazil." *Journal of Homosexuality* 11 (3–4): 137–53.
Gaspar, David Barry, and David P. Geggus. 1997. *A Turbulent Time: The French Revolution and the Greater Caribbean*. Bloomington: Indiana University Press.
Geertz, Clifford. 1973. *The Interpretation of Cultures*. New York: Basic Books.
Gellner, Ernest. 1983. *Nations and Nationalism*. Ithaca, NY: Cornell University Press.
Genovese, Eugene D. 1974[1972]. *Roll, Jordan, Roll*. New York: Vintage Books.
Georgia Writers' Project (Savannah Unit). 1940. *Drums and Shadows: Survival Studies among the Georgia Coastal Negroes*. Athens: University of Georgia Press.
Gerth, H. H., and C. Wright Mills, eds. 1946. *From Max Weber*. Trans. H. H. Gerth and C. Wright Mills. New York: Oxford University Press.
Ghosh, Amitav. 1992. *In an Antique Land*. New York: Vintage/Random House.
Giacomini, Sonia Maria. 1988. *Mulher e Escrava: Uma introdução histórica ao estudo da mulher negra no Brasil*. Petrópolis: Vozes.
Gilroy, Paul. 1993. *The Black Atlantic: Modernity and Double Consciousness*. Cambridge, MA: Harvard University Press.
Glick Schiller, N., L. Basch, and C. Blanc-Szanton, eds. 1992. *Towards a Transnational Perspective on Migration: Race, Class, Ethnicity, and Nationalism Reconsidered*. New York: New York Academy of Science.
Gonsalves de Mello, José Antônio, ed. 1988[1935]. *Estudos afro-brasileiros (Trabalhos apresentados ao Primeiro Congresso Afro-Brasileiro realizado no Recife, em 1934)*. Vol. 1. Recife, Brazil: Fundação Joaquim Nabuco/Editora Massangana.
————. 1988[1937]. *Novos estudos afro-brasileiros (Trabalhos apresentados ao Primeiro Congresso Afro-Brasileiro realizado no Recife, em 1934)*. Vol. 2. Recife, Brazil: Fundação Joaquim Nabuco/Editora Massangana.
Gordon, Robert W. 1931. "The Negro Spiritual." In Augustine T. Smythe et al, *The Carolina Low-Country*, pp. 189–222. New York: Macmillan.
Green, James N. 1999. *Beyond Carnival: Male Homosexuality in Twentieth-Century Brazil*. Chicago: University of Chicago Press.
Guha, Ranajit, ed. 1997. *A Subaltern Studies Reader, 1986–1995*. Minneapolis: University of Minnesota Press.
Gungwu, Wang. 1992[1991]. *China and the Chinese Overseas*. Singapore: Times Academic Press.
Hall, Gwendolyn Midlo. 1995 [1992]. *Africans in Colonial Louisiana: The Development of Afro-Creole Culture in the Eighteenth Century*. Baton Rouge: Louisiana State University Press.
Hall, Jacqueline Dowd. 1983. "'The Mind That Burns in Each Body': Women, Rape, and Racial Violence." In *Powers of Desire: The Politics of Sexuality*, ed. Ann Snitow, Christine Stansell, and Sharon Thompson, pp. 328–49. New York: Monthly Review Press.
Halter, Marilyn. 1993. *Between Race and Ethnicity: Cape-Verdean-American Immigrants, 1860–1965*. Urbana: University of Illinois Press.

Hannerz, Ulf. 1996. *Transnational Connections*. London: Routledge.

Harding, Rachel E. 2000. *A Refuge in Thunder: Candomble and Alternative Spaces of Blackness*. Bloomington: University of Indiana Press.

Hardwicke, C. Grafton, and Tweeddale Argyll. 1745. *A New General Collection of Voyages and Travels: Consisting of the Most Esteemed Relations, which have been hitherto published in any Language: Comprehending every Thing remarkable in its Kind, in Europe, Asia, Africa, and America . . .* Vol. 2. London: Thomas Astley.

———. 1746. *A New General Collection of Voyages and Travels: Consisting of the Most Esteemed Relations, which have been hitherto published in any Language: Comprehending every Thing remarkable in its Kind, in Europe, Asia, Africa, and America . . .* Vol. 3. London: Thomas Astley.

Harris, Michael D. 1999. "Departures and Returns: African Artists in the West." In *Transatlantic Dialogue: Contemporary Art in and out of Africa*, by Michael D. Harris, principal author, pp. 10–31. Chapel Hill: Ackland Art Museum, University of North Carolina.

Harvey, David. 1989. *The Condition of Postmodernity*. Cambridge, MA: Blackwell.

Healey, Mark Alan. 2000[1998]. "'The Sweet Matriarchy of Bahia': Ruth Landes' Ethnography of Race and Gender." *Dispositio/n* 23 (50):87–116.

Hebdige, Dick. 1979. *Subculture: The Meaning of Style*. London: Methuen.

Henfrey, Colin. 1981. "The Hungry Imagination: Social Formation, Popular Culture and Ideology in Bahia." In *The Logic of Poverty: The Case of the Brazilian Northeast*, ed. Simon Mitchell, pp. 58–108. London: Routledge and Kegan Paul.

Le Herissé, A. 1911. *L'Ancien Royaume du Dahomey: Moeurs, Réligion, Histoire*. Paris: Émile Larose, Libraire-Éditeur.

Herskovits, Melville J. 1938. *Dahomey: An Ancient West African Kingdom*. 2 vols. New York: J. J. Augustin.

———.1943. "The Negro in Bahia, Brazil: A Problem in Method." *American Sociological Review* 8(4): 394–402.

———. 1958[1941]. *The Myth of the Negro Past*. Boston: Beacon.

———. 1966[1945]. "Problem, Method and Theory in Afroamerican Studies." In *The New World Negro*, ed. Frances S. Herskovits, pp. 43–61. Bloomington: Indiana University Press.

———. 1966 [1953]. "The *Panan*, an African Rite of Transition." In *The New World Negro*, ed. Frances S. Herskovits, pp. 199–226. Bloomington: Indiana University Press/Minerva.

———. 1966[1955]. "The Social Organization of the Candomblé." In *The New World Negro*, ed. Frances S. Herskovits, pp. 226–47. Bloomington: Indiana University Press.

———. 1966[1956]. "Some Modes of Ethnographic Comparison." In *The New World Negro*, ed. Frances S. Herskovits, pp. 71–82. Bloomington: Indiana University Press.

———. 1966[1958]. "Some Economic Aspects of the Afrobahian Candomblé." In *The New World Negro*, ed. Frances S. Herskovits, pp. 248–66. Bloomington: Indiana University Press.

——. 1966[1930]. "The Negro in the New World: The Statement of a Problem." In *The New World Negro*, ed. Frances S. Herskovits, pp. 1–12. Bloomington: Indiana University Press.

——. 1966[1935]. "What Has Africa Given America?" In *The New World Negro*, ed. Frances S. Herskovits, pp. 168–74. Bloomington: Indiana University Press.

——. 1966[1937]. "African Gods and Catholic Saints in New World Negro Belief." In *The New World Negro*, ed. Frances S. Herskovits, pp. 321–29. Bloomington: Indiana University Press.

Herskovits, Melville J., and Frances S. Herskovits. 1934. *Rebel Destiny*. New York: McGraw-Hill.

——. 1976[1933]. *An Outline of Dahomean Religious Belief*. Millwood, NY: Kraus Reprint.

Herzfeld, Michael. 1997. *Cultural Intimacy: The Social Poetics of the Nation-State*. New York: Routledge.

Hethersett, A. L. 1944[pre-1896]. *Iwe Kika Ẹkẹrin Li Ede Yoruba*. Lagos: C.M.S. Bookshop; Exeter: James Townsend and Sons (Printers).

Heyward, Du Bose. 1931. "The Negro in the Low-Country." In Augustine T. Smythe et al., *The Carolina Low-Country*, pp. 169–87. New York: Macmillan.

Heywood, Linda M. 2000a. "Introduction." In *Central Africans and Cultural Transformations in the American Diaspora*, ed. Linda M. Heywood, pp. 1–18. Cambridge: Cambridge University Press.

——. 2000b. "Portuguese into African: The Eighteenth-Century Central African Background of Atlantic Creole Cultures." In *Central Africans and Cultural Transformations in the American Diaspora*, ed. Linda M. Heywood, pp. 91–113. Cambridge: Cambridge University Press.

Hobsbawm, Eric, and Terence Ranger, eds. 1983. *The Invention of Tradition*. Cambridge: Cambridge University Press.

Hoetink, H. 1973. *Slavery and Race Relations in the Americas*. New York: Harper and Row.

Holloway, Joseph E., ed. 1990. *Africanisms in American Culture* Bloomington: Indiana University Press.

Howe, M. A. DeWolfe. 1930. "The Song [Songs] of Charleston." *Atlantic Monthly*, July 5: 108–111.

Idowu, E. Bọlaji. 1963. *Olodumare: God in Yoruba Belief*. New York: Praeger.

Ignatiev, Noel. 1995. *How the Irish Became White*. New York: Routledge.

Jackson, Peter A. 2000. "Reading Rio from Bangkok: An Asianist Perspective on Brazil's Male Homosexual Cultures" (review article). *American Ethnologist* 27 (4):950–60.

Jackson, Walter. 1986. "Melville Herskovits and the Search for Afro-American Culture." In *Malinowski, Rivers, Benedict and Others*, ed. George Stocking, pp. 95–126. Madison: University of Wisconsin Press.

James, C.L.R. 2001[1938]. *The Black Jacobins: Toussaint L'Ouverture and the San Domingo Revolution*. London: Penguin.

Jameson, Fredric. 1991. *Postmodernism, or, The Cultural Logic of Late Capitalism*. Durham, NC: Duke University Press.

Johnson, Rev. Samuel. 1921[1897]. *The History of the Yorubas*. Lagos: C.S.S. Bookshops.

Kaké, Ibrahima B. 1982. "The Impact of Afro-Americans on French-Speaking Black Africans, 1919–1945." In *Global Dimensions of the African Diaspora*, ed. Joseph E. Harris, pp. 195–209. Washington, DC: Howard University Press.

Kaplan, Flora E. S., ed. 1997. *Queens, Queen Mothers, Priestesses, and Power: Case Studies in African Gender*. New York: New York Academy of Sciences.

Karasch, Mary C. 1987. *Slave Life in Rio de Janeiro 1808–1850*. Princeton, NJ: Princeton University Press.

Kasinitz, Philip. 1992. *Caribbean New York: Black Immigrants and the Politics of Race*. Ithaca, NY: Cornell University Press.

Kearney, M. 1995. "The Local and the Global: The Anthropology of Globalization and Transnationalism." *Annual Review of Anthropology* 24:547–65.

Kessler, Suzanne J., and Wendy McKenna. 1978. *Gender: An Ethnomethodological Approach*. New York: Wiley.

Kiti, Gabriel. 1929. "Rites funéraires usités chez les Alladanous et diverse tribus de race Goun ou Aïzo habitant la banlieue de Porto-Novo." *Les missions catholiques* [Bureaux de la Propagation de la Foi, Lyon] 3.097 (1 Nov.)

Kopytoff, Jean Herskovits. 1965. *A Preface to Modern Nigeria: The "Sierra Leoneans" in Yoruba, 1830–1890*. Madison: University of Wisconsin Press.

Kotkin, Joel. 1993. *Tribes*. New York: Random House.

Kramer, Fritz. 1993[1987]. *The Red Fez: Art and Spirit Possession in Africa*. Trans. Malcolm R. Green. London: Verso.

Kubik, Gerhard. 1979. *Angolan Traits in Black Music, Games and Dances of Brazil: A Study of African Cultural Extensions Overseas*. Lisbon: Junta de Investigações Científicas do Ultramar/Centro de Estudos de Antropologia Cultural.

Kulick, Don. 1998. *Travestí: Sex, Gender and Culture among Brazilian Transgendered Prostitutes*. Chicago: University of Chicago Press.

Labat, K. 1731. *Voyage du Chevalier des Marchais en Guinée, Isles Voisines, et a Cayenne, Fait en 1725, 1726 and 1727, enrichi d'un grand nombre de Cartes et de figures en Tailles douce[s]*. Vol. 2. Amsterdam: Aux dépens de la Compagnie.

Labouret, Henri, and Paul Rivet. 1929. *Le royaume d'Arda et son évangélisation au XVIIe siècle*. Paris: Institut d'Éthnologie.

Lachatañeré, Rómulo. 1992[1939]. *El Sistema Religioso de los Afrocubanos*. Havana: Editorial de Ciencias Sociales.

Lakoff, George, and Mark Johnson. 1980. *Metaphors We Live By*. Chicago: University of Chicago Press.

Landes, Ruth. 1940. "A Cult Matriarchate and Male Homosexuality." *Journal of Abnormal and Social Psychology* 35: 386–97.

———. 1947. The City of Women. New York: Macmillan.

———. 1973. *Pan-Africanism and Nationalism in West Africa, 1900–1945*. Oxford: Clarendon.

———. 1994[1947]. *The City of Women*. 2nd ed. Albuquerque: University of New Mexico Press.

Langley, J. Ayọdele. 1973. *Pan-Africanism and Nationalism in West Africa, 1900–1945*. Oxford: Clarendon.

Laotan, A. B. 1943. *The Torch Bearers, or Old Brazilian Colony in Lagos.* Lagos: Ifẹ-Olu Printing Works.

Larose, Serge. 1977. "The Meaning of Africa in Haitian Vodu." In *Symbols and Sentiments,* ed. Ioan Lewis, pp. 85–116. London: Academic Press.

Law, Robin. 1997. "Ethnicity and the Slave Trade: 'Lucumi' and 'Nago' as Ethnonyms in West Africa." *History in Africa* 24:1–16.

———. 1977. *The Oyo Empire c. 1600–c. 1836: A West African Imperialism in the Era of the Slave Trade.* Oxford: Clarendon.

———. 1990. "Constructing 'A Real National History': A Comparison of Edward Blyden and Samuel Johnson." In *Self-Assertion and Brokerage,* ed. P. F. Moraes Farias and Karen Barber, pp. 78–100. Birmingham University African Studies Series 2. Birmingham: Centre of West African Studies.

———. 1993[1984]. "How Truly Traditional Is Our Traditional History? The Case of Samuel Johnson and the Recording of Yoruba Oral Tradition." In *Pioneer, Patriot and Patriarch: Samuel Johnson and the Yoruba People,* ed. Toyin Falọla, pp. 47–63. Madison: African Studies Program, University of Wisconsin.

———. 1996. "Local Amateur Scholarship in the Construction of Yoruba Ethnicity, 1880–1914." In *Ethnicity in Africa: Roots, Meanings and Implications,* ed. Louise de la Gorgendiére, Kenneth King, and Sarah Vaughan, pp. 55–90. Edinburgh: Center of African Studies, University of Edinburgh.

Law, Robin, and Kristin Mann. 1999. "West Africa in the Atlantic Community: The Case of the Slave Coast." *William and Mary Quarterly,* 3rd ser., 56:307–34.

Leacock, Seth, and Ruth Leacock. 1975. *Spirits of the Deep.* Garden City, NY: Anchor/Double day.

Leão Teixeira, Maria Lina. 1987. "Lorogun—Identidades sexuais e poder no candomblé." In *Candomblé: Desvendando Identidades,* ed. Carlos Eugênio Marcondes de Moura, pp. 33–52. São Paulo: EMW Editores.

Lévi-Strauss, Claude. 1966[1962]. *The Savage Mind.* Chicago: University of Chicago Press.

Levine, Lawrence W. 1977. *Black Culture and Black Consciousness: Afro-American Folk Thought from Slavery to Freedom.* Oxford: Oxford University Press.

Lewis, I.M. 1989[1971]. *Ecstatic Religion: A Study of Shamanism and Spirit Possession.* 2nd ed. London: Routledge.

Lie, John. 1995. "From International Migration to Transnational Diaspora." *Contemporary Sociology* 24 (4):303–06.

Lima, Délcio Monteiro de. 1983. *Os Homoeróticos.* Rio de Janeiro: F. Alves.

Lima, Vivaldo da Costa. 1977[1971] *A família-de-santo nos Candomblés jejenagôs da Bahia.* Master's thesis, Federal University of Bahia.

———. 1976. "O conceito de 'nação' nos Candomblés da Bahia." *Afro-Àsia* 12:65–90.

———. 1987. "O Candomblé da Bahia na Década de 30." In *Cartas de Édison Carneiro a Artur Ramos,* ed. Waldir Freitas Oliveira and Vivaldo da Costa Lima, pp. 37–73. São Paulo: Corrupio.

Lincoln, C. Eric. 1973[1961]. *The Black Muslims in America,* Rev. ed. Boston: Beacon.

Lindsay, Lisa A. 1994. "'To Return to the Bosom of their Fatherland': Brazilian Immigrants in Nineteenth-Century Lagos." *Slavery and Abolition* 15 (1):22–50.

Linebaugh, Peter, and Marcus Rediker. 2000. *Many-Headed Hydra: Sailors, Slaves, Commoners, and the Hidden History of the Revolutionary Atlantic.* Boston: Beacon.

Lopes, Helena Theodoro, José Jorge Siqueira, and Mana Beatriz Nascimento. 1987. *Negro e cultura no Brasil: Pequena Enciclopédia da Cultura Brasileira.* Rio de Janeiro: UNIBRADE/UNESCO.

Lucas, J. Olumide. 1970. *Religions in West Africa and Ancient Egypt.* Apapa, Lagos State, Nigeria: Nigerian National Press.

Malinowski, Bronislaw. 1922. *Argonauts of the Western Pacific.* New York: Dutton.

Malkki, Liisa H. 1995. *Purity and Exile: Violence, Memory, and National Cosmology among Hutu Refugees in Tanzania.* Chicago: University of Chicago Press.

Mann, Kristin. 1985. *Marrying Well: Marriage, Status and Social Change among the Educated Elite in Colonial Lagos.* Cambridge: Cambridge University Press.

Manning, Patrick. 1982. *Slavery, Colonialism and Economic Growth in Dahomey, 1640–1960.* Cambridge: Cambridge University Press.

Matory, J. Lorand, 1986. "Vessels of Power: The Dialectical Symbolism of Power in Yoruba Religion and Polity." Master's Thesis, University of Chicago.

———. 1988. "Homens montados: Homossexualidade e simbolismo da possessão nas religiões afro-brasileiras." In *Escravidão e Invenção da Liberdade,* ed. João José Reis, pp. 215–31. São Paulo: Brasiliense.

———. 1991. "Sex and the Empire That Is No More: A Ritual History of Women's Power among the Oyo-Yoruba." Ph.D. diss., University of Chicago.

———. 1994a. "Rival Empires: Islam and the Religions of Spirit Possession among the Oyo-Yoruba." *American Ethnologist* 21:495–515.

———. 1994b. *Sex and the Empire That Is No More: Gender and the Politics of Metaphor in Oyo Yoruba Religion.* Minneapolis: University of Minnesota Press.

———. 1996. "Return, Race, and Religion in a Transatlantic Yorùbá Nation." Paper presented at the annual meeting of the African Studies Association, San Francisco, 23–26 Nov.

———. 1999a. "Afro-Atlantic Culture: On the Live Dialogue between Africa and the Americas." In *Africana: The Encyclopedia of the African and African American Experience,* ed. Henry Louis Gates and K. Anthony Appiah, pp. 36–44. New York: Basic/*Civitas* Books.

———. 1999b. "The English Professors of Brazil." *Comparative Studies in Society and History* 41 (1): 72–103.

———. 1999c. "Jeje: Repensando Nações e Transnacionalismo" *Mana: Estudos de Antropologia Social* 5 (1):57–80.

———. 1999d. "Man in the 'City of Women.'" Lecture presented at "From Local to Global: Rethinking Yoruba Religious Traditions for the Next Millenium," an international conference cosponsored by the Department of Religious Studies and the Department of African–New World Studies, Florida International University, 9–12 Dec.

————. 2000. "Surpassing 'Survival': On the Urbanity of 'Traditional Religion' in the Afro-Atlantic World". *Black Scholar* 30(3–4):36–43.

————. 2003. "Gendered Agendas: The Secrets Scholars Keep about Yoruba-Atlantic Religion." *Gender and History* 15:408–38.

————. Forthcoming a. "An African Empire in America: Texts, Migration, and the Trans-Atlantic Rise of the Lucumí Nation." In *Religion Outside the Institutions*, ed. Karen McCarthy Brown and Lynn Davidman. Princeton, NJ: Princeton University Press/Center for the Study of American Religion.

————. Forthcoming b *Sex and the Empire That Is No More: Gender and the Politics of Metaphor in Oyo Yoruba Religion.* 2nd ed. New York: Berghahn.

————. In preparation. *The Other African Americans: Black Ethnic Diversity in the United States.* New York: New York University Press.

Mattoso, Katia M. de Queiros. 1986[1979]. *To Be a Slave in Brazil, 1550–1888,* Trans. Arthur Goldhammer. New Brunswick, NJ: Rutgers University Press.

Maupoil, Bernard. 1988[1934–36] *La Géomancie à l'ancienne Côte des Esclaves.* Paris: Institut d'Éthnologie.

McLeod, John. 1820. *A Voyage to Africa with Some Account of the Manners and Customs of the Dahomian People.* London: John Murray.

Mead, Margaret. 1928. *Coming of Age in Samoa.* Boston: Beacon.

————. 1963[1935]. *Sex and Temperament in Three Primitive Societies.* New York: Morrow Quill.

Mendonça Teles, Gilberto, ed. 1986. *Vanguarda Européia e Modernismo Brasileiro: Apresentaçao dos principais poemas, manifestos, prefácios e conferências vanguardistas, de 1857 a 1972.* 10th ed. Rio de Janeiro: Editora Record.

Mercier, P. 1954. "The Fon of Dahomey." In *African Worlds*, ed. Daryll Forde, pp. 210–34. Plymouth, UK: International African Institute/Oxford University Press.

Métraux, Alfred. 1972 [1959]. *Voodoo in Haiti.* 2nd ed. Introduction by Sidney W. Mintz. New York:. Schocken Books.

Middleton, David, and Derek Edwards. 1990. "Conversational Remembering: A Social Psychological Approach." In *Collective Remembering*, ed. David Middleton and Derek Edwards, pp. 23–45. London: Sage.

Miller, Joseph C. 2000. "Central Africa during the Era of the Slave Trade." In *Central Africans and Cultural Transformations in the American Diaspora*, ed. Linda M. Heywood, pp. 21–69. Cambridge: Cambridge University Press.

Minority Rights Group. 1995. *No Longer Invisible: Afro-Latin Americans Today.* London: Minority Rights Publications.

Mintz, Sidney. 1985. *Sweetness and Power.* New York: Viking/Penguin.

————. 1998. "The Localization of Anthropological Practice: From Area Studies to Transnationalism." *Critique of Anthropology* 18 (2):117–33.

Mintz, Sidney W., and Richard Price. 1992 [1976]. *The Birth of African-American Culture: An Anthropological Perspective.* Boston: Beacon.

Mitchell, Michael. 1983. "Race, Legitimacy and the State in Brazil." Paper presented at the annual meeting of the Latin American Studies Association, Mexico City, 29 Sept.–1 Oct.

Moloney, Alfred C. 1889. "Cotton Interests, Foreign and Native, in Yoruba, and Generally in West Africa." *Journal of the Manchester Geographical Society* 5:265–76.

Moog, Vianna. 1993[1954]. *Bandeirantes e Pioneiros*, 18th ed. Rio de Janeiro: Civilização Brasileira.

Moreau de Saint-Méry, Médéric Louis Elie. 1958[1797]. *Description topographique, physique, civile, politique et historique de la partie française de l'Isle Saint-Domingue*. Paris: Société de l'Histoire des Colonies Françaises et Librairie Larose.

Moss, Leonard W., and Stephen C. Cappannari. 1982. "In Quest of the Black Virgin: She Is Black Because She Is Black." In *Mother Worship*, ed. James J. Preston, 53–74. Chapel Hill: University of North Carolina Press.

Mosse, George. 1985. *Nationalism and Sexuality*. Madison: University of Wisconsin Press.

Motta, Roberto. 1976. "Carneiro, Ruth Landes e os Candomblés Bantus." *Revista do Arquivo Público*, Recife 30 (32):3–218.

———. 1982. "Bandeira de Alairá: A festa de Xangô-São João e problemas do sincretismo." In *Bandeira de Alairá (Outros escritos sobre religião dos orixás)*, ed. Carlos Eugênio Marcondes de Moura, pp. 1–11. São Paulo: Nobel.

———. 1992. "The Churchifying of Candomblé: Priests, Anthropologists, and the Canonization of the African Religious Memory in Brazil." Paper presented at the panel Reafricanization and Urban Growth in Brazil, Latin American Studies Association (LASA), 17th International Congress, Los Angeles, September.

———. 1994. L'invention de l'Afrique dans le Candomblé du Brésil. *Storia, Antropologia e Scienze del Linguaggio* 9 (2–3): 65–85.

Mulira, Jesse Gaston. 1984. "*A History of the Mahi Peoples from 1774 to 1920*." Ph.D. diss., University of California, Los Angeles.

Murray, Stephen O. 1998. "Overview [West Africa]." In *Boy-Wives and Female Husbands*, ed. Stephen O. Murray and Will Roscoe, p. 91–109. New York: St. Martin's.

Nascimento, Luiz Cláudio do, and Cristiana Isidoro. 1988. *A Boa Morte em Cachoeira*. Cachoeira, Bahia, Brazil: Centro de Estudos, Pesquisa e Ação Sócio-Cultural de Cachoeira.

Nonini, Donald M., and Aihwa Ong. 1997. "Introduction: Chinese Transnationalism as an Alternative Modernity." In *Ungrounded Empires: The Cultural Politics of Modern Chinese Transnationalism*, ed. Donald M. Nonini and Aihwa Ong, pp. 3–33. New York: Routledge.

Norris, Robert. 1789. *Memoires of the Reign of Bossa Ahadee, King of Dahomy*. London: W. Lowndes.

Ogilby, John. 1670. *Africa: Being an Accurate Description of the Regions of Aegypt, Barbary, Lybia, and Billedulgerid, the Land of the Negroes, Guinee, Aethiopia, and the Abyssines . . . Collected and Translated from the Most Authentick Authors, and augmented by later observations* . . . London: Thomas Johnson.

Ogundipę-Leslie, 'Mọlara. 1985. "Women in Nigeria." In *Women in Nigeria Today*, ed. S. Bappa, J. Ibrahim, A. M. Imam, F.J.A. Kamara, H. Mahdi,

M. A. Modibbo, A. S. Mohammed, H. Mohammed, A. R. Mustapha, N. Perchonock, and R. I. Pittin, pp. 119–31. London: Zed Books.

Okediji, Moyọ. 1999. "Returnee Recollections: Transatlantic Transformations." In *Transatlantic Dialogue: Contemporary Art in and out of Africa*, by Michael D. Harris, principal author, pp. 32–51. Chapel Hill: Ackland Art Museum, University of North Carolina.

Okediji, O. O., and F. O. Okediji. 1966. "Marital Stability and Social Structure in an African City." *Nigerian Journal of Economic and Social Studies* 8 (1):151–63.

Okonkwo, Rina. 1985. *Heroes of West African Nationalism*. Enugu, Nigeria: Delta.

Olinto, Antônio. 1964. *Brasileiros na África*. Rio: Edições G.R.D.

———. 1980[1964]. *Brasileiros na África*. 2nd ed. São Paulo: G. R. Dorea; Brasília: Instituto Nacional do Livro.

Oliveira, Waldir Freitas, and Vivaldo da Costa Lima, eds. 1987. "Os Estudos Africanistas na Bahia dos Anos 30." In *Cartas de Édison Carneiro a Artur Ramos: De 4 de janeiro de 1936 a 6 de dezembro de 1938*, ed. Waldir Freitas Oliveira and Vivaldo da Costa Lima, pp. 23–35. São Paulo: Corrupio.

Olupọna, Jacob K. 1997. "Women's Rituals, Kingship and Power among the Ondo-Yoruba of Nigeria." In *Queens, Queen Mothers, Priestesses, and Power*, ed. Flora E. S. Kaplan, pp. 315–36. New York: New York Academy of Sciences.

Omari, Mikelle Smith. 1984. *From the Inside to the Outside: The Art and Ritual of Bahian Candomblé*. Los Angeles: Museum of Cultural History, University of California Monograph Series, 24.

Omu, Fred I. A. 1978. *Press and Politics in Nigeria, 1880–1937*. London: Longmans.

Ong, Aihwa. 1999. *Flexible Citizenship: The Cultural Logics of Transnationality*. Durham, NC: Duke University Press.

Ong, Aihwa, and Donald M. Nonini, eds. 1997. *Ungrounded Empires: The Cultural Politics of Modern Chinese Transnationalism*. New York: Routledge.

Ortiz, Fernando. 1921. "Los Cabildos Afro-Cubanos." *Revista Bimestre Cubana* 16 (1): 5–39.

———. 1973[1906]. *Los Negros Brujos*. Miami: Ediciones Universal.

———. 1995[1947]. *Cuban Counterpoint: Tobacco and Sugar*. Durham, NC: Duke University Press.

Ortiz, Renato. 1991[1978]. *A morte branca do feiticeiro negro: Umbanda e sociedade brasileira*. São Paulo: Editora Brasiliense.

Otero, Solimar. 2000. "Rethinking the Diaspora: Reflections on Africa and the Americas." Paper presented at the conference "Transcending Traditions: African, Afro-American and African Diaspora Studies in the 21st Century," African Studies Center and the Afro-American Studies Program, University of Pennsylvania, 20 Apr.

Oyewumi, Oyeronkẹ 1997. *The Invention of Women: Making an African Sense of Western Gender Discourses*. Minneapolis: University of Minnesota Press.

Palmié, Stephan. 2002. *Wizards and Scientists: Explorations in Afro-Cuban Modernity*. Durham, NC: Duke University Press.

Pamies, Alberto N. 1973. "Prólogo." In *Los Negros Brujos*, by Fernando Ortiz, pp. vii–xix. Miami: Ediciones Universal.

Pandey, Gyanendra. 1992. "In Defence of the Fragment: Writing about Hindu-Muslim Riots in India Today." In *A Subaltern Reader*, ed. Ranajit Guha, pp. 1–33. Minneapolis: University of Minnesota Press.

Parker, Andrew, Mary Russo, Doris Sommer, and Patricia Yaeger. 1992. *Nationalisms and Sexualities*. New York: Routledge.

Parker, Richard G. 1991. *Bodies, Pleasures, and Passions: Sexual Culture in Contemporary Brazil*. Boston: Beacon.

———. 1998. *Beneath the Equator: Cultures of Desire, Male Homosexuality, and Emerging Gay Communities in Brazil*. New York: Routledge.

Patterson, Orlando. 1991. *Freedom*, Vol. 1: *Freedom in the Making of Western Culture*. New York: Basic Books.

Peel, J.D.Y. 1993[1989]. "The Cultural Work of Yoruba Ethnogenesis." In *Pioneer, Patriot and Patriarch: Samuel Johnson and the Yoruba People*, ed. Toyin Falola, pp. 65–75. Madison: African Studies Program/University of Wisconsin–Madison.

———. 2000. *Religious Encounter and the Making of the Yoruba*. Bloomington: Indiana University Press.

———. 1995. "A Comparative Analysis of Ogun in Pre-colonial Yorubaland." Unpublished manuscript.

Peixoto, António da Costa. 1943–1944[1741]. *Obra Nova de Língua Geral de Mina*. Manuscrito da Biblioteca Pública de Èvora. Lisbon: Agência Geral das Colónias.

Pereira, Manuel Nunes. 1979[1947]. *A Casa das Minas: Contribuição ao Estudo das Sobrevivências do Culto dos Voduns, do Panteão Daomeano, no Estado do Maranhão, Brasil*. Petrópolis: Vozes.

Philips, John Edward. 1990. "The African Heritage of White America." In *Africanisms in American Culture*. ed. Joseph E. Holloway, pp. 225–39. Bloomington: Indiana University Press.

Pierson, Donald. 1942[1939]. "Negroes in Brazil". Ph.D. diss., University of Chicago.

Pietz, William. 1993. "Fetishism and Materialism: The Limits of Theory in Marx". In *Fetishism as Cultural Discourse*, ed. Emily Apter and William Pietz, pp. 119–51. Ithaca, NY: Cornell University Press.

———. 1995. "The Spirit of Civilization: Blood Sacrifice and Monetary debt." *Res* 28(Autumn):23–38.

Pires, Vicente Ferreira. 1957[1800]. *Viagem de Africa em o Reino de Dahomé*. São Paulo: Editora Nacional.

Pires, Regina Helena Gonçalves. 1992. "Curuzú: Caminhos e descaminhos na construção do si mesmo e do outro." Master's thesis, Federal University of Bahia.

Pliya, Jean. 1970. *Histoire Dahomey: Afrique Occidentale*. Issy les Moulineaux, France: Classiques Africains.

Polanyi, Karl. 1966. *Dahomey and the Slave Trade*. Seattle: University of Washington Press.

Potash, Betty. 1989. "Gender Relations in Sub-Saharan Africa." In *Gender and Anthropology*, ed. Sandra Morgen, pp. 189–227. Washington, DC: American Anthropological Association.

Prandi, Reginaldo. 1991. *Os Candomblés de São Paulo: A Velha Magia na Metró-pole Nova*. São Paulo: Editora Hucitec/Editora da Universidade de São Paulo.
Price, Richard. 1983. *First-Time: The Historical Vision of an Afro-American People*. Baltimore: Johns Hopkins University Press.
———. 1990. *Alabi's World*, Baltimore: Johns Hopkins University Press.
Price, Sally, and Richard Price. 1999. *Maroon Arts: Cultural Vitality in the African Diaspora*. Boston: Beacon.
Price-Mars, Jean. 1990[1928]. *So Spoke the Uncle* [Ainsi Parla l'Oncle]. Trans. Magdaline W. Shannon. Washington, DC: Three Continents Press.
Pries, Ludger, ed. 2001. *New Transnational Social Spaces: International Migration and Transnational Companies in the Early 21st Century*. London: Routledge
Puckett, Newbell Niles. 1969 [1926]. *Folk Beliefs of the Southern Negro*. New York: Dover.
Quénum, Maximilien. 1938[1931]. *Au Pays des Fons (Us et Coutumes du Dahomey)*. Paris: Larose.
Querino, Manuel. 1988[1938]. *Costumes Africanos no Brasil*. Recife, Brazil: Fundação Joaquim Nabuco–Editora Massangana.
Raboteau, Albert J. 1980 [1978]. *Slave Religion: The "Invisible Institution" in the Antebellum South*. Oxford: Oxford University Press.
Ramos, Arthur. 1940 [1934]. *O Negro Brasileiro: Etnografia religiosa*. Vol. 1. Augmented second ed. São Paulo: Companhia Editora Nacional.
———. 1942. *A Aculturação Negra no Brasil*. São Paulo: Companhia Editora Nacional.
———. 1946 [1937]. *As Culturas Negras no Novo Mundo*. O Negro Brasileiro. Vol. 3. 2nd ed. São Paulo: Companhia Editora Nacional.
Raphael, Allison. 1981. *Samba and Social Control: Popular Culture and Racial Democracy in Rio de Janeiro*. Ann Arbor, MI: University Microfilms.
Rassinoux, Jean. 1987. *Dictionnaire Français-Fon*. Saint-Étienne, France: Imprimerie Dumas.
Reid, Ira DeG. 1969[1939]. *The Negro Immigrant*. New York: Arno and the New York Times.
Reis, João José. 1983. "Magia Jeje na Bahia: A Invasão do Calundu do Pasto de Cachoeira, 1785." *Revista Brasileira de História* 8 (16):57–81.
———. 1986. "Nas Malhas do Poder Escravista: A Invasão do Candomblé do Accú na Bahia, 1829." *Religião e Sociedade* 13 (3):108–27.
———. 1987. *Rebelião escrava no Brasil: A história do levante dos malês, 1835*. 2nd ed. São Paulo: Editora Brasiliense.
Reisman, Karl. 1970. "Cultural and Linguistic Ambiguity in a West Indian Village." In *Afro-American Anthropology*, ed. Norman E. Whitten and John F. Szwed, pp. 129–144. New York: Free Press.
Renaud, [?], and A. Akindélé. 1939. "La Collectivité chez les Goun de l'ancien Royaume de Porto-Novo, 1938." *Coutumiers Juridiques de l'Afrique Occidentale Française* 3, série A, no. 10, 535–56.
Ribeiro, René. 1969. "Personality and the Psychosexual Adjustment of Afro-Brazilian Cult Members." *Journal de la Société des Américanistes* 58:109–20.
Rio, João do (aka Paulo Barreto). 1976[1906]. *As Religiões do Rio*. Rio de Janeiro: Nova Aguilar/Biblioteca Manancial.

Risério, Antônio, and Gilberto Gil. 1988. *O Poético e o Político e outros escritos*. Rio de Janeiro: Paz e Terra.

Roach, Joseph. 1996. *Cities of the Dead: Circum-Atlantic Performance*. New York: Columbia University Press.

Robotham, Don. 1998. "Transnationalism in the Caribbean: Formal and Informal." *American Ethnologist* 25 (2): 307–21.

Rodney, Walter. 1982[1972]. *How Europe Underdeveloped Africa*. Enugu, Nigeria: Ikenga.

Rodrigué, Emilio. 1991. *Gigante pela Própria Natureza*. São Paulo: Escuta.

Rodrigues, José Honório. 1965. *Brazil and Africa*. Trans. Richard A. Mazzara and Sam Hileman. Berkeley and Los Angeles: University of California Press.

Rodrigues, Raimundo Nina. 1935[1900/1896]. *O Animismo Fetishista dos Negros Bahianos*. Rio: Civilização Brasileira. Published as a series of magazine articles in 1896 (in *Revista Brasileira*); published as a single volume in French in 1900.

———. 1945[1905]. *Os Africanos no Brasil*. São Paulo: Companhia Editora Nacional.

———. 1988[1905]. *Os Africanos no Brasil*. 7th ed. Brasília: Editora Universidade de Brasília.

Roediger, David R. 1991. *The Wages of Whiteness: Race and the Making of the American Working Class*. London: Verso.

Rosengarten, Dale. 1987[1986]. *Row upon Row: Sea Grass Baskets of the South Carolina Low Country*. Columbia: McKissick, Museum, University of South Carolina.

———. 1997. "*Social Origins of the African-American Lowcountry Basket*." Ph.D. diss., Harvard University.

Rosenthal, Judy. 1998. *Possession, Ecstacy, and Law in Ewe Voodoo*. Charlottesville: University Press of Virginia.

Rotilu, Oyeyinka. 1932. "A Notable Ẹgba, Mr. Adeyẹmọ Alakija." *Nigerian Daily Times*, Dec., pp. 36–37.

Rouch, Jean. 1954. *Les Maîtres Fous*. Film produced under the auspices of the Centre national de la recherche scientifique and of the Institut français d'Afrique noire.

Rush, Dana. 1997. *Vodun Vortex: Accumulative Arts, Histories, and Religious Consciousness along Coastal Benin*. Ann Arbor, MI: University Microfilms.

———. 1999. "Eternal Potential: Chromolithographs in *Vodunland*." *African Arts* 32 (4):61–75, 94–96.

Russell-Wood, A.J.R. 1974. "Black and Mulatto Brotherhoods in Colonial Brazil: A Study in Collective Behavior." *Hispanic American Historical Review* 54 (4): 567–602.

Salvaing, Bernard. 1994. *Les Missionaires à la Rencontre de l'Afrique au XIXe Siècle (Côte des Esclaves et pays yoruba, 1840–1891)*. Paris: Edition L'Harmattan.

Santos, Deoscóredes dos (Mestre Didi). 1962. *Axé Opô Afonjá*. Rio de Janeiro: G. R. Dorea.

———. 1967. *West African Sacred Art and Rituals in Brazil*. Monograph produced with the assistance of the Institute of African Studies, University of Ibadan, Nigeria.

———. 1988[1962]. *História de um Terreiro Nagô*, 2nd ed. São Paulo: Max Limonad.

Santos, Jocélio Teles dos. 1995. *O Dono da terra: O Caboclo nos Candomblés da Bahia*. Salvador: Sarah Letras.

Sarracino [Magriñat], Rodolfo. 1988. *Los Que Volvieron a Africa*. Havana: Editorial de Ciencias Sociales (Sociologiá).

Sassen, Saskia. 1998. *Globalization and Its Discontents: Essays on the New Mobility of People and Money*. New York: New Press.

Schudson, Michael. 1997. "Lives, Laws, and Language: Commemorative versus Non-commemorative Forms of Effective Public Memory." *Communication Review* 2 (1): 3–17.

Schwartz, Stuart B. 1992. *Slaves, Peasants, and Rebels: Reconsidering Brazilian Slavery*. Urbana: University of Illinois Press.

Scott, David. 1991. "That Event, This Memory: Notes on the Anthropology of the African Diasporas in the New World." *Diaspora* 1 (3): 261–84.

Scott, James C. 1985. *Weapons of the Weak*. New Haven, CT: Yale University Press.

———. 1990. *Domination and the Arts of Resistance: Hidden Transcripts*. New Haven, CT: Yale University Press.

Segurola, R. P. B. 1968[1963]. *Dictionnaire Fon-Français* 2 vols. Porto-Novo, Benin (Dahomey): Centre Catéchétique de Porto-Novo.

Shaw, Rosalind, 2002. *Memories of the Slave Trade: Ritual and the Historical Imagination in Sierra Leone*. Chicago: University of Chicago Press.

Silverstein, Leni M. 1979. "Mãe de Todo Mundo: Modos de Sobrevivência nas Comunidades de Candomblé da Bahia." *Religião e Sociedade* 4: 143–69.

Simpson, George E. 1966. "Baptismal, 'Mourning,' and 'Building' Ceremonies of the Shouters in Trinidad." *Journal of American Folklore* 79: 537–50.

Skertchly, J. A. 1874. *Dahomey as It Is*. London: Chapman and Hall.

Skidmore, Thomas E. 1974. *Black into White: Race and Nationality in Brazilian Thought*. New York: Oxford University Press.

———. 1985. "Race and Class in Brazil: Historical Perspectives." In *Race, Class and Power in Brazil*, ed. Pierre-Michel Fontaine, pp. 11–24. Los Angeles: Center for Afro-American Studies, University of California.

Skinner, Elliot P. 1982. "The Dialectic between Diasporas and Homelands." In *Global Dimensions of the African Diaspora*, ed. Joseph E. Harris, pp. 17–45. Washington, DC: Howard University Press.

Smith, Guillaume. 1751. *Nouveau Voyage de Guinée*, 2 vols. Paris: Chez Durand [et] Pissot.

Smith, Robert C. 2001. "Comparing Local-Local Swedish and Mexican Transnational Life." In *New Transnational Social Spaces*, ed. Ludger Pries, pp. 37–58. London: Routledge.

Smith, William. 1967[1744]. *A New Voyage to Guinea*. London: Frank Cass.

Smythe, Augustine T., Herbert Ravenel Sass, Alfred Huger, Beatrice Ravenel, Thomas A. Waring, Archibald Rutledge, Josephine Pinckney, Caroline Pinckney Rutledge, DuBose Heyward, Katherine C. Hutson, and Robert W. Gordon. 1931. *The Carolina Low-Country*. New York: Macmillan.

Snelgrave, William. 1734. *A New Account of Some Parts of Guinea and the Slave Trade*. London: James, John and Paul Knapton.

Society for the Preservation of Spirituals. 1955. "The Society for the Preservation of Spirituals in Concert." 3rd series (recording). Columbia, SC: Society for the Preservation of Spirituals.

Sogbossi, Hippolyte Brice. 1996. "Aproximación al estudio de la tradición lingüístico-cultural de los arará en Jovellanos, Perico y Agramonte (Cuba)." Ph.D. diss., University of Havana.

Sommer, Doris. 1991. *Foundational Fictions: The National Romances of Latin America*. Berkeley and Los Angeles: University of California Press.

Stepan, Nancy Leys. 1991. *The Hour of Eugenics: Race, Gender, and Nation in Latin America*. Ithaca, NP: Cornell University Press.

Stoller, Paul. 1995. *Embodying Colonial Memories: Spirit Possession, Power and the Hauka in West Africa*. New York: Routledge

Stuckey, Sterling. 1987. *Slave Culture: Nationalist Theory and the Foundations of Black America*. New York: Oxford University Press.

Sweet, James H. 1996. "Male Homosexuality and Spiritism in the African Diaspora: The Legacies of a Link." *Journal of the History of Sexuality* 7 (21): 184–202.

Szanton Blanc, Cristina. 1997. "The Thoroughly Modern 'Asian': Capital, Culture, and Nation in Thailand and the Philippines." In *Ungrounded Empires: The Cultural Politics of Modern Chinese Transnationalism*, ed. Donald M. Nonini and Aihwa Ong, pp. 261–86. New York: Routledge.

Tedlock, Dennis, and Bruce Mannheim. 1995. "Introduction." In *The Dialogic Emergence of Culture*, ed. Dennis Tedlock and Bruce Mannheim, pp. 1–32. Urbana: University of Illinois Press.

Thomas, William I., and Florian Znaniecki. 1974[1918–20]. *The Polish Peasant in Europe and America*. New York: Octagon/Farrar, Straus and Giroux.

Thompson, Robert Farris. 1969. "Abatan: A Master Potter of the Egbado Yoruba." In *Tradition and Creativity in Tribal Art*, ed. Daniel Biebuyck, pp. 120–82. Berkeley and Los Angeles: University of California Press.

———. 1983. *Flash of the Spirit: African and Afro-American Art and Philosophy*. New York: Random House.

Thornton, John. 1992. *Africa and the Africans in the Making of the Atlantic World, 1400–1680*. Cambridge: Cambridge University Press.

———. 2000. "Religious and Ceremonial Life in the Kongo and Mbundu Areas, 1500–1700." In *Central Africans and Cultural Transformations in the American Diaspora*, ed. Linda M. Heywood, pp. 71–90. Cambridge: Cambridge University Press.

Trexler, Richard C. 1995. *Sex and Conquest: Gendered Violence, Political Order, and the European Conquest of the Americas*. Ithaca, NY: Cornell University Press.

Trocki, Carl A. 1997. "Boundaries and Transgressions: Chinese Enterprise in Eighteenth- and Nineteenth-Century Southeast Asia." In *Ungrounded Empires: The Cultural Politics of Modern Chinese Transnationalism*, ed. Donald M. Nonini and Aihwa Ong, pp. 61–85. New York: Routledge.

Trosky, Susan M., and Donna Olendorf. 1992. "Harris, Joel Chandler, 1848–1908." In *Contemporary Authors*, pp. 191–94. Detroit: Coale Research.

Trouillot, Michel-Rolph. 1995. *Silencing the Past: Power and the Production of History*. Boston: Beacon.

Tsing, Anna. 2000. "The Global Situation." *Cultural Anthropology* 15 (3): 327–60.

Turner, J. Michael. 1975 [1974]. *Les Brésiliens: The Impact of Former Brazilian Slaves upon Dahomey*. Ann Arbor, MI: University Microfilms.

———. 1985. "Brown into Black." In *Race, Class and Power in Brazil*, ed. Pierre-Michel Fontaine, pp. 73–94. Los Angeles: Center for Afro-American Studies, University of California.

Turner, Lorenzo D. 1942. "Some Contacts of Brazilian Ex-Slaves with Nigeria, West Africa." *Journal of Negro History* 27:55–67.

———. 1949. *Africanisms in the Gullah Dialect*. Chicago: University of Chicago Press.

Turner, Victor. 1969. *The Ritual Process: Structure and Anti-Structure*. Ithaca, NY: Cornell University Press.

———. 1983. "Carnaval in Rio: Dionysian Drama in an Industrializing Society." In *The Celebration of Society: Perspectives on Contemporary Cultural Performance*, ed. Frank Manning, pp. 103–24. Bowling Green, OH: Bowling Green University: London, Canada: Congress of Social and Humanistic Studies, University of Western Ontario.

Vass, Winifred Kellsberger. 1979. *The Bantu-Speaking Heritage of the United States*. Los Angeles: Center for Afro-American Studies, University of California.

Verdery, Katherine. 1998. "Transnationalism, Nationalism, Citizenship, and Property: Eastern Europe since 1989." *American Ethnologist* 25 (2):291–306.

Verger, Pierre. 1953. "Le Culte des Vodoun d'Abomey Aurait-Il Été Apporté à Saint-Louis de Maranhon par la Mère du Roi Ghézo?" *Mémoire de l'Institut Français d'Afrique Noire* 27:157–60.

———. 1966. *Le Fort St. Jean-Baptiste D'Ajuda*. Mémoire de l'Institut de Recherches Appliquées du Dahomey, no. 1. Porto Novo: Imprimerie Nationale.

———. 1970[1957]. *Notes sur le culte des Oriṣa et Vodun à Bahia, la Baie de tous les Saints, au Brésil et à l'ancienne Côte des Esclaves en Afrique*. Amsterdam: Swets and Zeitlinger.

———. 1976[1968]. *Trade Relations between the Bight of Benin and Bahia from the 17th to the 19th Century*. Trans. Evelyn Crawford Ibadan, Nigeria: Ibadan University Press.

———. 1980. "Primeiros Terreiros de Candomblé." In *Iconografia dos Deuses Africanos no Candomblé da Bahia*, by Carybé, Pierre Verger, and Waldeloir Rego. São Paulo: Editora Raízes Artes Gráficas.

———. 1981. *Orixás: Deuses Iorubás na África e no Novo Mundo*. Salvador, Brazil: Corrupio; São Paulo: Círculo do Livro.

———. ed. 1981[1850]. *Notícias da Bahia–1850*. Salvador, Bahia: Corrupio.

———. 1987[1968]. *Fluxo e refluxo do tráfico de escravos entre o Golfo do Benin e a Bahia de Todos os Santos dos Séculos XVII a XIX*. São Paulo: Corrupio.

Verger, Pierre. 1995. *Ewé: O uso das plantas na sociedade Iorubá*. São Paulo: Editora Schwarcz/Odebrecht/Companhia das Letras.

Verneau, René. 1890–91. *Les races humaines*. Paris: Librairie J.-B. Ballière et Fils.

Vlach, John Michael. 1990[1978]. *The Afro-American Tradition in Decorative Arts*. Athens and London: Brown Thrasher Books/University of Georgia Press.

Wafer, Jim. 1991. *The Taste of Blood: Spirit Possession in Brazilian Candomblé*. Philadelphia: University of Pennsylvania Press.

Wafer, Jim, and Hédimo Rodrigues Santana. 1990. "Africa in Brazil: Cultural Politics and the Candomblé Religion." *Folklore Forum* 23 (1/2): 98–114.

Wagley, Charles. 1963[1952]. "Introduction." In *Race and Class in Rural Brazil*, ed. Charles Wagley, pp. 7–15. New York: UNESCO/International Documents Service, Columbia University Press.

Wallerstein, Immanuel. 1990. "La construction des peuples: Racisme, nationalisme, ethnicité." In *Race, nation, classe*, ed. Étienne Balibar and Immanuel Wallerstein, pp. 95–116. Paris: Éditions la Découverte.

Watson, James L., ed. 1997. *Golden Arches East: McDonald's in East Asia*. Stanford, CA: Stanford University Press.

Werbner, Richard. 1998. "Smoke from the Barrel of a Gun: Postwars of the Dead, Memory and Reinscription in Zimbabwe." In *Memory and the Postcolony: African Anthropology and the Critique of Power*, ed. Richard Werbner, pp. 71–102. London: Zed.

White, Hayden. 1981. "The Value of Narrativity." In *On Narrative*, ed. W.J.T. Mitchell, pp. 1–23. Chicago: University of Chicago Press.

Wikan, Unni. 1977. "'Man Becomes Woman': Transsexualism in Oman as a Key to Gender Roles." *Man*, n.s., 12:304–19.

Williams, Eric. 1993[1944]. *Capitalism and Slavery*. London: André Deutsch.

Wimberly, Fayette. 1998. "The Expansion of Afro-Bahian Religious Practices in Nineteenth-Century Cachoeira." In *Afro-Brazilian Culture and Politics: Bahia, 1790s to 1990s*, ed. Hendrik Kraay, pp. 74–89. Armonk, NY: Sharpe.

Wolf, Eric R. 1982. *Europe and the People without History*. Berkeley and Los Angeles: University of California Press.

Wood, J. Buckley. 1881. "On the Inhabitants of Lagos: Their Character, Pursuits, and Languages." *Church Missionary Intelligence and Record*, n.s., 6: 683–91.

Wood, Peter H. 1974. *Black Majority*. New York: Knopf.

———. 1975. "'It Was a Negro Taught Them': A New Look at African Labor in Early South Carolina." In *Discovering Afro-America*, ed. Roger D. Abrahams and John F. Szwed, pp. 27–45. Leiden: Brill.

Yai, Olabiyi. 1992. "From Vodu to Mawu: Monotheism and History in the Fon Cultural Area." *SAPINA Newsletter* (A Bulletin of the Society for African Philosophy in North America) 4 (2–3):10–29.

Yelvington, Kevin A. 1997. "Herskovits' Jewishness." *History of Anthropology Newsletter* 27 (2): 3–9.

Index